Very Massive Stars in the Local Universe

Astrophysics and Space Science Library

EDITORIAL BOARD

Chairman

W. B. BURTON, *National Radio Astronomy Observatory, Charlottesville, Virginia, U.S.A. (bburton@nrao.edu); University of Leiden, The Netherlands* (burton@strw.leidenuniv.nl)

F. BERTOLA, *University of Padua, Italy*
C. J. CESARSKY, *Commission for Atomic Energy, Saclay, France*
P. EHRENFREUND, *Leiden University, The Netherlands*
O. ENGVOLD, *University of Oslo, Norway*
A. HECK, *Strasbourg Astronomical Observatory, France*
E. P. J. VAN DEN HEUVEL, *University of Amsterdam, The Netherlands*
V. M. KASPI, McGill *University, Montreal, Canada*
J. M. E. KUIJPERS, *University of Nijmegen, The Netherlands*
H. VAN DER LAAN, *University of Utrecht, The Netherlands*
P. G. MURDIN, *Institute of Astronomy, Cambridge, UK*
B. V. SOMOV, *Astronomical Institute, Moscow State University, Russia*
R. A. SUNYAEV, *Space Research Institute, Moscow, Russia*

More information about this series at
http://www.springer.com/series/5664

Jorick S. Vink
Editor

Very Massive Stars in the Local Universe

Editor
Jorick S. Vink
Armagh Observatory
Armagh
Ireland

ISSN 0067-0057 ISSN 2214-7985 (electronic)
ISBN 978-3-319-09595-0 ISBN 978-3-319-09596-7 (eBook)
DOI 10.1007/978-3-319-09596-7
Springer Cham Heidelberg New York Dordrecht London

Library of Congress Control Number: 2014953251

© Springer International Publishing Switzerland 2015
This work is subject to copyright. All rights are reserved by the Publisher, whether the whole or part of the material is concerned, specifically the rights of translation, reprinting, reuse of illustrations, recitation, broadcasting, reproduction on microfilms or in any other physical way, and transmission or information storage and retrieval, electronic adaptation, computer software, or by similar or dissimilar methodology now known or hereafter developed. Exempted from this legal reservation are brief excerpts in connection with reviews or scholarly analysis or material supplied specifically for the purpose of being entered and executed on a computer system, for exclusive use by the purchaser of the work. Duplication of this publication or parts thereof is permitted only under the provisions of the Copyright Law of the Publisher's location, in its current version, and permission for use must always be obtained from Springer. Permissions for use may be obtained through RightsLink at the Copyright Clearance Center. Violations are liable to prosecution under the respective Copyright Law.
The use of general descriptive names, registered names, trademarks, service marks, etc. in this publication does not imply, even in the absence of a specific statement, that such names are exempt from the relevant protective laws and regulations and therefore free for general use.
While the advice and information in this book are believed to be true and accurate at the date of publication, neither the authors nor the editors nor the publisher can accept any legal responsibility for any errors or omissions that may be made. The publisher makes no warranty, express or implied, with respect to the material contained herein.

Cover illustration: The young massive cluster R136 at the heart of the Tarantula nebula (30 Doradus) in the Large Magallanic CLoud (LMC). The region contains the most massive stars known in the local Universe up to 300 times the mass of the Sun.
Credit: NASA, ESA, and E. Sabbi (ESA/STScI)
Acknowledgment: R. O'Connell (University of Virginia) and the Wide Field Camera 3 Science Oversight Committee.

Printed on acid-free paper

Springer is part of Springer Science+Business Media (www.springer.com)

Preface

Over the last couple of years, evidence has been mounting for the existence of very massive stars (VMS) in the Local Universe up to 300 solar masses. With this paradigm shift in the stellar upper-mass limit, it seems timely to construct a textbook on the physics unique to VMS.

As the top-end of the stellar initial mass function is imprinted on the observed spectra of distant populations, such as the star-forming galaxies detected at the highest redshifts, this paradigm shift on the stellar upper-mass limit may have major implications far beyond the field of stellar physics.

For these reasons, we decided to work on a textbook that discusses the evidence for and against the existence of stars up to 300 solar masses, and that entails VMS formation, mass loss, evolution, and death.

The book in front of you may be considered a spin-off from a fruitful joint discussion (JD) at the 2012 IAU General Assembly in Beijing that was organized by the IAU Working Group on Massive Stars.

The textbook comprises seven in-depth chapters and an introduction on the role of VMS in the Universe. The book is intended to describe the status of the field and the physics specific to VMS, with sufficient background material to enable a graduate student or a researcher from a different area to enter this exciting new field of research.

I would like to take the opportunity to warmly thank the authors of the book and the people at Springer for their help with the editorial aspects. I would also like to thank the participants of the fruitful JD held in 2012. Finally, I would like to express my gratitude to all my collaborators, including post-docs and students over the years, without whom the science would have been far less enjoyable!

Armagh, Ireland　　　　　　　　　　　　　　　　　　　　　　　　　　Jorick S. Vink
April 2014

Contents

1 Very Massive Stars in the Local Universe 1
Jorick S. Vink
   1.1 Introduction 1
   1.2 The Role of Very Massive Stars in the Universe 2
   1.3 Definition of a Very Massive Star 3
   1.4 The Very Existence of Very Massive Stars 4
   1.5 The Evolution and Fate of Very Massive Stars 5
   References 6

2 Empirical Properties of Very Massive Stars 9
Fabrice Martins
   2.1 Historical Background and Definition 9
   2.2 Very Massive Single Stars 13
      2.2.1 Atmosphere Models and Determination of Stellar Parameters 13
      2.2.2 Uncertainties on the Luminosity 19
      2.2.3 Uncertainties in Evolutionary Tracks 22
      2.2.4 The Best Cases for Very Massive Single Stars 24
   2.3 Very Massive Stars in Binary Systems 33
      2.3.1 Massive Binaries and Dynamical Masses 33
      2.3.2 The Most Massive Binary Systems 36
   Summary and Conclusions 39
   References 40

3 The Formation of Very Massive Stars 43
Mark R. Krumholz
   3.1 Introduction 43
   3.2 The Formation of Very Massive Stars by Accretion 44
      3.2.1 Fragmentation 45
      3.2.2 Radiation Pressure 50

		3.2.3	Ionization Feedback	52
		3.2.4	Stellar Winds	55
	3.3	The Formation of Very Massive Stars by Collision		56
		3.3.1	Gas Accretion-Driven Collision Models	57
		3.3.2	Gas-Free Collision Models	60
		3.3.3	Stellar Evolution and Massive Star Mergers	62
	3.4	Observational Consequences and Tests		64
		3.4.1	The Shape of the Stellar Mass Function	64
		3.4.2	Environmental-Dependence of the Stellar Mass Function	66
		3.4.3	Companions to Massive Stars	67
	Conclusions and Summary: Does Star Formation Have an Upper Mass Limit?			69
	References			71
4	**Mass-Loss Rates of Very Massive Stars**			**77**
	Jorick S. Vink			
	4.1	Introduction		77
	4.2	O Stars with Optically Thin Winds		78
		4.2.1	Stellar Wind Equations	79
		4.2.2	CAK Solution	80
		4.2.3	Predictions Using a Monte Carlo Radiative Transfer Approach	81
		4.2.4	Line Acceleration Formalism $g(r)$ for Monte Carlo Use	83
	4.3	Wolf-Rayet Stars with Optically Thick Winds		85
		4.3.1	Wolf-Rayet (WR) Stars	85
		4.3.2	WR Wind Theory	85
		4.3.3	Hydrodynamic Optically Thick Wind Models	87
	4.4	VMS and the Transition Between Optically Thin and Thick Winds		88
		4.4.1	Analytic Derivation of Transition Mass-Loss Rate	89
		4.4.2	Models Close to the Eddington Limit	90
	4.5	Predictions for Low Metallicity Z and Pop III Stars		91
	4.6	Luminous Blue Variables		93
		4.6.1	What Is an LBV?	93
		4.6.2	Do LBVs Form Pseudo-photospheres?	94
		4.6.3	Winds During S Doradus Variations	95
		4.6.4	Super-Eddington Winds	96
	4.7	Observed Wind Parameters		97
		4.7.1	Ultraviolet P Cygni Resonance Lines	98
		4.7.2	The Hα Recombination Emission Line	98
		4.7.3	Radio and (Sub)millimetre Continuum Emission	99
	4.8	Wind Clumping		100
		4.8.1	Optically Thin Clumping ("Micro-clumping")	100
		4.8.2	The P v Problem	101

	4.8.3	Optically Thick Clumping ("Macro"-clumping)	102
	4.8.4	Quantifying the Number of Clumps	104
	4.8.5	Effects on Mass-Loss Predictions	106
Summary and Conclusion			107
References			108

5 Instabilities in the Envelopes and Winds of Very Massive Stars 113
Stanley P. Owocki

5.1	Background: VMS M-L Relation and the Eddington Limit	113
5.2	Mean Opacity Formulations	117
	5.2.1 Flux-Weighted Mean Opacity	117
	5.2.2 Planck Mean and its Dominance by Line Opacity	118
	5.2.3 Rosseland Opacity and Radiative Diffusion in Stellar Interior	119
5.3	Effect of Radiation Pressure on Stellar Envelope	120
	5.3.1 Mass-Luminosity Scaling for Radiative Envelope	120
	5.3.2 Virial Theorem and Stellar Binding Energy	122
	5.3.3 OPAL Opacity	122
	5.3.4 Envelope Inflation and the Iron Bump Eddington Limit	124
5.4	Basic Formalism for Envelope Instability and Mass Loss	126
	5.4.1 General Time-Dependent Conservation Equations	126
	5.4.2 Local Linear Analysis for "Strange-Mode" Instability of Hydrostatic Envelope	127
	5.4.3 General Equations for Steady, Spherically Symmetric Wind	129
5.5	Line-Driven Stellar Winds	130
	5.5.1 The CAK/Sobolev Model for Steady-State Winds	131
	5.5.2 Non-Sobolev Models of Wind Instability	135
	5.5.3 Clumping, Porosity and Vorosity: Implications for Mass Loss Rates	140
5.6	Continuum-Driven Mass Loss from Super-Eddington LBVs	143
	5.6.1 Lack of Self-Regulation for Continuum Driving	143
	5.6.2 Convective Instability of a Super-Eddington Interior	144
	5.6.3 Flow Stagnation from Photon Tiring	145
	5.6.4 Porosification of VMS Atmospheres by Stagnation and Instabilities	146
	5.6.5 Continuum-Driven Winds Regulated by Porous Opacity	148
	5.6.6 Simulation of Stagnation and Fallback Above the Tiring Limit	150
	5.6.7 LBV Eruptions: Enhanced Winds or Explosions?	151
Concluding Summary		152
References		153

6	**Evolution and Nucleosynthesis of Very Massive Stars**	157
	Raphael Hirschi	
	6.1 Introduction	157
	6.2 Stellar Evolution Models	159
	6.2.1 Stellar Structure Equations	159
	6.2.2 Mass Loss	161
	6.2.3 Rotation and Magnetic Mields	162
	6.3 General Properties and Early Evolution of VMS	168
	6.3.1 VMS Evolve Nearly Homogeneously	168
	6.3.2 Evolutionary Tracks	170
	6.3.3 Lifetimes and Mass-Luminosity Relation	175
	6.3.4 Mass Loss by Stellar Winds	176
	6.3.5 Mass Loss Rates and Proximity of the Eddington Limit	178
	6.3.6 Evolution of the Surface Velocity	181
	6.4 WR Stars from VMS	182
	6.5 Late Evolution and Pre-SN Properties of Very Massive Stars	185
	6.5.1 Advanced Phases, Final Masses and Masses of Carbon-Oxygen Cores	186
	6.5.2 Do VMS Produce PISNe?	189
	6.5.3 Supernova Types Produced by VMS	190
	6.5.4 GRBs from VMS?	190
	6.6 The Final Chemical Structure and Contribution to Galactic Chemical Evolution	192
	Summary and Conclusion	195
	References	196
7	**The Deaths of Very Massive Stars**	199
	Stan. E. Woosley and Alexander Heger	
	7.1 Introduction	199
	7.2 The Deaths of Stars 8–80 $M_\odot$	200
	7.2.1 Compactness as a Guide to Outcome	200
	7.2.2 8–30 $M_\odot$; Today's Supernovae and Element Factories	202
	7.2.3 Stars 30–80 $M_\odot$; Black Hole Progenitors	204
	7.2.4 Yesterday's Metal Poor Stars	205
	7.3 Pulsational Pair Instability Supernovae (80–150 $M_\odot$)	207
	7.3.1 Pulsationally Unstable Helium Stars	208
	7.3.2 Light Curves for Helium Stars	212
	7.3.3 Type II Pulsational Pair Instability Supernovae	213
	7.3.4 Nucleosynthesis	215
	7.4 150–260 $M_\odot$; Pair Instability Supernovae	216
	7.5 Above 260 $M_\odot$	217
	7.6 The Effects of Rotation	217
	7.6.1 Magnetar Powered Supernova Light Curves	219
	7.6.2 Gamma-Ray Bursts (GRBs)	221

		7.7	Final Comments	222
			References	224
8	**Observed Consequences of Preupernova Instability in Very Massive Stars**			227
	Nathan Smith			
	8.1	Introduction		227
	8.2	LBVs and Their Giant Eruptions		228
		8.2.1	Basic Observed Properties of LBVs	229
		8.2.2	The Evolutionary State of LBVs	232
		8.2.3	A Special Case: Eta Carinae	234
		8.2.4	Giant Eruptions: Diversity, Explosions, and Winds	236
	8.3	Very Luminous Supernovae		240
		8.3.1	Background	240
		8.3.2	Sources of Unusually High Luminosity	240
		8.3.3	Type IIn SLSNe	243
		8.3.4	CSM Mass Estimates for SLSNe IIn	245
		8.3.5	Connecting SNe IIn and LBVs	250
		8.3.6	Requirements for Pre-SN Eruptions and Implications	252
		8.3.7	Type Ic SLSNe and GRBs	253
	8.4	Detected Progenitors of Type IIn Supernovae		256
	8.5	Direct Detections of Pre-SN Eruptions		259
	8.6	Looking Forward (or Backward, Actually)		262
		References		263
Index				267

Chapter 1
Very Massive Stars in the Local Universe

Jorick S. Vink

Abstract Recent studies have claimed the existence of very massive stars (VMS) up to 300 $M_\odot$ in the local Universe. As this finding may represent a paradigm shift for the canonical stellar upper-mass limit of 150 $M_\odot$, it is timely to evaluate the physics specific to VMS, which is currently missing. For this reason, we decided to construct a book entailing both a discussion of the accuracy of VMS masses (Martins), as well as the physics of VMS formation (Krumholz), mass loss (Vink), instabilities (Owocki), evolution (Hirschi), and fate (theory – Woosley and Heger; observations – Smith).

1.1 Introduction

It has been thought for many years that very massive stars (VMS) with masses substantially larger than 100 $M_\odot$ may occur more frequently in the early Universe, some few hundred million years after the Big Bang. The reason for the expectation that the first few stellar generations would generally have been more massive is that there was less cooling during the formation process of these metal-poor objects than in today's metal-rich Universe (e.g. Bromm et al. 1999; Abel et al. 2002; Omukai and Palla 2003; Yoshida et al. 2004; Ohkubo et al. 2009).

Furthermore, as radiation-driven winds are thought to be weaker at the lower metal content of the early Universe (e.g. Kudritzki 2002; Vink and de Koter 2005; Krticka and Kubat 2006; Gräfener and Hamann 2008; Muijres et al. 2012), this could imply that the final masses of VMS in the early Universe would be almost equally high as their initial masses. This could then lead to the formation of 10^2–10^3 $M_\odot$ intermediate-mass black holes (IMBHs), with masses in between stellar mass black holes and supermassive black holes of order 10^5 $M_\odot$ in the centres of galaxies. IMBHs have been hypothesized to be the central engines of ultraluminous x-ray sources (ULXs). Moreover, in a high stellar mass – low mass loss – situation it might become possible to produce pair-instability supernovae (PISNe) in the initial

---

J.S. Vink (✉)
Armagh Observatory, Armagh, Ireland
e-mail: jsv@arm.ac.uk

mass range of 140–260 $M_\odot$ (Woosley et al. 2002; see also Fowler and Hoyle 1964; Barkat et al. 1967; Bond et al. 1984; Langer et al. 2007; Moriya et al. 2010; Pan et al. 2012; Dessart et al. 2013; Whalen et al. 2013). Such PISNs are very special as just one such explosion could potentially produce more metals than an entire initial mass function (IMF) below it (Langer 2012).

Interestingly, Crowther et al. (2010) re-analyzed the most massive hydrogen-and nitrogen-rich Wolf-Rayet (WNh) stars in the center of R136, the ionizing cluster of the Tarantula nebula in the Large Magellanic Cloud (LMC). The conclusion from their analysis was that stars usually assumed to be below the canonical stellar upper-mass limit of 150 $M_\odot$ (of e.g. Figer 2005), were actually found to be much more luminous (see also Hamann et al. 2006; Bestenlehner et al. 2011), with initial masses up to $\sim$200–300 $M_\odot$. As this finding may represent a paradigm shift for the canonical stellar upper-mass limit of 150 $M_\odot$, it is timely to discuss the status of the data as well as VMS theory.

Whilst textbooks and reviews have been devoted to the physics of canonical massive single and binary stars (Maeder 2009; Langer 2012) there is as yet no source that specifically addresses the physics unique to VMS. As such objects are in close proximity to the Eddington limit, this is likely to affect both their formation, and via their mass loss also their fates.

1.2 The Role of Very Massive Stars in the Universe

The first couple of stellar generations may be good candidates for the reionization of the Universe (e.g. Haehnelt et al. 2001; Barkana and Loeb 2001; Ciardi and Ferrara 2005; Fan et al. 2006) and their ionizing properties at very low metallicity (Z) may also be able to explain the extreme Lyα and He II emitting galaxies at high redshift (Malhotra and Rhoads 2002; Kudritzki 2002; Schaerer 2003; Stark et al. 2007; Ouchi et al. 2008).

Notwithstanding the role of the first stars, the interest in the current generation of massive stars has grown as well. Massive stars are important drivers for the evolution of galaxies, as the prime contributors to the chemical and energy input into the interstellar medium (ISM) through stellar winds and supernovae (SNe). A number of exciting developments have taken place in recent years, including the detection of long-duration gamma-ray bursts (GRBs) at redshifts of 9 (e.g. Tanvir et al. 2009), just a few hundred millions years after the Big Bang (Cucchiara et al. 2011). This provides convincing evidence that massive stars are able to form and die massive when the Universe was not yet enriched.

Very massive stars are usually found in and around young massive clusters, such as the Arches cluster in the Galactic centre and the local starburst region R136 in the LMC. Young clusters are also relevant for the unsolved problem of massive star formation. For decades it was a real challenge to form stars over 10–20 $M_\odot$, as radiation pressure on dust grains might halt and reverse the accretion flow onto the central object (e.g. Yorke and Kruegel 1977; Wolfire and Cassinelli 1987). Because of this issue, theorists have been creative in forming massive stars via competitive accretion and collisions in dense cluster environments (e.g., Bonnell et al. 1998). In more recent times several multi-D simulations have shown that massive stars might form via disk accretion after all (e.g., Krumholz et al. 2009; Kuiper et al. 2010). In the light of recent claims for the existence of VMS in dense clusters, however, the issue of forming VMS in extreme environments is discussed by Mark Krumholz in Chap. 3.

The fact that so many VMS are located within dense stellar clusters still allows for an intriguing scenario in which VMS may originate from collisions of smaller objects (e.g., Portegies Zwart et al. 1999; Gürkan et al. 2004), leading to the formation of VMS up to $1,000 M_\odot$ at the cluster center, which may produce IMBHs at the end of their lives, but only if VMS mass loss is not too severe (see Belkus et al. 2007; Yungelson et al. 2008; Glebbeek et al. 2009; Pauldrach et al. 2012; Yusof et al. 2013).

1.3 Definition of a Very Massive Star

One of the very first questions that arises when one prepares a book on VMS is what actually constitutes a "very" massive star. One may approach this in several different ways.

Theoretically, "normal" massive stars with masses above $\sim 8\,M_\odot$ are those that produce core-collapse SNe (Smartt et al. 2009), but what happens at the upper-mass end? Above a certain critical mass, one would expect the occurrence of PISNe, and ideally this could be the lower-mass limit for the definition of our VMS. However, in practice this number is not known a priori (due to mass loss), and therefore the initial and final masses are likely not the same. In other words, the initial main-sequence mass for PISN formation is model-dependent, and thus somewhat arbitrary. Furthermore, there is the complicating issue of pulsational pair-instability (PPI) at masses below those of full-fedged PISNe (e.g. Woosley et al. 2007). One could alternatively resort to the mass of the helium (He) core for which stars reach the conditions of electron/positron pair-formation instability. Heger showed this minimum mass to be $\sim 40\,M_\odot$ to encounter the PPI regime and $\sim 65\,M_\odot$ to enter the arena of the true PISNe (see also Chatzopoulos and Wheeler 2012).

Another definition could involve the spectroscopic transition between normal main-sequence O-type stars and hydrogen-rich Wolf-Rayet stars (of WNh type), which have also been shown to be core H burning main sequence objects. However, such a definition would be dependent on the mass-loss transition point between O-type and WNh stars, which is set by the transition luminosity (Vink and Gräfener 2012) and is expected to be Z dependent.

For these very reasons, we decided at the joint discussion meeting at the 2012 IAU GA in Beijing to follow a more pragmatic approach, defining stars to be *very* massive when their initial masses are $\simeq 100\,M_\odot$ (Vink et al. 2013).

1.4 The Very Existence of Very Massive Stars

With this definition, the question of whether *very* massive stars exist can easily be answered affirmatively, but the more relevant question during the joint discussion was whether the widely held "canonical" upper-mass limit of 150 $M_\odot$ has been superseded, as some part of the astronomical community had expressed some skepticism regarding very high masses in R136, in the light of an earlier spectacular claim for the existence of a 2,500 $M_\odot$ star R136 in the 30 Doradus region of the LMC (e.g. Cassinelli et al. 1981). Higher spatial resolution showed that R136 was actually not a single supermassive star, but it eventually revealed a young cluster containing several lower mass objects, including the current record holder R136a1.

Over the last few decades there has been a consensus of a 150 $M_\odot$ stellar upper mass limit (Weidner and Kroupa 2004; Figer 2005; Oey and Clarke 2005; Koen 2006), albeit the accuracy of these claims was surprisingly low (e.g. Massey 2011). Crowther et al. (2010) re-analyzed the VMS data in R136 claiming that the cluster hosts several stars with masses as high as 200–300 $M_\odot$. In addition they performed a sanity check on similar WNh objects in the Galactic starburst cluster NGC 3603. Although these objects were fainter than those in R136, the advantage was the available dynamical mass estimate by Schnurr et al. (2008) of the binary object NGC 3603-A1 with a primary mass of $116 \pm 31 M_\odot$. This was deemed important as the least model-dependent way to obtain stellar masses is through the analysis of the light-curves and radial velocities induced by binary motions (see Martins' Chap. 2).

It could still be argued that the luminosities derived by Crowther et al. are uncertain and that these central WNh stars might in reality involve multiple sources due to insufficient spatial resolution, especially considering that the highest resolution data of the young Galactic Arches cluster with the largest telescope (Keck) only has a limiting resolution of 50 mas, and given that R136 is 7 times more distant than the Arches cluster, the achievable resolution if the Arches cluster were in the LMC would mean that R136 would not be resolved. This suggests that we still cannot be 100 % certain that the bright WNh stars in R136 could not "break up" into lower-mass objects.

For this reason it was rather relevant that Bestenlehner et al. (2011) found an almost identical twin of R136a3 WNh star in 30 Doradus: VFTS 682. Its key relevance is that it is located in apparent isolation from the R136 cluster, and as a result the chance of line-of-sight contamination is insignificant in comparison to R136. The VFTS 682 object thus offered a second sanity check on the reliability of the luminosities of the R136 core stars. Bestenlehner et al. argued for a high luminosity of $\log(L/L_\odot) = 6.5$ with a present-day mass of 150 $M_\odot$ for VFTS 682, which implies an initial mass on the zero-age main sequence (ZAMS) higher than the canonical upper-mass limit.

In other words, although one cannot exclude the possibility that the object R136a1 claimed to be $\sim$300 $M_\odot$ in the R136 cluster might still "dissolve" when higher spatial resolution observations become available, the sanity checks involving binary dynamics and isolated objects make it quite convincing that stars with ZAMS masses at least up to 200 $M_\odot$ exist.

A more detailed overview of the masses of VMS and the upper end of the IMF will be described in Martins' Chap. 2.

1.5 The Evolution and Fate of Very Massive Stars

Very massive stars are thought to evolve almost chemically homogeneously (Hirschi's Chap. 6), implying that knowing the exact details of the mixing processes (e.g., rotation, magnetic fields) are less relevant in comparison to their canonical ~10–60 $M_\odot$ counterparts. Instead, the evolution and death of VMS is dominated by mass loss.

At some level it does not matter 'how' VMS became such massive objects. First of all we do not yet definitively know the formation mode of 'very' massive stars, and whether the formation involves disk accretion or coalescence of less massive objects. Secondly, there is a possibility that binary evolution already during early core hydrogen (H) burning resulted in the formation of massive blue stragglers (Schneider et al. 2014; de Mink et al. 2014), but the fate of these effectively single VMS will naturally be determined by single-star mass loss.

The existence of the Humphreys-Davidson (HD) limit at approximately solar metallicity tells us that VMS do not become red supergiants (RSG) but that they remain on the hot side of the Hertzsprung-Russell (HR) diagram as luminous O stars and Wolf-Rayet-type objects. For these hot stars the mass loss is thought to be driven by million of iron lines in a radiatively-driven wind, but what is not yet known is whether episodes of super-Eddington (Shaviv 1998), continuum-driven mass loss (such as may occur in Eta Carinae and other Luminous Blue Variable (LBV) star eruptions) may also play a role (see Vink's Chap. 4 and Owocki's Chap. 5). What is clear is that the Eddington Γ limit will play a dominant role in the mass-loss physics.

We should also note that the Eddington limit is relevant for another issue relating to VMS physics. When objects approach the Eddington limit, they may or may not *inflate* (Ishii et al. 1999; Petrovic et al. 2006), i.e. be subject to enormous radius and temperature changes (Gräfener et al. 2012). This implies that the temperatures and thus the ages of VMS are highly uncertain.

A final issue concerns the fate of VMS. In the traditional view, after core H-burning, VMS would become LBVs, remove large amounts of mass, exposing their bare-naked helium (He) cores, burn He for another 10^5 before giving rise to H-poor Type Ibc SNe (e.g. Conti 1976; Yoon et al. 2012; Georgy et al. 2012). However since 2006 there have been indications that some massive stars may explode prematurely as H-rich type II SNe already during the LBV phase (Kotak and Vink 2006; Gal-Yam et al. 2007; Mauerhan et al. 2013).

Might some of the most massive stars even produce PISNe? And how do PISNe compare to the general population of super-luminous SNe (SLSNe) that have recently been unveiled by Quimby et al. (2011), and are now seen out to high redshifts (Cooke et al. 2012)? Gal-Yam et al. (2009) discovered an intruiging optical transient with an observed light curve that fits the theoretical one calculated from pair-instability supernova with a He core mass around 100 $M_\odot$ (see also Kozyreva et al. 2014).

Even if the SLSNe turn out to be unrelated to PISNe as argued by Nicholl et al. (2013) and Inserra et al. (2013), we should note that alternative models such as magnetar models (e.g. Kasen and Bildsten 2010) would also involve rather massive stars, and if the high luminosity is not the result of a magnetar, but for instance due to mass loss, then the amounts of mass loss inferred for interacting type IIn SNe are so humongous (of order tens of solar masses; see Smith's Chap. 8) that they can only originate from VMS.

In summary, the evolution of VMS into the PISN and/or SLSNe regime can only be understood once we obtain a comprehensive framework regarding the evolution and physics of VMS. In this book, a number of experts discuss aspects of their research field relevant to VMS in the local Universe. In Chap. 2 Fabrice Martins discusses the observational data of VMS with a special emphasis on the luminosity and mass determinations of both single and binary VMS. The rest of the book is mostly theoretical. In Chap. 3, Mark Krumholz discusses the different formation modes of VMS. As mass loss is so dominant for the evolution and fate of VMS, the next topics involve the physics of both their stellar winds (Jorick Vink; Chap. 4) and instabilities (Stan Owocki; Chap. 5), before Raphael Hirschi discusses the evolution of VMS in Chap. 6. We finish with an overview of the possible theoretical outcomes in Chap. 7 by Woosley and Heger, and an overview of the observations of VMS fate by Nathan Smith in Chap. 8.

References

Abel, T., Bryan, G. L., & Norman, M. L. (2002). *Science, 295*, 93.
Barkana, R., & Loeb, A. (2001). *Physics Reports, 349*, 125.
Barkat, Z., Rakavy, G., & Sack, N. (1967). *Physical Review Letters, 18*, 379.
Belkus, H., Van Bever, J., & Vanbeveren, D. (2007). *Astrophysical Journal, 659*, 1576.
Bestenlehner, J. M., Vink, J. S., Gräfener, G., et al. (2011). *Astronomy and Astrophysics, 530*, L14.
Bond, J. R., Arnett, W. D., & Carr, B. J. (1984). *Astrophysical Journal, 280*, 825.
Bonnell, I. A., Bate, M. R., & Zinnecker, H. (1998). *Monthly Notices of the Royal Astronomical Society, 298*, 93.
Bromm, V., Coppi, P. S., & Larson, R. B. (1999). *Astrophysical Journal Letters, 527*, L5.
Cassinelli, J. P., Mathis, J. S., & Savage, B. D. (1981). *Science, 212*, 1497.
Chatzopoulos, E., & Wheeler, J. C. (2012). *Astrophysical Journal, 760*, 154.
Ciardi, B., & Ferrara, A. (2005). *Space Science Reviews, 116*, 625.
Conti, P. S. (1976). *Memoires of the Societe Royale des Sciences de Liege, 9*, 193.
Cooke, J., Sullivan, M., Gal-Yam, A., et al. (2012). *Nature, 491*, 228.
Crowther, P. A., Schnurr, O., Hirschi, R., et al. (2010). *Monthly Notices of the Royal Astronomical Society, 408*, 731.
Cucchiara, A., Cenko, S. B., Bloom, J. S., et al. (2011). *Astrophysical Journal, 743*, 154.

de Mink, S. E., Sana, H., Langer, N., Izzard, R. G., & Schneider, F. (2014). *Astrophysical Journal, 782*, 7.
Dessart, L., Waldman, R., Livne, E., Hillier, D. J., & Blondin, S. (2013). *Monthly Notices of the Royal Astronomical Society, 428*, 3227.
Fan, X., Carilli, C. L., & Keating, B. (2006). *Annual Review of Astronomy and Astrophysics, 44*, 415.
Figer, D. F. (2005). *Nature, 434*, 192.
Fowler, W. A., & Hoyle, F. (1964). *Astrophysical Journal Supplement Series, 9*, 201.
Gal-Yam, A., Leonard, D. C., Fox, D. B., et al. (2007). *Astrophysical Journal, 656*, 372.
Gal-Yam, A., Mazzali, P., Ofek, E. O., et al. (2009). *Nature, 462*, 624.
Georgy, C., Ekström, S., Meynet, G., et al. (2012). *Astronomy and Astrophysics, 542*, A29.
Glebbeek, E., Gaburov, E., de Mink, S. E., et al. (2009). *Astronomy and Astrophysics, 497*, 255.
Gräfener, G., & Hamann, W.-R. (2008). *Astronomy and Astrophysics, 482*, 945.
Gräfener, G., Owocki, S. P., & Vink, J.S. (2012). *Astronomy and Astrophysics, 538*, 40.
Gürkan, M. A., Freitag, M., & Rasio, F. A. (2004). *Astrophysical Journal, 604*, 632.
Haehnelt, M. G., Madau, P., Kudritzki, R., & Haardt, F. (2001). *Astrophysical Journal, 549L*, 151.
Hamann, W.-R., Gräfener, G., & Liermann, A. (2006). *Astronomy and Astrophysics, 457*, 1015.
Inserra, C., Smartt, S. J., Jerkstrand, A., et al. (2013). *Astrophysical Journal, 770*, 128.
Ishii, M., Ueno, M., & Kato, M. (1999). *Publications of the Astronomical Society of Japan, 51*, 417.
Kasen, D., & Bildsten, L. (2010). *Astrophysical Journal, 717*, 245.
Koen, C. (2006). *Monthly Notices of the Royal Astronomical Society, 365*, 590.
Kotak, R., & Vink, J. S. (2006). *Astronomy and Astrophysics, 460*, L5.
Kozyreva, A., Blinnikov, S., Langer, N., & Yoon, S.-C. (2014). *Astronomy and Astrophysics, 565*, 70.
Krticka, J., & Kubat, J. (2006). *Astronomy and Astrophysics, 446*, 1039.
Krumholz, M. R., Klein, R. I., McKee, C. F., et al. (2009). *Science, 323*, 754.
Kudritzki, R.-P. (2002). *Astrophysical Journal, 577*, 389.
Kuiper, R., Klahr, H., Beuther, H., & Henning, T. (2010). *Astrophysical Journal, 722*, 1556.
Langer, N. (2012). *Annual Review of Astronomy and Astrophysics, 50*, 107.
Langer, N., Norman, C. A., de Koter, A., et al. (2007). *Astronomy and Astrophysics, 475*, L19.
Maeder, A. (2009). *Astronomy and astrophysics library*. Berlin/Heidelberg: Springer.
Malhotra, S., & Rhoads, J. E. (2002). *Astrophysical Journal, 565L*, 71.
Massey, P. (2011). *American Shetland Pony Club, 440*, 29.
Mauerhan, J. C., Smith, N., Filippenko, A., et al. (2013). *Monthly Notices of the Royal Astronomical Society, 430*, 1801.
Moriya, T., Tominaga, N., Tanaka, M., Maeda, K., & Nomoto, K. (2010). *Astrophysical Journal, 717*, 83.
Muijres, L., Vink, J. S., de Koter, A., Hirschi, R., Langer, N., & Yoon, S.-C. (2012). *Astronomy and Astrophysics, 546*, 42.
Nicholl, M., Smartt, S. J., Jerkstrand, A., et al. (2013). *Nature, 502*, 346.
Oey, M. S., & Clarke, C. J. (2005). *Astrophysical Journal, 620*, 43.
Ohkubo, T., Nomoto, K., Umeda, H., Yoshida, N., & Tsuruta, S. (2009). *Astrophysical Journal, 706*, 1184.
Omukai, K., & Palla, F. (2003). *Astrophysical Journal, 589*, 677.
Ouchi, M., Shimasaku, K., Akiyama, M., et al. (2008). *Astrophysical Journal Supplement Series, 176*, 301O.
Pan, T., Loeb, A., & Kasen, D. (2012). *Monthly Notices of the Royal Astronomical Society, 423*, 2203.
Pauldrach, A. W. A., Vanbeveren, D., & Hoffmann, T. L. (2012). *Astronomy and Astrophysics, 538*, 75.
Petrovic, J., Pols, O., & Langer, N. (2006). *Astronomy and Astrophysics, 450*, 219.
Portegies Zwart, S. F., Makino, J., McMillan, S. L. W., & Hut, P. (1999). *Astronomy and Astrophysics, 348*, 117.

Quimby, R. M., Kulkarni, S. R., Kasliwal, M. M., et al. (2011). *Nature, 474*, 487.
Schaerer, D. (2003). *Astronomy and Astrophysics, 397*, 527.
Schneider, F. R. N., Izzard, R. G., de Mink, S. E., et al. (2014). *Astrophysical Journal, 780*, 117.
Schnurr, O., Moffat, A. F. J., St-Louis, N., et al. (2008). *Monthly Notices of the Royal Astronomical Society, 389*, 806.
Shaviv, N. J. (1998). *Astrophysical Journal, 494*, 193.
Smartt, S. J., Eldridge, J. J., Crockett, R. M., & Maund, J. R. (2009). *Monthly Notices of the Royal Astronomical Society, 395*, 1409.
Stark, D. P., Ellis, R. S., Richard, J., Kneib, J.-P., Smith, G. P., Santos, M. R. (2007). *Astrophysical Journal, 663*, 10.
Tanvir, N. R., Fox, D. B., Levan, A. J., et al. (2009). *Nature, 461*, 1254.
Vink, J. S., & de Koter, A. (2005). *Astronomy and Astrophysics, 442* 587.
Vink, J. S., & Gräfener, G. (2012). *Astrophysical Journal, 751*, 34.
Vink, J. S., Heger, A., Krumholz, M. R., Puls, J., et al. (2013). *History in Africa*. arXiv1302.2021
Weidner, C., & Kroupa, P. (2004). *Monthly Notices of the Royal Astronomical Society, 348*, 187.
Whalen, D. J., Even, W., & Frey, L. H. (2013). *Astrophysical Journal, 777*, 110.
Wolfire, M. G., & Cassinelli, J. P. (1987). *Astrophysical Journal, 319*, 850.
Woosley, S. E., Heger, A., & Weaver, T. A. (2002). *Reviews of Modern Physics, 74*, 1015.
Woosley, S. E., Blinnikov, S., & Heger, A. (2007). *Nature, 450*, 390.
Yoon, S.-C., Gräfener, G., Vink, J. S., Kozyreva, A., & Izzard, R. G. (2012). *Astronomy and Astrophysics, 544*, L11.
Yorke, H. W., & Kruegel, E. (1977). *Astronomy and Astrophysics, 54*, 183.
Yoshida, N., Bromm, V., Hernquist, L. (2004). *Astrophysical Journal, 605*, 579.
Yungelson, L. R., van den Heuvel, E. P. J., Vink, J. S., et al. (2008). *Astronomy and Astrophysics, 477*, 223.
Yusof, N., Hirschi, R., Meynet, G., et al. (2013). *Monthly Notices of the Royal Astronomical Society, 433*, 1114.

Chapter 2
Empirical Properties of Very Massive Stars

Fabrice Martins

Abstract In this chapter we present the properties of the most massive stars known by the end of 2013. We start with a summary of historical claims for stars with masses in excess of several hundreds, even thousands of solar masses. We then describe how we determine masses for single stars. We focus on the estimates of luminosities and on the related uncertainties. We also highlight the limitations of evolutionary models used to convert luminosities into masses. The most luminous single stars in the Galaxy and the Magellanic Clouds are subsequently presented. The uncertainties on their mass determinations are described. Finally, we present binary stars. After recalling some basics of binary analysis, we present the most massive binary systems and the estimates of their dynamical masses.

2.1 Historical Background and Definition

Massive stars are usually defined as stars with masses higher than $8\,M_\odot$. In the standard picture of single star evolution, these objects end their lives as core collapse supernovae. Unlike lower mass stars, they go beyond the core carbon burning phase, and produce many of the elements heavier than oxygen. In particular, they are major producers of α elements. Consequently, massive stars are key players for the chemical evolution of the interstellar medium and of galaxies: the fresh material produced in their cores is transported to the surface and subsequently released in the immediate surroundings by powerful stellar winds (and the final supernova explosion). The origin of these winds is rooted in the high luminosity of massive stars (several 10^4 to a few 10^6 times the solar luminosity). Photons are easily absorbed by the lines of metals present in the upper layers, which produces a strong radiative acceleration sufficient to overcome gravity and to accelerate significant amounts of material up to speeds of several thousands of km s^{-1} (Castor et al. 1975). Mass loss rates of 10^{-9}–$10^{-4}\,M_\odot$ year^{-1} are commonly observed in various types of massive stars. Another property typical of massive stars is their high effective

---

F. Martins (✉)
LUPM, Université Montpellier 2, CNRS, Place Eugène Bataillon, F-34095 Montpellier, France
e-mail: fabrice.martins@univ-montp2.fr

temperature. T_{eff} exceeds 25,000 K on the main sequence. In the latest phases of evolution, massive stars can be very cool (about 3,500 K in the red supergiant phase) but also very hot (100,000 K in some Wolf-Rayet stars) depending on the initial mass. A direct consequence of the high temperature is the production of strong ionizing fluxes which create HII regions.

Spectrally, massive stars appear as O and early B (i.e. earlier than B3) stars on the main sequence. Once they evolve, they become supergiants of all sorts: blue supergiants (spectral type O, B and A), yellow supergiants (spectral type F and G) and red supergiants (spectral type K and M). The most massive stars develop very strong winds which produce emission lines in the spectra: these objects are Wolf-Rayet stars. The strong mass loss of WR stars peels them off, unravelling deep layers of chemically enriched material. WN stars correspond to objects showing the products of hydrogen burning (dominated by nitrogen), while WC (and possibly WO) stars have chemical compositions typical of helium burning (where carbon is the main element). For stars more massive than about 25 $M_\odot$, there exists a temperature threshold (the Humphreys-Davidson limit, see Humphreys and Davidson 1994) below which stars are expected to become unstable, the ratio of their luminosity to the Eddington luminosity reaching unity. This temperature limit is higher for higher initial masses. Stars close to the Humphreys-Davidson limit are usually Luminous Blue Variable objects (such as the famous η Car).

From the above properties, massive stars are defined as stars with masses higher than 8 $M_\odot$. But the question of the upper limit on the mass of stars is not settled. For some time, it was thought that the first stars formed just after the big-bang had masses well in excess of 100 $M_\odot$, and likely of 1,000 $M_\odot$ (Bromm et al. 1999). The reason was a lack of important molecular cooling channels, favouring a large Jeans mass. Recent advances in the physics of low metallicity star formation have shown that masses of a few tens "only" could be obtained if feedback effects are taken into account (Hosokawa et al. 2011). At the same time, 3D hydrodynamical simulations of massive star formation at solar metallicity have been able to create objects with $M > 40 M_\odot$ through accretion (Krumholz et al. 2009), a process long thought to be inefficient for massive stars because of the strong radiative pressure.

Observationally, the existence of an upper mass limit for stars has always been debated actively. The method most often used to tackle this question relies on massive clusters. The idea is to determine the mass function of such clusters and to look for the most massive component. The mass function is extrapolated until there is only one star in the highest mass bin. The mass of this star is the maximum stellar mass expected in the cluster (M_{max}). M_{max} is subsequently compared to the mass of the most massive component observed in the cluster (M^{obs}_{max}). If $M^{obs}_{max} < M_{max}$ (and if all massive stars in the cluster are young enough not to have exploded as supernovae), then the lack of stars in the mass range M^{obs}_{max}–M_{max} is attributed to an upper mass cut-off in the mass function. Weidner and Kroupa (2004) used this method to infer an upper mass limit of about 150 $M_\odot$. Their analysis relied on the young cluster R136. Their conclusions were based on the results of Massey and Hunter (1998) who obtained masses of about 140–155 $M_\odot$ for the most massive members. We will see later that the most massive members of R136 may actually be more massive than 150 $M_\odot$, which could slightly change the conclusions of Weidner and Kroupa. Following the same method, Oey and Clarke (2005) determined the mass of the most massive member of a cluster as a function of the number of cluster components and of the upper mass cut-off. Using both R136 and a collection of OB associations, they confirmed that an upper mass limit between 120 and 200 $M_\odot$ should exist to explain the maximum masses observed. The studies of Weidner and Kroupa (2004) and Oey and Clarke (2005) rely on a statistical sampling of the mass function, without taking into account any physical effects that might alter the formation of massive stars. The subsequent work of Weidner and Kroupa (2006) and Weidner et al. (2010) show that a random sampling of the initial mass function may not be the best way of investigating the relation between the cluster mass and the mass of its most massive component. Feedback effects once the first massive stars are formed might be important, stopping the formation of objects in the mass bin M_{max}^{obs}–M_{max}. This could explain that in the Arches cluster no star with masses in excess of 130 $M_\odot$ is observed while according to Figer (2005), there should be 18 of them. Therefore, very massive stars are important not only for stellar physics, but also for star formation and the interplay between stars and the interstellar medium, both locally and on galactic scales. The reminder of this chapter focuses on the search for these objects, and their physical properties.

Many of the stars with masses claimed to be higher than 100 $M_\odot$ are located in the Magellanic Clouds. The most striking example is certainly that of R136, the core of the 30 Doradus giant HII region in the Large Magellanic Clouc (LMC) where the metallicity is about half the solar metallicity. Using photographic plates, Feitzinger et al. (1980) showed that R136 is made of three components (a, b and c) separated by ~ 1". R136a is the brightest and bluest object. Based on its effective temperature (50,000 < T_{eff} < 55,000 K) and bolometric luminosity ($3.1 \times 10^7 L_\odot$), Feitzinger et al. (1980) estimated a lower and upper mass limit of 250 and 1,000 $M_\odot$ respectively. This made R136a the most massive star at that time. Cassinelli et al. (1981) obtained an ultraviolet spectrum of R136a with the *International Ultraviolet Explorer* telescope and confirmed the hot and luminous nature of R136a: they obtained a temperature of 60,000 K and a luminosity close to $10^8 L_\odot$. Cassinelli et al. (1981) compared the morphology of the UV spectrum to that of known early O and WN stars. They concluded that the large terminal velocity deduced from the blueward extension of the P-Cygni profiles (3,400 km s^{-1}) and the shape of the CIV 1550 line (with a non zero flux in the blue part of the profile) was incompatible with a collection of known massive stars following a standard mass distribution. The authors favoured the solution of a unique object to explain these signatures. This object should have a mass of about 2,500 $M_\odot$ and a mass loss rate of $10^{-3.5 \pm 1.0}$ $M_\odot$ year^{-1}, far in excess of any other known O or Wolf-Rayet star. A similar mass was estimated for the progenitor of the peculiar supernova SN 1961V in the galaxy NGC 1058 (Utrobin 1984). The width of the maximum emission peak in the light curve of the supernova together with the bright magnitude of the progenitor were only reproduced by hydrodynamical models with masses of the order 2,000 $M_\odot$.

The high luminosity is often the first criterion to argue for very massive objects. We will return to this at length in Sect. 2.2. Other example of claims for very massive stars based on luminosity in the Magellanic Clouds exist. Humphreys (1983) reported on the brightest stars in the Local Group. She listed the brightest blue and red supergiants of six galaxies (Milky Way, SMC, LMC, M33, NGC 6822 and IC 1613). In the Magellanic Clouds, several stars reached bolometric magnitudes of -11 (LMC) and -10 (SMC), corresponding to luminosities in excess of $10^6 L_\odot$ and thus masses larger than $100 M_\odot$. Kudritzki et al. (1989) studied the most massive cluster in the SMC, NGC 346, and estimated a mass of $113^{+40}_{-29} M_\odot$ for the brightest component, NGC 346-1.

Since all of the above examples of very massive stars are located in the Magellanic Clouds, the question of distance and crowding rises. On the good side, distance is rather well constrained, so that luminosity estimates are usually more robust than in the Galaxy. On the other hand, the Magellanic Clouds are much further away than any Galactic region: they are more difficult to resolve. This problem turned out to be a key in the understanding of very massive objects. Weigelt and Baier (1985) used speckle interferometry to re-observe R136. They achieved a spatial resolution of 0.09" which broke the R136a object into 8 components, all located within 1". The three brightest members – R136a1, a2 and a3 – have similar magnitudes and are separated by a few tenths of arcseconds (Fig. 2.1, right). The $1,000 M_\odot$ star in R136a had long lived. Figure 2.1 illustrates the improvements in the imaging capabilities between the study of Feitzinger et al. (1980) and the recent paper by Crowther et al. (2010). Taking advantage of the developments in photometric observations (CCDs, adaptive optics) and image analysis (deconvolution), Heydari-Malayeri et al. showed in a series of papers that several of the claimed very massive stars were

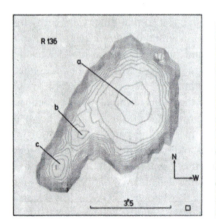
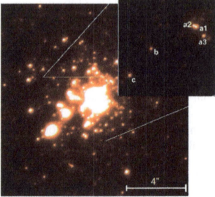

Fig. 2.1 R136 cluster in the giant HII region 30 Doradus (Large Magellanic Cloud). *Left panel*: photographic observation of Feitzinger et al. (1980). *Right panel*: adaptive Optics observations presented in Fig. 1 of Crowther et al. (2010, MNRAS, 408, 731 "The R136 star cluster hosts several stars whose individual masses greatly exceed the accepted 150Msolar stellar mass limit") (Reproduced with permission)

in fact multiple objects. HDE 268743, one of the bright LMC blue supergiant listed by Humphreys (1983), was first decomposed into 6 components (Heydari-Malayeri et al. 1988) before further observations with AO systems revealed not 6 but 12 stars. The mass of the most massive objects was estimated to be $\sim 50\,M_\odot$ (compared to more than $100\,M_\odot$ for the single initial object). Similarly, NGC 346-1 turned out to be made of at least three components (Heydari-Malayeri and Hutsemekers 1991), its mass shrinking from 113 (single object) to $58\,M_\odot$ (most massive resolved component). Two additional Magellanic Clouds very bright stars (Sk 157 and HDE 269936) were resolved into at least ten components by Heydari-Malayeri et al. (1989).

Spatial resolution is thus crucial to understand the nature of very massive stars. But mass estimates also rely on a number of models, for both stellar interiors and stellar atmospheres. The case of the Pistol star in the Galactic Center is an example of the effects of model improvements on mass estimates. Figer et al. (1998) studied that peculiar object using infrared spectroscopy and atmosphere models. Pistol is a late type massive star, probably of the class of Luminous Blue Variables. Figer et al. obtained a temperature of 14,000–20,000 K. Form that, they produced synthetic spectral energy distributions and fitted the observed near infrared photometry. They obtained luminosities between 4×10^6 and $1.5 \times 10^7\,L_\odot$, corresponding to masses in the range 200–250 $M_\odot$. Najarro et al. (2009) revisited the Pistol star with modern atmosphere models including the effects of metals on the atmospheric structure and synthetic spectra. They revised the temperature (11,800 K) and luminosity (1.6×10^6) of the Pistol star, with the consequence of a lower mass estimate: $100\,M_\odot$.

In the following sections of this chapter, we will illustrate how masses of single and binary stars can be determined. We will focus on the methods to constrain the stellar parameters, and especially the luminosity. We will highlight the assumptions of the analysis and raise the main sources of uncertainties.

2.2 Very Massive Single Stars

In this section we will describe how we determine the properties of single massive stars. We will see how an initial mass can be derived from the luminosity, and we will describe the related uncertainties. We will also explain how the present mass of stars can be obtained from the determination of gravity.

2.2.1 Atmosphere Models and Determination of Stellar Parameters

Two types of masses are usually determined for massive single stars. The "evolutionary" mass is the most often quoted. It is based on the luminosity of the star and its comparison to predictions of evolutionary calculations which provide a direct relation between L and the initial mass of the star. This is done in the Hertzsprung-Russell diagram. An estimate of the effective temperature of the star is required. The second mass is the "spectroscopic" mass. It is obtained from the determination of the surface gravity and its radius. From the definition of the gravity g, one has:

$$M = gR^2/G \tag{2.1}$$

where G is the constant of gravitation. The radius R is usually obtained from the estimate of both the effective temperature and of the luminosity, since by definition:

$$L = 4\pi R^2 \sigma T_{eff}^4 \tag{2.2}$$

with σ the Stefan-Boltzmann constant. Thus, the spectroscopic mass requires the knowledge of one more fundamental parameter (the surface gravity) compared to the evolutionary mass. For both masses, the determination of the effective temperature and of the luminosity are necessary.

The most important parameter to constrain is the effective temperature since once it is known the luminosity can be relatively easily determined. Atmosphere models are necessary to estimate T_{eff}. They predict the shape of the flux emitted by the star which can be compared to observations, either photometry or spectroscopy. The spectral energy distribution (SED) and lines strength depend sensitively on the effective temperature (and also to a lesser extent on other parameters). Iterations between models and observations allow to find the best models, and consequently the best temperature, to account for the properties of the star.

For low and intermediate mass stars, the effective temperature can be obtained from optical photometry. Since the peak of the SED is located around the visible wavelength range, a change in T_{eff} is mirrored by a change of optical colors (e.g. B-V or V-I, Bessell et al. 1998). For (very) massive stars, optical colors are almost insensitive to T_{eff}. Their high effective temperature shifts the SED peak to the ultraviolet wavelength range. The visible range is located is in the Rayleigh-Jeans tail of the flux distribution where the slope of the SED barely depends on T_{eff}. In principle, UV colors could be used to constrain the temperature of massive stars. But the UV range is dominated by metallic lines the strength of which depends on several parameters (metallicity, mass loss rate, microturbulence). Consequently, another way has to be found to estimate T_{eff}. The ionization balance is the standard diagnostic: for higher T_{eff} the ionization is higher and consequently lines from more ionized elements are stronger. Classically, for O stars, the ratio of HeI to HeII lines is used: HeII lines are stronger (weaker) and HeI lines weaker (stronger) at higher (lower) T_{eff}. Synthetic spectra are compared to observed He lines: if a good match is achieved, the effective temperature used to compute the synthetic spectrum is assigned to the star. He lines are the best temperature indicators between 30,000 and 45,000 K. Above (below), HeI (HeII) lines disappear. Alternative diagnostics have to be used. In the high temperature range, more relevant for very massive stars, nitrogen lines can replace helium lines (Rivero González et al. 2012). Their behaviour is more complex than helium lines and uncertainties on $T_{\rm eff}$ determinations are larger.

From above, we see that the determination of the effective temperature of a (very) massive star requires the use of synthetic spectroscopy and atmosphere models. Atmosphere models are meant to reproduce the level populations of all the ionization states of all elements present in the atmosphere of a star, as well as the shape and intensity of the associated radiation field. Their ultimate goal, for spectroscopic analysis of stars, is to predict the flux emitted at the top of the atmosphere so that it can be compared to observational data. An atmosphere model should account for the radial stratification of: the temperature, the density, velocity fields (if present), the specific intensity and opacities. The latter are directly related to level populations. Ideally, such models should be time-dependent and computed in 3D geometry. In practice, the current generation of atmosphere models for massive stars is far from this. The reasons are the following:

- Atmosphere models for massive stars have to be calculated in non-LTE (non Local Thermodynamic Equilibrium). This means that we cannot assume that the flux distribution at each point in the atmosphere is a blackbody. This assumption is not even valid locally. It would be relevant if radiation and matter were coupled only by collisional processes and were at equilibrium. But the very strong radiation field coming from the interior of the star prevents this situation from happening. Radiative processes are much more important than collisional ones. The populations of atomic levels are governed by radiative (de)population and cannot be estimated by the Saha-Boltzmann equation. Instead, the balance between all populating and depopulating routes from and to lower and higher energy levels has to be evaluated. For instance, to estimate the population of the first energy level above the ground state of an element, we have to know the rate at which electrons from the ground level are pushed into the first level, the rate of the inverse transition (from the first level to the ground state) and similarly from all transitions for levels beyond the second level to/from the first level. The computational cost is thus much larger than if the LTE approximation could be applied.

- Massive stars emit strong stellar winds. Consequently, their atmosphere are extended, with sizes typically between a few tens and up to one thousand times their stellar radius. Spherical effects are important and the assumption of a thin atmosphere (plane-parallel assumption) cannot be applied. A spherical geometry has to be adopted. In addition, and more importantly, the winds of massive stars are accelerated. Starting from a quasi static situation at the bottom of the atmosphere, material reaches velocities of several thousands of km s^{-1} above ten stellar radii. Doppler shifts are thus induced, which complicates the radiative transfer calculations. A photon emitted at the bottom of the atmosphere can travel freely throughout the entire atmosphere and be absorbed by a Doppler shifted line only in the upper atmosphere. Non local interaction between light and matter are thus possible.

- For realistic models, as many elements as possible have to be included. This is not only important to predict realistic spectra with numerous lines from elements heavier than hydrogen and helium. It is also crucial to correctly reproduce the physical conditions in the atmosphere. Indeed, having more elements implies additional sources of opacities which can affect the solution of the rate equations and consequently the entire atmosphere structure. The effects of elements heavier than hydrogen and helium on atmosphere models are known as line-blanketing effects.

The combination of these three ingredients makes atmosphere models for massive stars complex. For a reasonable treatment of non-LTE and line-blanketing effects, they are restricted to stationary and 1D computations. There are currently three numerical codes specifically devoted to the study of massive stars: CMFGEN (Hillier and Miller 1998), FASTWIND (Puls et al. 2005) and POWR (Hamann and Gräfener 2004). They all account for the three key ingredients described above. A fourth one (TLUSTY, Lanz and Hubeny 2003) assumes the plane-parallel configuration and is thus only adapted to spectroscopy in the photosphere of massive stars. All these models are computed for a given set of input parameters. The effective temperature, the luminosity (or the stellar radius), the surface gravity, the chemical composition and the wind parameters (mass loss rate and terminal velocity) are the main ones. The computation of the atmospheric structure is performed for this set of parameters. Once obtained, a formal solution of the radiative transfer equation is usually done to produce the emergent spectrum. It is this spectrum that is subsequently compared to observations. If a good match is obtained, the parameters of the models are considered to be the physical parameters of the star. CMFGEN and POWR are better suited to the analysis of VMS since as we will see below, VMS usually appear as Wolf-Rayet stars with numerous emission lines from H, He but also metals.

As explained above, the effective temperature is the first parameter to constrain. Once it is obtained, the luminosity determination can be made. There are usually two ways to proceed: either spectro-photometric data covering a large wavelength region exist and the SED can be fitted, or only photometry in a narrow wavelength region is available, and bolometric corrections have to be used. SED fitting is illustrated in the left panel of Fig. 2.2. Atmosphere models provide the spectral energy distribution at the stellar surface for the set of input parameters. This flux is scaled by the ratio $(R/d)^2$ where R is the stellar radius and d the distance of the target star. The resulting flux is then compared to the observational spectro-photometric data. In Fig. 2.2 we see the effect of a change of 0.1 dex on the luminosity. Optical and infrared photometry has been used together with flux calibrated UV spectra to build the observed SED. The red model, corresponding to $\log \frac{L}{L_\odot} = 5.2$, best reproduces these data. The right panel of Fig. 2.2 illustrates an important limitation of the determination of luminosities for massive stars: the knowledge of distances and their uncertainties. An error of only 10 % on the distance is equivalent to an error of about 0.1 dex on the luminosity. Distance is usually the main contributor to the uncertainty on the luminosity of Galactic objects.

2 Empirical Properties of Very Massive Stars

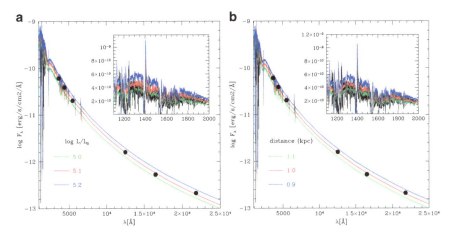

Fig. 2.2 Determination of the luminosity from the spectral energy distribution. The *black line* and *dots* are data for star HD 188209. *Left panel*: the colored lines are three models with luminosities differing by 0.1 dex. *Right panel*: the colored lines are the same model for three different distances (From Martins et al. (in prep))

If a sufficient number of spectro-photometric data is available, SED fitting can be performed rather safely (see Sect. 2.2.2 for limitations). Often, only photometry in the optical or the infrared range can be obtained. This is the case for stars located behind large amounts of extinction. In that case, luminosity has to be determined differently. This is done through an estimate of the bolometric correction. In the following, we will assume that only K-band photometry is available. The first step is to estimate the absolute K band magnitude

$$K = mK - A(K) - DM \quad (2.3)$$

where mK is the observed magnitude, $A(K)$ the amount of extinction in the K band and DM the distance modulus (DM = 5 × log(d)−5, d being the distance). As before, a good knowledge of the distance is required. We see that an estimate of the extinction is necessary too. We will come back to this issue in Sect. 2.2.2. In the second step, we need to add to the absolute K-band magnitude a correction to take into account the fact that we observe only a small fraction of the entire flux. This bolometric correction is computed from atmosphere models, and calibrated against effective temperature. For instance, Martins and Plez (2006) give

$$BC(K) = 28.80 - 7.24 \times log(T_{\text{eff}}) \quad (2.4)$$

The K-band bolometric correction is thus by definition the total bolometric magnitude minus the K-band absolute magnitude. Said differently, with K and BC(K), we have the total bolometric magnitude and thus the luminosity of the star:

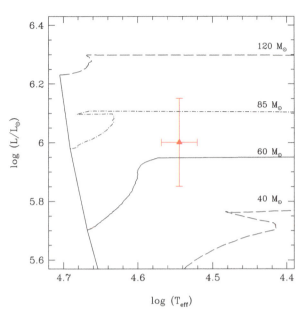

Fig. 2.3 Hertzsprung-Russell diagram illustrating the determination of evolutionary masses. Evolutionary tracks are from Meynet and Maeder (2003). The red square is for a star with $T_{\rm eff} = 35,000 \pm 2,000$ K and $\log \frac{L}{L_\odot} = 6.0 \pm 0.15$. The derived initial mass is 68^{+25}_{-14} M$_\odot$, while the estimated present-day mass is 50^{+18}_{-9} M$_\odot$

$$\log \frac{L}{L_\odot} = -0.4 \times (K + BC(K) - M_\odot^{bol}) \qquad (2.5)$$

where $M_\odot^{bol}$ is the bolometric magnitude of the Sun. Both methods (SED fitting or bolometric corrections) rely on atmosphere models and thus depend on the assumptions they are built on.

Once the effective temperature and luminosity are constrained, the evolutionary mass can be determined. Figure 2.3 shows a classical Hertzsprung-Russell diagram with the position of a bright O supergiant shown by the red symbol. From interpolation between the evolutionary tracks, we can estimate a present-day mass of 50^{+18}_{-9} M$_\odot$ and an initial mass of 68^{+25}_{-14} M$_\odot$. The difference between both estimates is due to mass loss through stellar winds. These masses are the evolutionary masses introduced at the beginning of this section. They depend on the way the interpolation between tracks is done and most importantly on the tracks themselves (Sect. 2.2.3). From Fig. 2.3, we see that very massive stars are objects with luminosities larger than one million times the Sun's luminosity.

The second mass estimate that can be given for single massive stars is the spectroscopic mass. As explained above, it is obtained from the surface gravity and an estimate of the stellar radius. The surface gravity, $\log g$, is obtained from the fit of Balmer, Paschen or Brackett lines in the optical/infrared range. Their width is sensitive to pressure broadening, especially Stark broadening. Stark broadening corresponds to a perturbation of the energy levels due to the electric field created by neighbouring charged particles. Broadening is thus stronger in denser environments, and consequently in stars with larger surface gravity. Hence, the width and strength of lines sensitive to Stark broadening effects are good estimates of log g. The most commonly used spectral diagnostics of surface gravity are Hβ and Hγ in the visible range. In the infrared, Brγ and Br10 are the best indicators. Synthetic spectra computed for a given log g are directly compared to the observed line profile. An accuracy of 0.1 dex on log g is usually achieved. This corresponds to an uncertainty of about 25 % on the stellar mass (without taking into account any error on the stellar radius).

The spectroscopic mass is more difficult to obtain than the evolutionary mass since it requires the observation of photospheric hydrogen lines. Such lines are sensitive to mass loss rate. When the wind strength becomes large, emission starts to fill the underlying photospheric profile leaving usually a pure emission profile, preventing the determination of log g. Unfortunately, this is often the case for very massive stars (see Sect. 2.2.4) which are very luminous and consequently have strong stellar winds.

The evolutionary mass and the spectroscopic mass should be consistent. However, as first pointed out by Herrero et al. (1992), the former are often systematically larger than the latter. Improvements in both stellar evolution and atmosphere models have reduced this discrepancy (e.g. Mokiem et al. 2007; Bouret et al. 2013), but the problem is still present for a number of stars. At present, the reason(s) for this difference is (are) not clear. Studies of binary systems tend to indicate that the evolutionary masses and the dynamical masses obtained from orbital solutions (see Sect. 2.3) are in good agreement below 30–50 $M_\odot$. Above, that limit, no clear conclusion can be drawn (Burkholder et al. 1997; Weidner and Vink 2010; Massey et al. 2012). The general conclusion is that there are at least two types of mass estimates for massive stars and that currently, no preference should be given to any of them.

In this section, we have presented the mass determinations for massive stars. For very massive objects, the evolutionary masses are usually quoted because of the shortcomings of the surface gravity determination. Evolutionary masses rely heavily on luminosity estimates and on the relation between luminosity and mass. In the next two sections, we will present the uncertainties related to both of them.

2.2.2 Uncertainties on the Luminosity

We now focus on the errors that enter the determination of the stellar luminosity. We assume we are dealing with single stars. In case of multiplicity, the determinations of L are obviously overestimated by an amount which depends on the number of companions and their relative brightness.

We have seen above that the luminosity of a star could be obtained from the monochromatic magnitude, a bolometric correction, an estimate of the extinction and of the distance (see Eqs. 2.3–2.5). This method is useful when not enough data are available to fit the entire SED. This is often the case for objects hidden behind large amounts of extinction. We have used this set of equations to compute the

Table 2.1 Effect of various uncertainties on luminosity and mass estimate[a]

Error	$\Delta \log \frac{L}{L_\odot}$	ΔM_{init} [$M_\odot$]	$\Delta M/M$ [%]
$\Delta m = 0.1$	0.04	7	9.0
$\Delta A = 0.2$	0.08	13	16.7
$\Delta d = 0.1$	0.09	15	19.2
$\Delta T_{\rm eff} = 3,000$	0.11	18	23.1
Combination of above errors	0.17	28	35.9
$\Delta d = 0.25$	0.22	36	46.1

[a]Calculations are for a star at 8 kpc, behind an extinction of 3 mag, with an observed magnitude of 11.2 and an effective temperature of 35,000 K. Photometry is taken in the K band. The luminosity and evolutionary mass of such a star would be $\log \frac{L}{L_\odot} = 6.06$ and 78 $M_\odot$ (initial mass). The mass is estimated using the evolutionary tracks of Meynet and Maeder (2003)

effects of uncertainties of several quantities on the derived luminosity. The results are summarized in Table 2.1. We have considered the case of a star with $\log \frac{L}{L_\odot}$ ~6.0. We have assumed it was observed in the K band. The distance and extinction are consistent with a position in the Galactic Center, a place where many massive stars are found. We have assumed typical errors on the magnitude, extinction, distance and effective temperature. The latter directly affects the uncertainty on the bolometric correction (Eq. 2.4). From Table 2.1, we see that the largest error budget is due to the uncertainties on the effective temperature and distance when the latter is poorly constrained. When combining all the sources of uncertainty, we get an error in the luminosity of about 0.17 dex (for a 10 % uncertainty on the distance).

In Table 2.1, we also provide estimates of the variations in mass estimates due to the above uncertainties. The evolutionary tracks of Meynet and Maeder (2003) have been used in this test case. The parameters we have chosen are those of a star with an initial mass of about 78 $M_\odot$ according to the Meynet and Maeder tracks. The individual errors induce changes of the initial mass between 7 and 18 $M_\odot$. The combined effects correspond to an uncertainty of 28 $M_\odot$, or 35 %. If the distance is poorly known (e.g. 25 % error, see last line of Table 2.1), the uncertainty can even reach almost 50 %.

Another source of uncertainty not taken into account in the above estimates is the shape of the extinction law. In Table 2.1 we have assumed a K-band extinction of 3.0. However, depending on the extinction law, the stellar flux will be redenned differently and for the same star, different values of the extinction can be obtained. As a consequence, the luminosity estimate will be affected. An illustration of this effect is given in the left panel of Fig. 2.4. Observational data (in black) for the Galactic star WR 18 are compared to a model with $\log \frac{L}{L_\odot} = 5.3$ (colored lines). In the UV, the extinction of Seaton (1979) is adopted. In the optical/infrared, three different extinction laws have been used: Howarth (1983) (green), Rieke and Lebofsky (1985) (red) and Nishiyama et al. (2009) (blue). All models assume E(B-V) = 0.9 and $R_V = 3.2$. The UV part of the SED is correctly reproduced, except in the region around 3,000 Å. The optical and infrared flux are different depending on the extinction curve. The laws of Howarth (1983) and Rieke and Lebofsky (1985)

2 Empirical Properties of Very Massive Stars

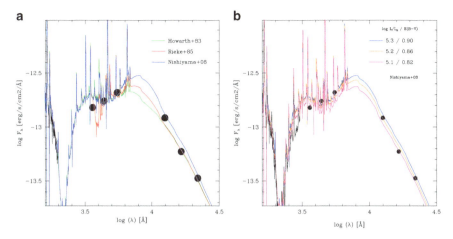

Fig. 2.4 Extinction law and luminosity determination. *Left panel*: effect of various extinction laws on the luminosity determination. The data are for the Galactic star WR 18 (*black line* and *dots*). The colored lines are models redenned with different extinction laws in the optical/infrared range. In the UV, the extinction law is that of Seaton (1979). *Right panel*: illustration of the luminosity tuning necessary to fit the SED with the extinction law of Nishiyama et al. (2009)

are relatively similar in the infrared but differ in the optical. The Nishiyama et al. law leads to a larger flux in the infrared compared to the other two laws. If we were to adopt the Nishiyama et al. law, we would need to reduce the luminosity to reproduce the infrared part of the SED. This is shown in the right panel of Fig. 2.4. The model with $\log \frac{L}{L_\odot} = 5.3$ and E(B-V) = 0.9 is shown in blue. Two additional models are shown: one with $\log \frac{L}{L_\odot} = 5.2$ and E(B-V) = 0.86 (orange) and one with $\log \frac{L}{L_\odot} = 5.1$ and E(B-V) = 0.82 (magenta). A luminosity intermediate between that of the two new models better reproduces the SED. Hence, using the Nishiyama et al. extinction leads to a downward revision of the luminosity by ~ 0.15 dex. At luminosities of $\sim 2 \times 10^5$ $L_\odot$, such a change corresponds to a reduction of the initial mass by 4 $M_\odot$, or 12 %.

In Table 2.1 we finally show the influence of the uncertainty on the effective temperature. It affects the bolometric correction and consequently the luminosity. A typical error of 3,000 K on T_{eff} (usually found for infrared studies) corresponds to a change in luminosity by 0.11 dex, and thus to an uncertainty of about 20–25 % on the initial mass. This estimate does not take into account the uncertainty in the relation between bolometric correction and temperature. It only accounts for the effect of T_{eff} on BC for the relation given in Eq. 2.4. Different model atmospheres provide slightly different calibrations of bolometric corrections. This adds another source of error in the mass estimate.

In conclusion, various uncertainties in the quantities involved in the luminosity determination lead to a typical error of about 0.15–0.20 dex on $\log \frac{L}{L_\odot}$. If the distance is poorly known, this uncertainty is larger. This translates into an error on the estimate of the initial mass of the order 10–50 %.

2.2.3 Uncertainties in Evolutionary Tracks

In the previous section, we have seen how the luminosity determination was affected by uncertainties in various observational quantities. We now focus on the uncertainties involved in the interpretation of the determined luminosity. As explained previously, the determination of the initial or present mass relies on comparison with evolutionary tracks (see Sect. 2.2.1). Such models rely on different assumptions to take into account the physical processes of stellar evolution. Consequently, they produce different outputs. In the following we will compare the public tracks of Brott et al. (2011) and Ekström et al. (2012). We refer to Martins and Palacios (2013) for a detailed comparison of various tracks.

Figure 2.5 shows the evolutionary tracks for Galactic stars from these two public grids of models. The tracks from Brott et al. (2011) are only available up to $60\,M_\odot$, so we do not show the higher mass models of Ekström et al. (2012). There are many differences between both sets of tracks (see Martins and Palacios 2013). The most important one for the sake of mass determination is the very different luminosities for a given mass. Looking at the $40\,M_\odot$ tracks, we see that the Ekstroem et al. tracks are about 0.2 dex more luminous than the Brott et al. tracks beyond the main

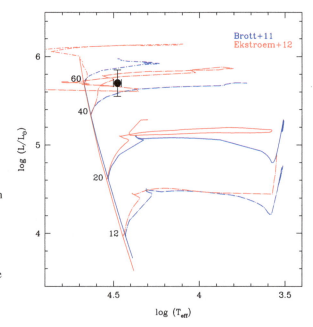

Fig. 2.5 Hertzsprung-Russell diagram with the evolutionary tracks of Brott et al. (2011) (*blue*) and Ekström et al. (2012) (*red*). The Brott et al. tracks are for an initial rotational velocity of 300 km s^{-1} while those of Ekstroem et al. assume a ratio of initial to critical rotation of 0.4

sequence. This offset is smaller at lower masses, and larger at higher masses. The direct consequence is that a lower initial mass is needed by the Ekstroem et al. tracks to reproduce the observed luminosity of a star. The black dot in Fig. 2.5 is an artificial star with $T_{\rm eff} = 30{,}000$ K and $\log \frac{L}{L_\odot} = 5.7$. Using the Ekstroem et al. tracks, one would find an initial mass of $37\,M_\odot$. For the Brott et al. tracks, the initial mass would be $48\,M_\odot$. The difference is of the order of 25 %.

The origin for the differences between the predictions of evolutionary tracks are manifold. One of the key effect known to affect the luminosity of a star is the amount of overshooting. The size of the convective core of massive stars is usually defined by the Schwarzschild criterion. But when this criterion applies to the velocity gradient of the convective region, and not to the velocity itself. It means that before reaching a zero velocity, the material transported by convection in the core travels a certain distance above the radius defined by the Schwarzschild criterion. This distance is not well known and is usually calibrated as a function of the local pressure scale height. The effect of this overshooting region is to bring to the core fresh material. This material is burnt and thus contributes to the luminosity of the star. Consequently, the larger the overshooting, the higher the luminosity of the star.

Another cause of luminosity increase is the effect of rotation. The physical reason is the same as for overshooting. When a star rotates, mixing of material takes place inside the star. Consequently, there is more material available for burning in the stellar core compared to a non-rotating star. The effect is an increase of the luminosity (Meynet and Maeder 2000). Quantitatively, the increase is between 0.1 and 0.3 dex depending on the initial mass for a moderate rotation of $300\,{\rm km\,s^{-1}}$ on the main sequence. Here again, the consequence for mass determinations is that lower masses are determined when tracks with rotation are used. In the case of the tracks of Ekström et al. (2012), and for the same example as above, a mass of $49\,M_\odot$ would be derived without rotation, compared to $37\,M_\odot$ with rotation. For very massive stars, the effects of rotation might be limited. The reason is the large mass loss rates at very high luminosities, causing efficient braking. However, the effects of rotation have not (yet) been investigated in the very high mass range.

Other parameters affecting the shape of evolutionary tracks are mass loss rate and metallicity. The former impact the evolution of the stars by removing material through stellar winds. Depending on the strength of the winds, the mass of a star at a given time will be different. Since mass and luminosity are directly related, a star with a strong mass loss will have a lower luminosity than a star with a lower mass loss (Meynet et al. 1994). Mass loss rates used in evolutionary calculations come from various sources. Some are empirical, some are theoretical. There are uncertainties associated with mass loss rates, but they are not straightforward to quantify. Clumping is known to affect the mass loss rate determinations in O-type stars (Bouret et al. 2005; Fullerton et al. 2006) but a good handle of its properties is missing. For cool massive stars, there is a wide spread in the mass loss rates of red supergiants (Mauron and Josselin 2011). Stellar winds are also weaker at lower metallicity (at least for hot massive stars). Their dependence is rather well constrained in the range $0.5 < Z/Z_\odot < 1.0$, but beyond, mass loss rates are based on extrapolations. Hence, evolutionary calculations, which adopt general (empirical or theoretical) prescriptions, can be adapted to explain the averaged properties of massive stars, but may fail to explain individual objects. They should thus be used with care.

2.2.4 The Best Cases for Very Massive Single Stars

After raising the sources of uncertainty in the determination of the mass of the most massive stars, we now turn to the presentation of the best cases. Since the stellar initial mass function (IMF) is a power law of the mass (at least above $\sim 1\,M_\odot$), massive stars are very rare. Consequently, we expect to find them more predominantly in clusters or association hosting a large number of stars. Adopting a standard Salpeter IMF, a cluster should have a mass in excess of a few $10^3\,M_\odot$ in order to host at least one star with mass in excess of $100\,M_\odot$. Thus, very massive stars have to be searched in massive clusters. In addition, massive stars live only a few million years, typically 2–3 Myr for objects above $100\,M_\odot$ (e.g. Yusof et al. 2013). Hence, VMS can only be found in young massive clusters. The best places to look for them would be young super star clusters. Observed in various types of galaxies, these objects have estimated masses in excess of $10^4\,M_\odot$, some of them reaching several $10^5\,M_\odot$ (Mengel et al. 2002; Bastian et al. 2006). However, none of these super star cluster is known in the local group, preventing current generation of telescopes and instruments to resolve their components individually.

In the Galaxy and its immediate vicinity, where individual stars can be observed, the best place to look for VMS is thus in young massive clusters. Although there has been a lot of improvement in the last decade in the detection of such objects (Figer et al. 2006; Chené et al. 2013), only a few are massive and young enough to be able to host VMS. In the Galaxy, the Arches and NGC3603 clusters are so far the two best candidates. In the Magellanic Clouds, R136 in the giant HII region 30 Doradus (see Sect. 2.1) is another interesting cluster. Beyond these three cases, known clusters are either too old or not massive enough. In the following, we will describe the evidence for the presence of VMS in the Arches, NGC3603 and R136 clusters. We will also highlight a couple of presumably isolated stars with large luminosities.

The Arches Cluster

The Arches cluster is located in the center of the Galaxy. It was discovered in the late 1990s through infrared imaging (Cotera et al. 1996; Figer et al. 1999). First thought to have a top heavy IMF (Figer et al. 2002) it is now considered to host a classical mass function (Espinoza et al. 2009). It hosts 13 Wolf-Rayet stars and several tens of O stars. All the Wolf-Rayet stars are of spectral WN7-9h. They are very luminous and their properties are consistent with those core-H burning objects

(Martins et al. 2008). Their luminosities, estimated using Eqs. 2.3 and 2.5, is larger than $10^6 L_\odot$, indicating that they are more massive than $80 M_\odot$. Martins et al. (2008) adopted a short distance (7.6 kpc) and a low extinction ($A_K = 2.8$) to obtain these luminosities. Revisiting the properties of two of the Arches WNh stars with a larger distance and extinction, Crowther et al. (2010) derived luminosities larger by about 0.25 dex. Consequently, they reported masses in excess of $160 M_\odot$. This illustrates the importance of extinction and distance in the estimates of the initial mass of VMS, as described in Sect. 2.2.2.

In Fig. 2.6 we present new HR diagrams of the Arches WNh stars. Table 2.2 provides the initial mass estimates depending on various assumptions. In this table, we have selected star F12 of Figer et al. (2002) as a test case. F12 is the hottest object in the HR diagram. Figure 2.6 and Table 2.2 are meant to further illustrate the role of various observational parameters on mass determinations. The upper left panel shows the effect of photometry. A change of 0.1 magnitude in the K-band photometry results in a change of about 10 % in the initial mass. The effect of extinction is similar. In our estimates, we have assumed a constant value of E(H-K) for all stars. Depending on the extinction law adopted, the ratio of selective to total absorption in the infrared ($R_K = A_K/E(H-K)$) is different. We have used the values of Rieke and Lebofsky (1985) ($R_K = 1.78$), Espinoza et al. (2009) ($R_K = 1.61$) and Nishiyama et al. (2009) ($R_K = 1.44$) to obtain the K-band extinction A_K. For the same observed color (H-K), the extinction can differ by 0.8 magnitude depending on the extinction law. The mass of star F12 varies between 93 and $138 M_\odot$ (35 %

Table 2.2 Effects of photometry, extinction and distance uncertainty on the initial mass estimate[a] for star F12 in the Arches cluster

mK / A_K / d(kpc)	M^a_{init} [$M_\odot$]
10.99[b]/3.1[c]/8.0	111
10.88[d]/3.1[c]/8.0	120
10.88[d]/3.06[d]/8.0	117
10.88[d] / 3.38[d]/8.0	138
10.88[d]/2.74[d]/8.0	93
10.99/2.74/7.6[e]	78
10.88/3.38/8.3[e]	146

[a] The mass is estimated using the evolutionary tracks of Meynet and Maeder (2003)
[b] Photometry is from Figer et al. (1999) and Espinoza et al. (2009)
[c] Extinction is from Stolte et al. (2002)
[d] Extinction is computed assuming E(H-K) = 1.9 and using the redenning laws of Rieke and Lebofsky (1985), Espinoza et al. (2009) and Nishiyama et al. (2009)
[e] Distances are from Eisenhauer et al. (2005) and Gillessen et al. (2009)

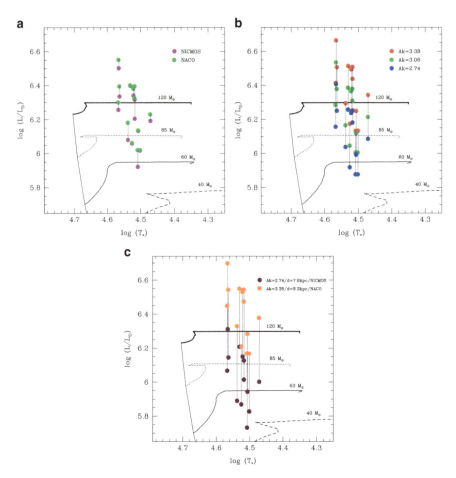

Fig. 2.6 HR diagram of the Arches cluster. The data points correspond to the WNh stars. The evolutionary tracks are from Meynet and Maeder (2003) and have Z = 0.02. *Upper left panel*: effect of photometry. The luminosity of the stars is computed using both the HST/NICMOS and VLT/NACO photometry (Figer et al. 1999; Espinoza et al. 2009). The distance is set to 8.0 kpc and effective temperatures are from Martins et al. (2008). *Upper right panel*: effect of extinction. An average value of E(H-K) = 1.9 is adopted and is transformed into A_K using the relations of Rieke and Lebofsky (1985), Nishiyama et al. (2009) and Espinoza et al. (2009). *Bottom panel*: extreme values of the Arches luminosities obtained by combining the faintest (brightest) photometry (Figer et al. 1999; Espinoza et al. 2009), lowest (highest) extinction (Rieke and Lebofsky 1985; Nishiyama et al. 2009) and shortest (largest) distances (Eisenhauer et al. 2005; Gillessen et al. 2009)

variation). In the bottom panel of Fig. 2.6, we show the most extreme variation in luminosity expected for the Arches star. For star F12, the initial mass can be as low as 78 M$_\odot$ or as large as 146 M$_\odot$ depending on the combination of photometry, extinction and distance adopted.[1] There is almost a factor of 2 in the estimated initial mass.

From the bottom panel of Fig. 2.6 we can conclude that according to the evolutionary tracks of Meynet and Maeder (2003), the most massive stars in the Arches cluster likely have masses higher than 80–90 $M_\odot$. Under the most favourable assumptions, the most massive objects can reach 150–200 $M_\odot$. This is the initial masses of stars with log $\frac{L}{L_\odot} > 6.5$ (the group of stars around log $\frac{L}{L_\odot} = 6.55$ corresponds to initial masses of 160 $M_\odot$). The truth probably lies somewhere in between these extreme values.

The above mass estimates have been obtained using the evolutionary tracks of Meynet and Maeder (2003). The grid of Ekström et al. (2012) is a revised version of these tracks, with a lower metal content (Z = 0.014 versus 0.02). Yusof et al. (2013) also published models for very massive stars at solar metallicity (Z = 0.014). Interestingly, none of the evolutionary calculations of Ekstroem et al. nor of Yusof et al. is able to reproduce the luminosity and temperature of the Arches stars. Their tracks for masses above 100 $M_\odot$ remain close to the zero age main sequence or even turn rapidly towards the blue part of the HR diagram. They never reach temperatures lower than about 40,000 K. Crowther et al. (2010) pointed out that metallicity in the Galactic center may be slightly super solar (Najarro et al. 2009). Evolutionary models with Z = 0.02 might be more relevant for the Arches cluster. In any case, the behaviour of evolutionary tracks at very high masses indicates that large uncertainties exist in the predicted temperature and luminosity of very massive stars. This should be kept in mind when quoting initial masses derived by means of evolutionary calculations.

R136 in 30 Doradus

R136 was once thought to be a single object of mass ~1,000 $M_\odot$, as described in Sect. 2.1. As shown in Fig. 2.1, three components are resolved by adaptive optics observations. Two other bright objects (R136b and R136c) are located next to the R136a stars. Contrary to the case of the stars in the Arches cluster, extinction is not the main limitation of the luminosity and mass determination. Crowding is at least as important. It is only when the *Hubble Space Telescope* started its operation that spectroscopic data could be obtained. Even with HST, a1 and a2 are difficult to separate.

de Koter et al. (1997) performed an analysis of the UV spectra of a1 and a3. Their results are shown by the green points in Fig. 2.7. They determined effective temperatures of about 45,000 K and luminosities close to $2.0 \times 10^6 L_\odot$, corresponding to initial masses between 100 and 120 $M_\odot$. Crowther and Dessart

---

[1]For masses above 120 $M_\odot$, we simply linearly extrapolate the luminosities and initial masses from the grid of Meynet and Maeder (2003), taking the 85 and 120 $M_\odot$ tracks as reference to estimate the dependence of mass on luminosity.

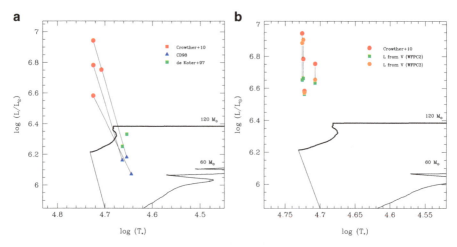

Fig. 2.7 Hertzsprung-Russell for the stars in R136. The evolutionary tracks at Z = 0.008 are from Meynet and Maeder (2005). The positions of the stars a1, a2, a3 and c are shown in *red* according to the analysis of Crowther et al. (2010). *Left*: the *green squares* are stars a1 and a3 from de Koter et al. (1997). The *blue triangles* are stars a1, a2 and a3 from Crowther and Dessart (1998). *Right*: the *green squares* correspond to luminosity estimates based on HST/WFPC2 photometry (Hunter et al. 1995), effective temperatures and extinction from Crowther et al. (2010), and bolometric corrections from Martins and Plez (2006). The orange hexagons are based on HST/WFPC3 photometry from De Marchi et al. (2011). The effective temperatures of stars a1, a2 and a3 have been shifted by 200 K for clarity

(1998) used improved model atmospheres to re-analyze a1 and a3, and included also a2. They extended their study to the optical range. They found very similar effective temperatures and luminosities 0.1–0.2 dex lower. A major re-investigation of the properties of the R136a components was performed by Crowther et al. (2010). They used atmosphere models including many metals, thus correctly taking into account the line-blanketing effects. In addition, they used the K-band spectra and photometry obtained by Schnurr et al. (2009), taking advantage of the high spatial resolution collected with the adaptive optics demonstrator *MAD* on the VLT (Campbell et al. 2010). Due to the inclusion of line-blanketing in atmosphere models, new, hotter effective temperatures were obtained ($T_{\mathrm{eff}} = 53,000$ K). The luminosities were significantly revised upward, reaching $10^{6.6-6.9}$ L$_\odot$ (see Fig. 2.7, left panel). Consequently, much larger initial masses were estimated: 320 M$_\odot$ for a1, 240 M$_\odot$ for a2, 165 M$_\odot$ for a3 and 220 M$_\odot$ for c.

There are several reasons for the revised luminosities: change of effective temperature and new bolometric corrections, different extinction, and use of different wavelength bands. de Koter et al. (1997) and to a lower extent Crowther and Dessart (1998) used atmosphere models with a limited treatment of line-blanketing. Crowther et al. (2010) included many metals in their models. As is well known, metals affect the determination of effective temperatures and bolometric corrections (Martins et al. 2005). Let us take the example of star R136a3. de Koter et al. (1997)

2 Empirical Properties of Very Massive Stars

Table 2.3 Luminosity estimates for the R136a stars using WFPC2 and WFPC3 photometry[a]

Star	T_{eff}	mV(WFPC2)	A_V	MV(WFPC2)	BC(V)	$\log \frac{L}{L_\odot}$
	[K]	mV(WFPC3)		MV(WFPC3)		
R136a1	53,000	12.84	1.80	−7.41	−4.48	6.65
		12.28		−7.97		6.88
R136a2	53,000	12.96	1.92	−7.41	−4.48	6.66
		12.34		−8.03		6.90
R136a3	53,000	13.01	1.72	−7.16	−4.48	6.56
		12.97		−7.20		6.57
R136c	51,000	13.47	2.48	−7.46	−4.37	6.63
		13.43		−7.50		6.65

[a] A distance modulus of 18.45 is adopted. Bolometric corrections are from Martins and Plez (2006). Photometry is from Hunter et al. (1995) for HST/WPFC2 and De Marchi et al. (2011) for HST/WPFC3. Extinction is from Crowther et al. (2010)

used an absolute visual magnitude MV = −6.73 and obtained $T_{\text{eff}} = 45,000$ K. According to the calibration of Vacca et al. (1996) based on models without line-blanketing, the corresponding bolometric correction is −4.17. Using Eq. 2.5 (for the optical range, see Martins and Plez 2006) we obtain $\log \frac{L}{L_\odot} = 6.26$, in perfect agreement with the value of 6.25 determined by de Koter et al. (1997). If we were using the lower effective temperature of Crowther and Dessart (1998) – $T_{\text{eff}} = 40,000$ K – one would get $\log \frac{L}{L_\odot} = 6.11$, close to 6.07 as derived by Crowther and Dessart (1998). If we now adopt the effective temperature derived by Crowther et al. (2010) ($T_{\text{eff}} = 53,000$ K) and use modern calibrations of bolometric corrections including the effects of line-blanketing (Martins and Plez 2006), we obtain BC = −4.48 and $\log \frac{L}{L_\odot} = 6.38$. Hence, better atmosphere models and temperature determinations contribute to an increase of about 0.2–0.3 dex in luminosity. But compared to the value obtained by Crowther et al. (2010) ($\log \frac{L}{L_\odot} = 6.58$) there is another 0.2 dex to explain. de Koter et al. (1997) adopted an extinction $A_V = 1.15$, using E(B-V) = 0.35 and a standard value of the ratio of selective to total extinction. Crowther et al. (2010) used the extinction determined by Fitzpatrick and Savage (1984) and obtained $A_V = 1.72$ for star a3. If we use this extinction, the absolute magnitude reaches −7.24 which, together with BC = −4.8 leads to $\log \frac{L}{L_\odot} = 6.59$, similar to Crowther et al. (2010). For star a3, the combination of new effective temperature, bolometric corrections, and a larger extinction explains the revised luminosity and thus initial mass. The results are summarized in Table 2.3 and Fig. 2.7 (right panel). Table 2.3 includes the optical photometry of Hunter et al. (1995) (HST/WFPC2) and that of De Marchi et al. (2011) (HST/WFPC3). Both are similar for star a3.

If we now turn to star R136c, the new stellar parameters together with optical HST photometry lead to $\log \frac{L}{L_\odot} = 6.65$. This is lower than the value of Crowther et al. (2010) by 0.1 dex, but still within the error bars. We also note that the WFPC2 and WFPC3 results are consistent. This is not true for stars a1 and a2. Using the

Table 2.4 Initial mass estimates for the R136a stars. $M_{C10}^{initial}$ and $M_{C10}^{current}$ are initial and current masses obtained from evolutionary tracks by Crowther et al. (2010). M_{G11} are upper limits from the homogeneous M-L relations of Gräfener et al. (2011) using a hydrogen mass fraction of 0.7 and the luminosities based on K-band, WFPC2 and WFPC3 photometry (see Table 2.3), assuming the distance and extinction of Crowther et al. (2010). All values are in $M_\odot$

Star	$M_{C10}^{initial}$	$M_{C10}^{current}$	M_{G11}^{K}	M_{G11}^{WFPC2}	M_{G11}^{WFPC3}
R136a1	320^{+100}_{-40}	265^{+80}_{-35}	372	230	336
R136a2	240^{+45}_{-45}	195^{+35}_{-35}	285	234	347
R136a3	165^{+30}_{-30}	135^{+25}_{-20}	206	200	203
R136c	220^{+55}_{-45}	175^{+40}_{-35}	271	223	231

WFPC2 magnitudes, we obtain $\log \frac{L}{L_\odot} = 6.65$ (6.66) for a1 (a2), while using WFPC3 photometry, one gets $\log \frac{L}{L_\odot} = 6.88$ (6.90) for a1 (a2). These values have to be compared to $\log \frac{L}{L_\odot} = 6.94$ and 6.78 obtained by Crowther et al. (2010). For both stars, WFPC2 photometry leads to lower luminosities than the K-band analysis of Crowther et al. (2010). The difference is significant for star a1. Luminosities are much higher if recent optical photometry is used. The results are now consistent for a1 (optical/K-band) while the optically derived luminosity of a2 is higher than the K-band luminosity. The changes in optical photometry might be due to crowding, stars a1 and a2 being separated by only 0.1".

In Table 2.4 we have gathered different mass estimates. The initial and current masses of Crowther et al. (2010) are shown in the first two columns. They are based on the dedicated evolutionary tracks presented in the study of Crowther et al. Masses can also be obtained from mass-luminosity relations. Gräfener et al. (2011) presented such relations for completely mixed stars. In that case, there is no gradient of molecular weight inside the star and the mass is maximum for a given luminosity. We have used their relation for a hydrogen mass fraction of 0.7 (slightly higher than the observed values) to provide *upper limits* on the current masses of the R136 stars. We have used the luminosities of Crowther et al. (2010) and those given in Table 2.3, based on WFPC2 and WPFC3 photometry. The results are shown in the last three columns of Table 2.4. The upper limits are consistent with the current mass estimates of Crowther et al. (2010). The impact of luminosity on the derived mass is clearly visible from Table 2.4: R136 a1 has an upper mass limit between 230 and 372 $M_\odot$ depending on its luminosity estimate.

The main conclusions regarding the revised R136 luminosities (and consequently masses) can be summarized as follows: (1) line-blanketed models lead to new effective temperatures and bolometric corrections which in turn contribute to an increase in luminosity; (2) higher visual extinction provides an additional source of luminosity increase; (3) photometric precision is important, especially in crowded regions. These three factors all affect the determination of luminosities, and thus initial masses, for stars in R136. Future studies of this region with high spatial resolution will certainly shed new light on the initial mass and nature of the R136 very massive components.

NGC 3603

NGC 3603 is a Galactic massive cluster dominated by three bright objects named A, B and C. The former is a known binary and will be described in Sect. 2.3. Stars B and C have been analyzed by Crowther and Dessart (1998) and revisited by Crowther et al. (2010). They are Wolf-Rayet stars of type WN6h.

Crowther and Dessart (1998) obtained $\log \frac{L}{L_\odot} = 6.16$ and 6.06 for stars B and C respectively. They did not use luminosities to infer stellar masses, but instead relied on the wind properties (Kudritzki et al. 1992). According to the radiatively driven wind theory, the wind terminal velocity is directly related to the escape velocity, which is itself related to the surface gravity. With the effective temperature and luminosity obtained from the spectroscopic analysis, the radius is known. Calculations of the terminal velocity in the framework of the radiatively driven wind theory and comparison to the observed velocities provides an estimate of the present stellar mass. Crowther and Dessart (1998) obtained M = 89 ± 12 (62 ± 8) for star B (C).

Crowther et al. (2010) used the more classical transformation of luminosities into stellar masses in the HR diagram. As for R136, they used atmosphere models including a proper treatment of line-blanketing effects. Consequently, the effective temperatures they derive are higher. They also take into account a larger extinction ($A_V \sim 4.7$ versus 3.8 for Crowther and Dessart) and a shorter distance (7.6 kpc versus 10.0 kpc, corresponding to distances modulus of 14.4 and 15.0 respectively). The effect of extinction and distance act in different directions, but the former is larger, so that in addition to the increased luminosity due to higher $T_{\rm eff}$, the luminosities obtained by Crowther et al. (2010) are higher than those of Crowther and Dessart (1998): $\log \frac{L}{L_\odot} = 6.46$ and 6.35 for B and C respectively. The present day masses (113 $M_\odot$ for both stars) and initial masses (166 ± 20 and $137^{+17}_{-14} M_\odot$) are higher than the wind masses quoted above. They were obtained using non-rotating tracks. Since evolutionary models including rotation have higher luminosities, lower initial masses would be determined if they were used.

The determination of the masses of the brightest objects in NGC 3603 illustrates once again the role of extinction and of better stellar parameters. But it also highlights the importance of distances. For the Arches cluster and R136, they were quite well constrained. For NGC 3603, the uncertainty is larger. Assuming the physical parameters of Crowther et al. (2010) for star B, but using a distance of 10.0 kpc (instead of 7.6) would lead to $\log \frac{L}{L_\odot} = 6.70$, 0.24 dex higher than reported by Crowther et al. The initial masses would then exceed 200 $M_\odot$. Accurate parallaxes hopefully provided by the *Gaia* mission will help to refine the mass estimates of NGC3603 B and C.

Other Candidates

There are a few stars that can be considered as candidates to have masses in excess of 100 $M_\odot$. Most of them are located in the Galactic Center. The first one is the Pistol star (Figer et al. 1998). As we have described in Sect. 2.1, it was once thought to have a luminosity between 1.5 and $4.0 \times 10^6 \, L_\odot$, corresponding to a mass larger than $200 \, M_\odot$. But the luminosity was revised by Najarro et al. (2009) so that the current initial mass of Pistol is thought to be closer to $100 \, M_\odot$. In addition, Martayan et al. (2011) reported the presence of a faint companion, calling for an additional downward revision of the mass of the brightest member.

If the Pistol star has lost its status of most luminous star in the Galactic Center, other objects have attracted attention in the last years. Barniske et al. (2008) studied two WN stars located between the Arches and central cluster, in relative isolation. Using 2MASS near-infrared and Spitzer mid-infrared photometry, they fitted the SED of these stars and determined $\log \frac{L}{L_\odot} = 6.3 \pm 0.3$ and 6.5 ± 0.2 for stars WR102c and WR102ka respectively. They adopted the extinction law of Rieke and Lebofsky (1985) and Lutz (1999), as compiled by Moneti et al. (2001). Keeping in mind the limitations described in Sect. 2.2.4, these two stars are comparable to the most luminous WNh stars in the Arches cluster, and could have initial masses between 100 and $150 \, M_\odot$.

Hamann et al. (2006) studied most of the Galactic WN stars and obtained luminosities larger than $2.0 \times 10^6 \, L_\odot$ for a few objects (WR24, WR82, WR85, WR131, WR147, WR158). All are H-rich WN stars, like all luminous stars in young massive clusters presented above. This seems to be a common characteristics of all suspected very massive stars: they have the appearance of WN stars, but contain a large hydrogen fraction. They are thus most likely core-H burning objects, still on the main sequence. Their Wolf-Rayet appearance is due to their high luminosity and consequently their strong winds.

Finally, Bestenlehner et al. (2011) reported on VFTS-682, a H-rich WN star located 29 pc away (in projection) from R136. Bestenlehner et al. rely on optical spectroscopy and optical-near/mid infrared photometry to constrain the stellar parameters. They determine an effective temperature of $54,500 \pm 3,000$ K. The luminosity estimate depends on the assumptions made regarding the extinction law. With a standard $R_V = 3.1$, they obtain $\log \frac{L}{L_\odot} = 5.7 \pm 0.2$. But the resulting SED does not match the 3.6 and 4.5 μm *Spitzer* photometry. To do so, the authors use a modified extinction law with $R_V = 4.7$. The SED is better reproduced and the luminosity increases to $\log \frac{L}{L_\odot} = 6.5 \pm 0.2$. The *Spitzer* 5.8 and 8.0 μm photometry remains unfitted, which is attributed to mid infrared excess possibly caused by circumstellar material. Bestenlehner et al. also argue that differences in the near infrared measurements between IRSF and 2MASS translates into an additional uncertainty of 0.1 dex on the luminosity. They favour the highest luminosity, which would correspond to a mass of about $150 \, M_\odot$. A luminosity of $10^{5.7} \, L_\odot$ would correspond to an initial mass of about $40 \, M_\odot$ (see Sect. 2.2.2).

In conclusion, there are several stars that can be considered good candidates for a VMS status. The best cases are located in the Aches, NGC 3603 and R136 clusters. However, we have highlighted the uncertainties affecting their luminosity determination and consequently their evolutionary masses. If masses in excess of $100 \, M_\odot$ are likely, values higher than $200 \, M_\odot$ are still subject to discussion and should be confirmed by new analysis.

2.3 Very Massive Stars in Binary Systems

In this section we focus on binary systems containing massive stars. After presenting the various types of binary stars, we recall some relations relating physical to observed properties of binary systems. Finally, we present the best cases for binary systems containing very massive stars.

2.3.1 Massive Binaries and Dynamical Masses

Types of Binaries

Mason et al. (1998) showed that two main types of binaries are detected: visual binaries and spectroscopic systems. The former are seen in imaging surveys. The two components of the systems are observed and astrometric studies covering several epochs can provide the orbital parameters, especially the separation and period. Visual binaries have periods larger than 100 years. The majority actually have periods of 10^4–10^6 years (corresponding to separations of the order of 0.1–10 pc). The reason is that massive binaries are located at large distances. Even with the best spatial resolution achievable today, only the systems with the widest separation can be detected. The other category of binaries is observed by spectroscopy. Due to the orbital motion of the two components around the center of mass, spectral lines are Doppler shifted. Radial velocities can be measured. If a periodicity is found in the RV curve, this is a strong indication of binarity. Spectroscopic binaries have short periods: from about 1 day to a few years, with a peak between 3 days and 1 month (separations of the order of a few tenths to one AU). This is again an observational bias: to detect significant variations of radial velocities the components have to be relatively close to each other, which implies short periods. Spectroscopic binaries cannot be spatially resolved. Hence, the spectrum collected by observations is a composite of the spectra of both components. Depending on the line strengths and the luminosity of the components, the spectral signatures of only one star or of both components can be observed. In the former (latter) case, the system is classified as SB1 (SB2). Spectroscopic binaries can sometimes experience eclipses if the inclination of the system is favourable. In that situation, the components periodically pass in front of each other, creating dips in the light curve. Such systems are the best to constrain masses as we will see below. Binary stars with periods between a few years and a century currently escape detection.

In the following we focus on spectroscopic binaries since all the suspected very massive stars in multiple systems discovered so far belong to this category.

Orbital Elements and Dynamical Masses

The first step to study a spectroscopic binary is to measure the radial velocities from spectral lines. The most widely used method consists in fitting Gaussian profiles to the observed lines. Such profiles are not able to account for the real shape of O stars lines, but it is a good approximation for their core, at least in the optical range. This is sufficient to measure a radial velocity (RV). Several lines are usually used and the final RV is the average of the individual RV measured on each line. An important limitation of this method is that it assumes that lines are formed in the photosphere at zero velocity with respect to the star's center of mass. For absorption lines of O stars, this is a good approximation. For stars with strong winds such as WNh stars, the observed lines will be mainly formed in the wind. Hence, they will have a non zero velocity due to the wind outflow. This can introduce a systematic offset in the radial velocity curve. Hopefully, the important quantity to constrain the mass of binary components is the amplitude of the RV curve, which is less sensitive to the above limitation. A way to avoid this problem is to use a cross-correlation method in which the spectrum of a single star with a spectral type similar to the components of the system is used to obtain RV.

The vast majority of spectroscopic binaries are studied using optical spectra. Hydrogen and helium lines are the main indicators, especially when the Gaussian fit method is employed. For the very massive stars we will describe below, nitrogen lines are also present and are included in the analysis. Near-infrared spectra are becoming available and are well suited to investigate the binarity of stars hidden behind several magnitudes of extinction. The K-band is the most commonly used spectral range in the near-infrared. Here again, hydrogen and helium lines are present. The typical width of absorption lines in O stars is 10–20 km s^{-1}. In order to correctly resolve them, at least ten points should be obtained throughout the line profile. This implies a minimum spectral resolution of $\sim$3,000. The higher the spectral resolution, the smaller the error on the RV measurement. For stars with strong winds, lines are broad and high spectral resolution is not required.

Once the radial velocity measurements are obtained, a period search can be performed using time series analysis. If a signal is detected, radial velocities can be phased to create the type of curves presented in Fig. 2.8. In a SB2 system, the RV changes of both components are seen. They are anti-correlated. In a SB1 system, only the radial velocity variations of the brightest star are seen. According to Kepler's laws, the semi-amplitude of the RV variations (K) is related to the orbital elements as follows:

$$K = (2\pi G)^{1/3} \left(\frac{M}{T}\right)^{1/3} \frac{q}{(1+q)^{2/3}} \frac{\sin i}{\sqrt{1-e^2}} \quad (2.6)$$

where G is the constant of gravitation, M the mass of the star, T the period, $q = M_2/M_1$ the ratio of the secondary to primary mass, i the inclination of the system and e its eccentricity. For a SB1 system, we can reprocess this relation to obtain the so-called mass function

$$f(m) = \frac{M \sin^3 i}{(1+q)^2} = \frac{K^3}{2\pi G}(1-e^2)^{3/2} T \quad (2.7)$$

This is not sufficient to constrain the mass of the system unless assumptions on the mass ratio are made. On the contrary, for a SB2 system we can use Kepler first law to derive

$$M_1 \sin^3 i = \frac{T}{2\pi G}(K_1 + K_2)K_2(1 - e^2)^{3/2} \qquad (2.8)$$

$$M_2 \sin^3 i = \frac{T}{2\pi G}(K_1 + K_2)K_1(1 - e^2)^{3/2} \qquad (2.9)$$

from which we immediately see that the mass ratio M_1/M_2 is the inverse of the RV amplitude ratio K_2/K_1.

With these expressions, it is possible to fit the observed RV curves and to estimate lower limits on the components masses. These masses are called "dynamical masses". The shape of the RV curve provides a first guess of the eccentricity. For a circular orbit, it should be a perfect sine curve. For more eccentric orbits, the RV curve becomes asymmetric with a maximum absolute value of the radial velocity concentrated in a short period of time corresponding to periastron passage. In Fig. 2.8 the observations are consistent with a zero eccentricity. The ratio of the semi-amplitude K_1/K_2 is the inverse of the mass ratio. Consequently, a look at the RV curve already provides information on the mass ratio of the system: for equal mass binaries, the amplitude of the radial velocity variations of the primary and of the secondary stars are the same. From the above equations, it appears that the determination of $M\sin^3 i$ is essentially model independent. The main limitations come from the measurement of the radial velocities.

To obtain dynamical masses and not just lower limits, one needs to determine the inclination of the system. This is possible for eclipsing binaries. A photometric

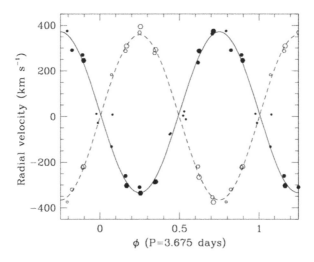

Fig. 2.8 Radial velocity curve of the binary system WR20a. *Filled (open) symbols* are for the primary (secondary) (From Rauw et al. (2004). Reproduced with permission)

monitoring over several periods provides the light curve, i.e. the evolution of the system's magnitude as a function of time. For eclipsing binaries, minima are observed in the light curve. They correspond to the physical situation where one star passes in front of the other and (partly) blocks its light. The shape of the light curve in the eclipse phases can be quite complex depending on the geometry of the system. For stars relatively separated and experiencing complete eclipses, a plateau is observed in the minimum of the light curve. If the eclipse is only partial, no plateau is seen. For close binaries, interactions between components affect the shape of the stars and their light. Tidal forces cause elongations along the binary axis. Light of one component can be reflected and/or heat the surface of the companion, making more complex the shape of the eclipses in the light curve. In the most extreme cases of contact binaries, material is exchanged between both stars which affects the system's photometry. All these processes render the fit of the light curve more difficult and more mode-dependent than the RV curve. Even in the simplest case of detached, non-interacting binaries, a limb darkening model has to be adopted to reproduce the eclipses. In addition, assumptions on the luminosity ratio are usually necessary to obtain a complete solution. Under these assumptions, it is possible to constrain the inclination of the system and thus, from Eq. 2.9, the dynamical masses.

In order to refine the analysis of RV and light curves, a better knowledge of the physical properties of the stars is useful. The method of spectral disentangling consists in the separation of the individual spectra from the combined, observed spectrum of the system. The principle is rather simple: subtract the spectrum of one star from the combined spectrum. In practice, this method requires again a good knowledge of the luminosity ratio of both components in order to evaluate their respective contribution. The radial velocity amplitudes have to be large enough so that the two spectra are well separated. A good sampling of the orbit is necessary. Under these conditions, it is possible to extract the individual spectra and perform spectral classification. A spectroscopic analysis of each star can be attempted, providing additional stellar parameters. Like for single stars, evolutionary and spectroscopic masses can be obtained and directly compared to dynamical masses (e.g. Mahy et al. 2011).

Provided that eclipses are observed, it is thus possible to obtain dynamical masses for massive binaries. The analysis of the RV curve is affected mainly by uncertainties in the measurements of radial velocities, while the interpretation of the light curve is much more model dependent. With these limitations in mind, we now present a few example of massive binary systems potentially hosting VMS.

2.3.2 The Most Massive Binary Systems

Now that the basic ingredients of mass determination in binary systems have been recalled, we turn to the presentation of some of the most interesting systems potentially hosting very massive stars.

WR20a

WR20a is located in the vicinity of the massive cluster Westerlund 2. Rauw et al. (2004) presented the optical spectrum of the system and first radial velocity measurements. They used Gaussian profiles to fit the main emission lines. The main difficulty they encountered was that the two components have the same spectral type: WN6ha. Hence, a period search was performed in the absolute difference of radial velocities ($| RV_1 - RV_2 |$). The preferred value of the period was 3.675 d, with a second possible value at 4.419 d. Assuming a zero eccentricity orbit (from the shape of the RV curve), Rauw et al. (2004) determined minimum masses of 70.7 ± 4.0 and $68.8 \pm 3.8\,M_\odot$ for the primary and secondary.

At the same time, Bonanos et al. (2004) collected I-band photometry for WR20a. The lightcurve showed clear minima, revealing the eclipsing nature of the system. They revised the period determination: 3.686 d. Performing a fit of the light curve, Bonanos et al. (2004) determined an inclination of $74°.5 \pm 2°.0$. Combined with the radial velocity measurements of Rauw et al. (2004), Bonanos et al. concluded that the current masses of the two components are 83.0 ± 5.0 and $82.0 \pm 5.0\,M_\odot$. The very similar masses are consistent with both stars having the same spectral type, and with the absence of strong asymmetry in the light curve (all eclipses have almost the same depth). Rauw et al. (2005) subsequently determined the stellar and wind parameters of the system and obtained an effective temperature of 43,000 K and a luminosity of $1.15 \times 10^6\,L_\odot$ for each component. The effective temperature is consistent with that assumed by Bonanos et al. (2004) for their light curve fit. Using the evolutionary tracks of Meynet and Maeder (2003), the stellar parameters corresponds to present mass of $71\,M_\odot$ and an initial mass of $84\,M_\odot$. Given the uncertainties related to evolutionary tracks described in Sect. 2.2.3, the agreement is good.

Although not a very massive star per see, WR20a is one of the binary star with the best determination of dynamical masses.

NGC 3603 A1

In Sect. 2.2.4 we have shown that NGC 3603 contained two stars (B and C) with high luminosities and good candidates for having initial masses in excess of $100\,M_\odot$. A third object lies in the core of NGC 3603: star A1. Using unresolved spectroscopy, Moffat and Niemela (1984) showed that object HD 97950, which includes all three stars A1, B and C, was variable in radial velocity. A period of 3.77 days could be identified. Subsequent investigation of NGC 3603 with HST by Moffat et al. (2004) revealed photometric variability with the same period. With the improved spatial resolution, the origin of the variability could be attributed to star A1 which was then classified as a double-line eclipsing binary (SB2). Moffat et al. (2004) performed a series of light curve fits assuming various values for the mass ratio of both components. They found that an inclination of 71° best reproduced their light curve. This inclination was obtained for a mass ratio between 0.5 and 2.0 and for an effective temperature of the secondary (primary) of 43,000 K (46,000 K).

Schnurr et al. (2008) used adaptive optics integral field spectroscopy in the K-band to monitor the radial velocity variations of A1. They showed that Brγ and HeII 2.189 μm were both in emission and were double peaked, with one peak stronger than the other. They concluded that the system was most likely formed by a WN6ha star and a less luminous O star. They determined radial velocities from Gaussian fits to the emission lines. Measurements of the secondary radial velocity variations were difficult when lines from the primary and secondary were blended and dominated by the primary's emission. This resulted in large uncertainties on the secondary RV curve and consequently on the dynamical mass estimates. Schnurr et al. adopted the inclination and period of Moffat et al. (2004) as well as a zero eccentricity and calculated an orbital solution. They obtained M = 116 ± 31 M$_\odot$ and M = 89 ± 16 M$_\odot$ for the primary and secondary respectively. The mass ratio is thus 1.3 ± 0.3.

Crowther et al. (2010) performed a spectroscopic study of the integrated spectrum of A1 at maximum separation of the two components. They found $T_{\text{eff}} = 42{,}000$ K for the primary and $T_{\text{eff}} = 40{,}000$ K for the secondary. This is lower than the values adopted by Moffat et al. (2004) for their light curve solution. Revisiting the inclination determination with these new temperatures would be useful. Crowther et al. also constrained the luminosity of the components and used non-rotating evolutionary tracks to estimate the current mass of the stars: 120^{+26}_{-17} and 92^{+16}_{-15} M$_\odot$. These values are in very good agreement with the dynamical masses. But here again, current masses determined with evolutionary tracks including rotation would have been lower.

R144 and R145

R144 and R145 are two H-rich WN stars located in the LMC. Moffat (1989) reported variability in R145 and tentatively derived a periodicity of 25.4 days, indicating a binary nature. Schnurr et al. (2009) monitored the spectroscopic variability of R145 and confirmed that most lines display changes in both shape and position. They observed mainly one set of line, thus one of the components. Its spectral is similar to other massive binaries: WN6h. Using a method based on the removal of the averaged primary spectrum in each individual spectra, they identified weak HeII features typical of O stars. Hence, they concluded that the secondary should be a less massive object than the primary. This was confirmed by estimates of the visible flux level and the study of the strength of the companion HeII lines in comparison with template spectra of O stars. The primary should be about 3 times more massive than the secondary.

In addition to spectroscopic observations, Schnurr et al. (2009) also obtained simultaneous polarimetric data in which they clearly detected Zeeman signatures in the Q and U profiles. Interestingly, these profiles also showed variability. A clear periodicity of 159 d could be identified. Using this period, they obtained an orbital solution for the radial velocity curve yielding $Msin^3 i = 116 \pm 33\,M_\odot$ ($48 \pm 20\,M_\odot$) for the primary (secondary). The mass ratio of 2.4 ± 1.2 was in reasonable agreement with that determined from the visible continuum flux. In a further step, Schnurr et al. (2009) performed a combined fit of the radial velocity and polarimetric curves and obtained $i = 38° \pm 9°$ (inclination of the system). With such an inclination, the primary and secondary mass should be larger than 300 and $125\,M_\odot$. These masses are inconsistent with the systems brightness though, questioning the inclination determination.

R144 was also monitored by Schnurr et al. (2009), but no periodicity was detected. Sana et al. (2013) presented new spectroscopic observations. They detected line shifts in NIII, NIV and NV lines. Interestingly, the NIII and NV shifts are anti-correlated, pointing towards a different origin for both types of lines. Sana et al. (2013) concluded that the optical spectra are the composite of the spectra of two types of stars: one WN5-6h mainly contributing to NV lines, and one WN6-7h star dominating the NIII features. A period search indicated a variability on a timescale of 2–12 months, without better constraint. Based on the visual photometry and assumptions regarding the bolometric correction of the components, Sana et al. (2013) estimated $\log \frac{L}{L_\odot} \sim 6.8$. This places R144 among the potential very massive stars.

For both R144 and R145, additional spectroscopic and photometric observations are required to better characterize the system's component. In particular, spectral disentangling would be helpful to constrain the nature of these massive systems.

In conclusion, there are relatively few binary systems potentially hosting VMS. The best cases would benefit for additional combined RV and light curve analysis (when eclipses are present) to refine the dynamical mass estimates. NGC 3603 A1 appears to be the best candidate.

Summary and Conclusions
In this chapter we have presented the observational evidence for the existence of very massive stars. Mass estimates of single stars are mainly obtained from the conversion of luminosities to evolutionary masses. We have highlighted the uncertainties in the determination of luminosities: crowding, accurate photometry, distance, extinction, atmosphere models all contribute to render uncertain luminosity estimates. The other source of error comes from evolutionary tracks. Different calculations produce different outputs depending on the assumptions they are built on. Even if the luminosity was perfectly well constrained, its transformation to masses relies on the predictions of evolutionary models. With these limitations in mind, there are several stars that can be considered as good candidates for a VMS status. They are manily located in the massive young clusters NGC 3603, R136 and the Arches. In R136, the brightest members may reach initial masses higher than $200\,M_\odot$.

(continued)

In the other two clusters, masses between 150 and 200 $M_\odot$ are not excluded. A few binary systems may also host stars with masses in excess of 100 $M_\odot$. NGC 3603 A1 seems to be the best candidate, with a M $\sim$ 115 $M_\odot$ primary star, but a better analysis of the light curve is needed to refine the analysis.

In conclusion, very massive stars may be present in our immediate vicinity. They usually look like WN5-9h stars, i.e. hydrogen rich mid to late WN stars. Super star clusters – the best places to look for VMS – being impossible to resolve with the current generation of instruments, these local VMS have to be re-observed and re-analyzed in order to minimize the uncertainties involved in their mass determination. This is important to understand the upper end of the initial mass function and the formation process of massive stars in general.

Acknowledgements The author thanks Paul Crowther for discussions on the mass determination of the R136 stars.

References

Barniske, A., Oskinova, L. M., & Hamann, W.-R. (2008). *Astronomy and Astrophysics, 486*, 971.
Bastian, N., Saglia, R. P., Goudfrooij, P., et al. (2006). *Astronomy and Astrophysics, 448*, 881.
Bessell, M. S., Castelli, F., & Plez, B. (1998). *Astronomy and Astrophysics, 333*, 231.
Bestenlehner, J. M., Vink, J. S., Gräfener, G., et al. (2011). *Astronomy and Astrophysics, 530*, L14.
Bonanos, A. Z., Stanek, K. Z., Udalski, A., et al. (2004). *Astrophysical Journal Letters, 611*, L33.
Bouret, J.-C., Lanz, T., & Hillier, D. J. (2005). *Astronomy and Astrophysics, 438*, 301.
Bouret, J.-C., Lanz, T., Martins, F., et al. (2013). *Astronomy and Astrophysics, 555*, A1.
Bromm, V., Coppi, P. S., & Larson, R. B. (1999). *Astrophysical Journal Letters, 527*, L5.
Brott, I., de Mink, S. E., Cantiello, M., et al. (2011). *Astronomy and Astrophysics, 530*, A115.
Burkholder, V., Massey, P., & Morrell, N. (1997). *Astrophysical Journal, 490*, 328.
Campbell, M. A., Evans, C. J., Mackey, A. D., et al. (2010). *Monthly Notices of the Royal Astronomical Society, 405*, 421.
Cassinelli, J. P., Mathis, J. S., & Savage, B. D. (1981). *Science, 212*, 1497.
Castor, J. I., Abbott, D. C., & Klein, R. I. (1975). *Astrophysical Journal, 195*, 157.
Chené, A.-N., Borissova, J., Bonatto, C., et al. (2013). *Astronomy and Astrophysics, 549*, A98.
Cotera, A. S., Erickson, E. F., Colgan, S. W. J., et al. (1996). *Astrophysical Journal, 461*, 750.
Crowther, P. A., & Dessart, L. (1998). *Monthly Notices of the Royal Astronomical Society, 296*, 622.
Crowther, P. A., Schnurr, O., Hirschi, R., et al. (2010). *Monthly Notices of the Royal Astronomical Society, 408*, 731.
de Koter, A., Heap, S. R., Hubeny, I. (1997). *Astrophysical Journal, 477*, 792.
De Marchi, G., Paresce, F., Panagia, N., et al. (2011). *Astrophysical Journal, 739*, 27.
Eisenhauer, F., Genzel, R., Alexander, T., et al. (2005). *Astrophysical Journal, 628*, 246.
Ekström, S., Georgy, C., Eggenberger, P., et al. (2012). *Astronomy and Astrophysics, 537*, A146.
Espinoza, P., Selman, F. J., & Melnick, J. (2009). *Astronomy and Astrophysics, 501*, 563.
Feitzinger, J. V., Schlosser, W., Schmidt-Kaler, T., & Winkler, C. (1980). *Astronomy and Astrophysics, 84*, 50.
Figer, D. F. (2005). *Nature, 434*, 192.
Figer, D. F., Kim, S. S., Morris, M., et al. (1999). *Astrophysical Journal, 525*, 750.

Figer, D. F., MacKenty, J. W., Robberto, M., et al. (2006). *Astrophysical Journal, 643*, 1166.
Figer, D. F., Najarro, F., Gilmore, D., et al. (2002). *Astrophysical Journal, 581*, 258.
Figer, D. F., Najarro, F., Morris, M., et al. (1998). *Astrophysical Journal, 506*, 384.
Fitzpatrick, E. L., & Savage, B. D. (1984). *Astrophysical Journal, 279*, 578.
Fullerton, A. W., Massa, D. L., Prinja, R. K. (2006). *Astrophysical Journal, 637*, 1025.
Gillessen, S., Eisenhauer, F., Trippe, S., et al. (2009). *Astrophysical Journal, 692*, 1075.
Gräfener, G., Vink, J. S., de Koter, A., & Langer, N. (2011). *Astronomy and Astrophysics, 535*, A56.
Hamann, W.-R., & Gräfener, G. (2004). *Astronomy and Astrophysics, 427*, 697.
Hamann, W.-R., Gräfener, G., Liermann, A. (2006). *Astronomy and Astrophysics, 457*, 1015.
Herrero, A., Kudritzki, R. P., Vilchez, J. M., et al. (1992). *Astronomy and Astrophysics, 261*, 209.
Heydari-Malayeri, M., & Hutsemekers, D. (1991). *Astronomy and Astrophysics, 243*, 401.
Heydari-Malayeri, M., Remy, M., & Magain, P. (1988). *Astronomy and Astrophysics, 201*, L41.
Heydari-Malayeri, M., Remy, M., Magain, P. (1989). *Astronomy and Astrophysics, 222*, 41.
Hillier, D. J., & Miller, D. L. (1998). *Astrophysical Journal, 496*, 407.
Hosokawa, T., Omukai, K., Yoshida, N., & Yorke, H. W. (2011). *Science, 334*, 1250.
Howarth, I. D. (1983). *Monthly Notices of the Royal Astronomical Society, 203*, 301.
Humphreys, R. M. (1983). *Astrophysical Journal, 269*, 335.
Humphreys, R. M., & Davidson, K. (1994). *Publications of the Astronomical Society of the Pacific, 106*, 1025.
Hunter, D. A., Shaya, E. J., Holtzman, J. A., et al. (1995). *Astrophysical Journal, 448*, 179
Krumholz, M. R., Klein, R. I., McKee, C. F., Offner, S. S. R., & Cunningham, A. J. (2009). *Science, 323*, 754.
Kudritzki, R. P., Cabanne, M. L., Husfeld, D., et al. (1989). *Astronomy and Astrophysics, 226*, 235.
Kudritzki, R.-P., Hummer, D. G., Pauldrach, A. W. A., et al. (1992). *Astronomy and Astrophysics, 257*, 655.
Lanz, T., & Hubeny, I. (2003). *Astrophysical Journal Supplement, 146*, 417.
Lutz, D. (1999). In P. Cox & M. Kessler (Eds.), *ESA special publication* (Vol. 427, p. 623). The universe as seen by ISO.
Mahy, L., Martins, F., Machado, C., Donati, J.-F., Bouret, J.-C. (2011). *Astronomy and Astrophysics, 533*, A9.
Martayan, C., Blomme, R., Le Bouquin, J.-B., et al. (2011). In C. Neiner, G. Wade, G. Meynet, & G. Peters, (Eds.), *IAU symposium* (Vol. 272, pp. 616–617). IAU Symposium.
Martins, F., Hillier, D. J., Paumard, T., et al. (2008). *Astronomy and Astrophysics, 478*, 219.
Martins, F., & Palacios, A. (2013). *Astronomy and Astrophysics, 560*, A16.
Martins, F., & Plez, B. (2006). *Astronomy and Astrophysics, 457*, 637.
Martins, F., Schaerer, D., Hillier, D. J. (2005). *Astronomy and Astrophysics, 436*, 1049.
Mason, B. D., Gies, D. R., Hartkopf, W. I., et al. (1998). *Astronomical Journal, 115*, 821.
Massey, P., & Hunter, D. A. (1998). *Astrophysical Journal, 493*, 180.
Massey, P., Morrell, N. I., Neugent, K. F., et al. (2012). *Astrophysical Journal, 748*, 96.
Mauron, N., & Josselin, E. (2011). *Astronomy and Astrophysics, 526*, A156.
Mengel, S., Lehnert, M. D., Thatte, N., Genzel, R. (2002). *Astronomy and Astrophysics, 383*, 137.
Meynet, G., & Maeder, A. (2000). *Astronomy and Astrophysics, 361*, 101.
Meynet, G., & Maeder, A. (2003). *Astronomy and Astrophysics, 404*, 975.
Meynet, G., & Maeder, A. (2005). *Astronomy and Astrophysics, 429*, 581.
Meynet, G., Maeder, A., Schaller, G., Schaerer, D., Charbonnel, C. (1994). *Astronomy and Astrophysics Supplement, 103*, 97.
Moffat, A. F. J. (1989). *Astrophysical Journal, 347*, 373.
Moffat, A. F. J. & Niemela, V. S. (1984). *Astrophysical Journal, 284*, 631.
Moffat, A. F. J., Poitras, V., Marchenko, S. V., et al. (2004). *Astronomical Journal, 128*, 2854.
Mokiem, M. R., de Koter, A., Evans, C. J., et al. (2007). *Astronomy and Astrophysics, 465*, 1003.
Moneti, A., Stolovy, S., Blommaert, J. A. D. L., Figer, D. F., Najarro, F. (2001). *Astronomy and Astrophysics, 366*, 106.

Najarro, F., Figer, D. F., Hillier, D. J., Geballe, T. R., Kudritzki, R. P. (2009). *Astrophysical Journal, 691*, 1816.
Nishiyama, S., Tamura, M., Hatano, H., et al. (2009). *Astrophysical Journal, 696*, 1407.
Oey, M. S., & Clarke, C. J. (2005). *Astrophysical Journal Letters, 620*, L43.
Puls, J., Urbaneja, M. A., Venero, R., et al. (2005). *Astronomy and Astrophysics, 435*, 669.
Rauw, G., Crowther, P. A., De Becker, M., et al. (2005). *Astronomy and Astrophysics, 432*, 985.
Rauw, G., De Becker, M., Nazé, Y., et al. (2004). *Astronomy and Astrophysics, 420*, L9.
Rieke, G. H., & Lebofsky, M. J. (1985). *Astrophysical Journal, 288*, 618.
Rivero González, J. G., Puls, J., Massey, P., Najarro, F. (2012). *Astronomy and Astrophysics, 543*, A95.
Sana, H., van Boeckel, T., Tramper, F., et al. (2013). *Monthly Notices of the Royal Astronomical Society, 432*, L26.
Schnurr, O., Casoli, J., Chené, A.-N., Moffat, A. F. J., St-Louis, N. (2008). *Monthly Notices of the Royal Astronomical Society, 389*, L38.
Schnurr, O., Chené, A.-N., Casoli, J., Moffat, A. F. J., & St-Louis, N. (2009). *Monthly Notices of the Royal Astronomical Society, 397*, 2049.
Seaton, M. J. (1979). *Monthly Notices of the Royal Astronomical Society, 187*, 73.
Stolte, A., Grebel, E. K., Brandner, W., & Figer, D. F. (2002). *Astronomy and Astrophysics, 394*, 459.
Utrobin, V. P. (1984). *Astrophysics and Space Science, 98*, 115.
Vacca, W. D., Garmany, C. D., & Shull, J. M. (1996). *Astrophysical Journal, 460*, 914.
Weidner, C., & Kroupa, P. (2004). *Monthly Notices of the Royal Astronomical Society, 348*, 187.
Weidner, C., & Kroupa, P. (2006). *Monthly Notices of the Royal Astronomical Society, 365*, 1333.
Weidner, C., Kroupa, P., & Bonnell, I. A. D. (2010). *Monthly Notices of the Royal Astronomical Society, 401*, 275.
Weidner, C., & Vink, J. S. (2010). *Astronomy and Astrophysics, 524*, A98.
Weigelt, G., & Baier, G. (1985). *Astronomy and Astrophysics, 150*, L18.
Yusof, N., Hirschi, R., Meynet, G., et al. (2013). *Monthly Notices of the Royal Astronomical Society, 433*, 1114.

Chapter 3
The Formation of Very Massive Stars

Mark R. Krumholz

Abstract In this chapter I review theoretical models for the formation of very massive stars. After a brief overview of some relevant observations, I spend the bulk of the chapter describing two possible routes to the formation of very massive stars: formation via gas accretion, and formation via collisions between smaller stars. For direct accretion, I discuss the problems of how interstellar gas may be prevented from fragmenting so that it is available for incorporation into a single very massive star, and I discuss the problems presented for massive star formation by feedback in the form of radiation pressure, photoionization, and stellar winds. For collision, I discuss several mechanisms by which stars might be induced to collide, and I discuss what sorts of environments are required to enable each of these mechanisms to function. I then compare the direct accretion and collision scenarios, and discuss possible observational signatures that could be used to distinguish between them. Finally, I come to the question of whether the process of star formation sets any upper limits on the masses of stars that can form.

3.1 Introduction

The mechanism by which the most massive stars form, and whether there is an upper limit to the mass of star that this mechanism can produce, has been a problem in astrophysics since the pioneering works of Larson and Starrfield (1971) and Kahn (1974). These authors focused on the physical mechanisms that might inhibit accretion onto stars as they accreted interstellar matter, and we will return to this topic below. However, a more modern approach to the problem of very massive stars requires placing them in the context of a broader theory of the stellar initial mass function (IMF).

The IMF is characterized by a peak in the range 0.1–1 $M_\odot$, and a powerlaw tail at higher masses of the form $dn/d\ln m \propto m^\Gamma$ with $\Gamma \approx -1.35$ (Bastian et al. 2010, and references therein). However, the mass to which this simple powerlaw

---

Mark R. Krumholz (✉)
Department of Astronomy & Astrophysics, University of California, Santa Cruz, CA 95064 USA
e-mail: mkrumhol@ucsc.edu

extends is not very well-determined. It is not possible to measure the IMF for field stars to very high masses due to uncertainties in star formation histories and the limited number of very massive stars available in the field. Measurements of the high-mass end of the IMF in young clusters must target very massive systems in order to achieve strong statistical significance, and such clusters are rare and thus distant. This creates significant problems with confusion. The limited studies that are available suggest that the a powerlaw with $\Gamma \approx -1.35$ remains a reasonable description of the IMF out to masses of $\sim 100\, M_\odot$ or more (e.g. Massey and Hunter 1998; Kim et al. 2006; Espinoza et al. 2009). However, it is by no means implausible that there are hidden features lurking in the IMF at the highest masses. Indeed some analyses of the IMF have claimed to detect an upper cutoff (see the Chapter by F. Martins in this volume for a critical review).

This observational question of whether the most massive stars are simply the "tip of the iceberg" of the normal IMF, or whether they represent a fundamentally distinct population, animates the theoretical question about how such stars form. The two dominant models for how massive stars form are formation by accretion of interstellar material, i.e. the same mechanism by which stars of low mass form, and formation by collisions between lower mass stars, which would represent a very different formation mechanism from the bulk of the stellar population.[1] In the remainder of this chapter, I review each of these models in turn (Sects. 3.2 and 3.3), pointing out its strengths, weaknesses, and areas of incompleteness. I then discuss the observable predictions made by each of these models, and which might be used to discriminate between them (Sect. 3.4). Finally, I summarize and return to the question first raised by Larson and Starrfield (1971) and Kahn (1974): is there an upper mass limit for star formation, and if so, why (section "Conclusions and Summary: Does Star Formation Have an Upper Mass Limit?")?

3.2 The Formation of Very Massive Stars by Accretion

The great majority of stars form via the collapse of cold, gravitationally-unstable, molecular gas, and the subsequent accretion of cold gas onto the protostellar seeds that the collapse produces (McKee and Ostriker 2007, and references therein). There are numerous competing models for the origin of the observed $\Gamma \approx -1.35$ slope (e.g. Bonnell et al. 2001; Padoan and Nordlund 2002; Padoan et al. 2007; Hennebelle and Chabrier 2008, 2009, 2013; Krumholz et al. 2011, 2012; Hopkins 2012), but in essentially all of these models, the massive end of this tail is populated

---

[1]Mergers between two members of a tight binary as a result of the growth of stellar radii during main sequence or post-main sequence evolution, or as a result of secular interactions in hierarchical triples, is a third possible mechanism by which massive stars can and probably do gain mass (Sana et al. 2012; de Mink et al. 2013; Moeckel and Bonnell 2013). However, I do not discuss this possibility further, because it provides at most a factor of 2 increase in stellar mass.

by stars forming in rare, high-density regions that provide at least the potential for large mass reservoirs to be accreted onto protostellar seeds at high rates. Some but not all of these models identify the regions that give rise to massive stars with observed "cores": compact ($\sim$0.01 pc), dense ($>10^5$ molecules cm^{-3}) regions of gas, the largest of which can have masses large enough to form very massive stars (e.g.,. Beuther and Schilke 2004; Beuther et al. 2005; Bontemps et al. 2010). None of these models predict that there is an upper limit to the masses of either cores or stars, and there is no observational evidence of a truncation either. Thus, it would seem that there is no barrier in terms of mass supply to the formation of very massive stars via the same accretion processes that give rise to the remainder of the IMF. However, the fact that there is a large supply of mass available does not guarantee that it can actually accrete onto a single object and thereby produce a very massive star. There are four major challenges to getting the available interstellar mass into a star, which we discuss below: fragmentation, radiation pressure, photoionization, and stellar winds. I discuss each of these challenges in turn in the remainder of this section.

3.2.1 *Fragmentation*

The first challenge, fragmentation, can be stated very simply. When gravitationally-unstable media collapse, they tend to produce objects with a characteristic mass comparable to the Jeans mass,

$$M_J = \frac{\pi}{6} \frac{c_s^3}{\sqrt{G^3 \rho}} = 0.5 \left(\frac{T}{10\,\text{K}}\right)^{3/2} \left(\frac{n}{10^4\,\text{cm}^{-3}}\right)^{1/2} M_\odot, \qquad (3.1)$$

where c_s is the sound speed, ρ is the gas density, T is the gas temperature, and n is the gas number density. The temperature and density values to which I have scaled in the above equation are typical values in star-forming regions. Clearly, a massive star is an object whose mass is far larger than the Jeans mass of the interstellar gas from which it is forming. Why, then, does this gas not fragment into numerous small stars rather than forming a single large one? Indeed, hydrodynamic simulations of the collapse of compact, massive regions such as the observed massive cores show that they generally fail to produce massive stars (Dobbs et al. 2005), and larger-scale simulations of star cluster formation appear to produce mass functions that are better described by truncated powerlaws than pure powerlaws (e.g. Maschberger et al. 2010), and where the formation of the most massive stars is limited by "fragmentation-induced starvation" (Peters et al. 2010b; Girichidis et al. 2012).

While these results might seem to present a serious challenge to the idea that massive stars form by accretion, they are mostly based on simulations that include no physics other than hydrodynamics and gravity. More recent simulations including a wider range of physical processes suggest that the fragmentation problem is much less severe than was once believed. Fragmentation is reduced by two primary effects: radiation feedback and magnetic fields.

Radiation feedback works to reduce fragmentation by heating the gas, raising its pressure and thus its Jeans mass (cf. Eq. 3.1). Although massive stars can of course produce a tremendous amount of heating, the more important effect from the standpoint of suppressing fragmentation is the early feedback provided by low mass stars, whose luminosities are dominated by accretion rather than internal energy generation. Krumholz (2006) first pointed out the importance of this effect, showing that even a $1\,M_\odot$ star accreting at the relatively high rates expected in the dense regions where massive stars form could radiate strongly enough to raise the gas temperature by a factor of a few at distances of $\approx 1,000$ AU. Since the minimum fragment mass is roughly the Jeans mass, and this varies as temperature to the $3/2$ power (Eq. 3.1), this effect raises the minimum mass required for gas to fragment by a factor of ≈ 10. Subsequent radiation-hydrodynamic simulations by a number of authors (Krumholz et al. 2007, 2010, 2011; Bate 2009, 2012; Offner et al. 2009b) have confirmed that radiation feedback dramatically suppresses fragmentation compared to the results obtained in purely hydrodynamic models. Krumholz and McKee (2008) argue that this effect will efficiently suppress fragmentation in regions of high column density, allowing massive stars to form without their masses being limited by fragmentation. In contrast, Peters et al. (2010b) find that fragmentation limits the growth of massive stars even when heating by direct stellar photons is included, but their simulations do not include the dust-reprocessed radiation field that is likely more important for regulating fragmentation, and are limited to regions of much lower density than the typical environment of massive star formation.

Magnetic fields limit fragmentation in two ways. First, they remove angular momentum. In a collapsing cloud, the densest regions collapse fastest, and as the gas falls inward it attempts to rotate faster and faster in order to conserve angular momentum. When the collapsing gas is threaded by a magnetic field, however, the resulting differential rotation between inner collapsing regions and outer ones that have not yet begun to collapse twists the magnetic field lines. The twist produces a magnetic tension force that transfer angular momentum from the inner to the outer regions, a process known as magnetic braking. Formally, for an axisymmetric flow, one can show (e.g., Stahler and Palla 2005) that the time rate of change of the angular momentum of a fluid element at a distance ϖ from the rotation axis due to magnetic forces is given by

$$\frac{\partial J}{\partial t} = \frac{1}{4\pi}\left[B_\varpi \frac{\partial}{\partial \varpi}(\varpi B_\phi) + \varpi B_z \frac{\partial}{\partial z} B_\phi\right] \tag{3.2}$$

where $\mathbf{B}$ is the magnetic field, and we have used cylindrical coordinates such that the components of $\mathbf{B}$ are (B_ϖ, B_ϕ, B_z). Thus in general if the toroidal (ϕ) component of the magnetic field varies with either radial or vertical position, and the field also has a poloidal (ϖ or z) component, there will be a magnetic torque that alters the angular momentum of the gas. For the types of magnetic field configurations produced by collapse, the net effect is to transport angular momentum outward. This process inhibits the formation of rotationally-flattened structures (e.g. accretion disks). This is significant from the standpoint of fragmentation, because rotational flattening raises the density of the gas as it approaches the star, and dense, rotationally-flattened structures are vulnerable to the Toomre instability (see below), in which the self-gravity of a flattened rotating structure overcomes support from thermal pressure and angular momentum, leading to fragmentation and collapse.

Second, magnetic fields provide extra pressure support that prevents regions from collapsing unless their magnetic flux to mass ratios are below a critical value

$$\left(\frac{\Phi}{M}\right)_{\text{crit}} = (4\pi^2 G)^{1/2}. \quad (3.3)$$

Regions with masses small enough such that $\Phi/M < (\Phi/M)_{\text{crit}}$ are said to be magnetically sub-critical, meaning that they do not have enough mass to overcome magnetic pressure support and collapse. Observations indicate that star-forming cores, over a wide range of size and density scales, tend to have flux to mass ratios that are roughly uniformly distributed from 0 up to $(\Phi/M)_{\text{crit}}$ (Crutcher 2012, and references therein). Thus the median core is magnetically supercritical, and is able to collapse despite magnetic support. However, gravity overcomes magnetic support only by a factor of ~2. If the flux to mass ratio is at all non-uniform, this implies that there may be significant amounts of mass contained in regions that are too magnetized to collapse. Simulations of massive protostellar cores by Hennebelle et al. (2011) find that, for realistic levels of magnetization, the number of fragments is reduced by a factor of ~2 compared to a purely hydrodynamic calculation.

More recently, Commerçon et al. (2011) and Myers et al. (2013) have studied the collapse of massive cores using both radiative feedback and magnetic fields, and the effects amplify one another. At early times, the extra magnetic braking provided by magnetic fields removes angular momentum and channels material to the center faster. This tends to raise the accretion rate and thus the luminosity, making radiative heating more effective. Moreover, radiative and magnetic suppression of fragmentation are complementary in that they operate in different regions. Radiation suppresses fragmentation within ≈1,000 AU of a forming star, as found by Krumholz (2006) and subsequent radiation-hydrodynamic simulations, but becomes ineffective at larger radii. However, the regions more than ≈1,000 AU from a forming star are precisely those that are mostly likely to be magnetically sub-critical, and thus magnetic fields are able to suppress fragmentation in these regions. Because each mechanism operates where the other is weakest, the combination of the two reduces fragmentation much more efficiently than one might naively guess. Figure 3.1 shows an illustration of this effect: a simulation with magnetic fields and radiation shows almost no fragmentation, while ones with either alone both experience some fragmentation, though still less than in a purely hydrodynamic case. Based on these simulations, Myers et al. (2013) conclude that compact, dense regions such as the observed massive cores are likely to form single star systems, rather than fragment strongly.

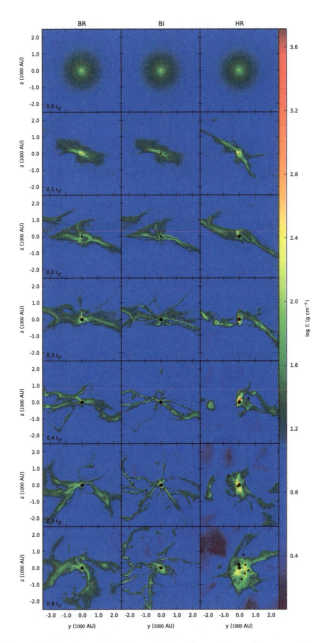

Fig. 3.1 Column densities from simulations of the collapse of massive protostellar cores (Myers et al. 2013). The *left column* (BR) shows simulations including both magnetic fields and radiative feedback. The *middle column* (MI) uses magnetic fields but no radiation, while the *right column* (HR) uses radiation but has not magnetic field. Rows show snapshots at uniformly-spaced times, from the initial state to 0.6 core free-fall times. The region shown is the central 5,000 AU around the most massive star. Colors show column density, and *black circles* show stars, with the size of the circle indicating the stellar mass. The initial magnetic field is oriented vertically in this projection See Myers et al. (2013) for more details

While this would seem to settle the question of whether fragmentation might limit stellar masses, it is worth noting that there is one final possible fragmentation mechanism that has not yet been evaluated via simulations. Kratter and Matzner (2006) point out that the disks around massive stars are likely to be gravitationally-unstable. Gravitational stability for a pressure-supported disk can be characterized by the (Toomre 1964) parameter,

$$Q = \frac{\kappa_{ep} c_s}{\pi G \Sigma}, \tag{3.4}$$

where κ_{ep} is the epicyclic frequency (equal to the angular frequency of the orbit for a Keplerian disk), c_s is the gas sound speed, and Σ is the gas surface density. Values of $Q < 1$ indicate instability of the disk to axisymmetric perturbations, and non-axisymmetric perturbations begin to appear at $Q \approx 1-2$. Depending on the properties of the disk, these instabilities can run away and cause the disk to fragment into point masses. For a steady disk with dimensionless viscosity α (Shakura and Sunyaev 1973), the accretion rate through the disk is (e.g., Kratter et al. 2010)

$$\dot{M} = \frac{3\alpha c_s^3}{GQ} = 1.5 \times 10^{-4} \frac{\alpha}{Q} \left(\frac{T}{100\,\mathrm{K}} \right)^{3/2} M_\odot\,\mathrm{yr}^{-1}, \tag{3.5}$$

where the numerical evaluation for the sound speed uses $c_s = \sqrt{k_B T/\mu}$ and the mean particle mass $\mu = 3.9 \times 10^{-24}$ g, appropriate for fully molecular gas of standard cosmic composition. Local instabilities such as the magnetorotational instability in the disk cannot produce $\alpha > 1$, and the disk cannot be gravitationally stable if $Q < 1$, so the accretion rate through a gravitationally-stable disk where angular momentum is transported primarily by local instabilities is strictly limited from above. Accretion rates of $10^{-4}\,M_\odot$ year^{-1} in such disks are possible only if the temperature is ≈ 100 K.

This means that there is a race between stellar heating and accretion. Forming a very massive star via accretion in a time less than its main sequence lifetime of a few Myr requires extremely high accretion rates – $\sim 10^{-3}\,M_\odot$ year^{-1} for a $>100\,M_\odot$ star. However, such high accretion rates tend to be more than a disk can process without going unstable and fragmenting, unless radiation from the central star can heat the disk up, allowing it to transport mass more quickly while remaining stable. However, this process of heating to allow more mass through has a limit: once the temperature required to stabilize the disk exceeds the dust sublimation temperature, it will not be easy to heat the disk further, and this may result in an instability so violent that the disk fragments entirely, halting further accretion. Kratter and Matzner (2006) estimate that this could limit stellar masses of $\sim 120\,M_\odot$. Simulations thus far have not probed this possibility, as no 3D simulations have reached such high stellar masses. However, we caution that Kratter and Matzner's scenario did not consider the effects of magnetic fields, which limit the disk radius and help stabilize it against fragmentation, or the effects of molecular opacity in the gas, which can provide coupling to the stellar radiation field and a means to heat the disk to temperatures above the dust sublimation temperature (Kuiper and Yorke 2013).

3.2.2 Radiation Pressure

The second potential difficulty in forming massive stars via accretion is the radiation pressure problem, first pointed out by Larson and Starrfield (1971) and Kahn (1974). The problem can be understood very simply: the inward gravitational force per unit mass exerted by a star of mass M and luminosity L on circumstellar material with specific opacity κ located at a distance r is $f_{\text{grav}} = GM/r^2$, while the outward radiative force $f_{\text{rad}} = \kappa L/(4\pi r^2 c)$. Since the radial dependence is the same, the net force will be inward only if

$$\frac{L}{M} < \frac{4\pi G c}{\kappa} = 2{,}500 \left(\frac{\kappa}{5\,\text{cm}^2\,\text{g}^{-1}}\right)^{-1} \frac{L_\odot}{M_\odot}. \tag{3.6}$$

All stars above $\sim 20\,M_\odot$ have $L/M > 2{,}500\,L_\odot/M_\odot$, so the question naturally arises: why doesn't radiation pressure expel circumstellar material and prevent stars from growing to masses substantially larger than $\sim 20\,M_\odot$?

The choice of opacity κ to use in evaluating this limit is somewhat subtle, because the dominant opacity source for circumstellar material will be dust that is mixed with the gas, which provides a highly non-gray opacity that will vary with position as starlight passes through the dust and is reprocessed by absorption and re-emission. Thus there is no single value of κ that can be used in the equation above, and for an accurate result one must first compute the radiation field that results from the interaction of stellar photons with circumstellar dust, and then ask how the resulting radiation force compares to gravity. Nonetheless, detailed one-dimensional calculations by Wolfire and Cassinelli (1986, 1987), Preibisch et al. (1995), and Suttner et al. (1999), including effects such as grain growth and drift relative to the gas, nonetheless confirm that radiation pressure is sufficient to halt accretion onto massive protostars at masses of $\sim 20\,M_\odot$ for Milky Way dust abundances.

However, spherical symmetry is likely to be a very poor assumption for this problem, and a number of authors point out that relaxing it might reduce or eliminate the radiation pressure problem. The central idea behind these models is that the optically thick circumstellar matter around a rapidly-accreting protostar is capable of reshaping the radiation field emitted by the star, and making it non-spherical. If the radiation can be beamed, then the radiation force can be weaker than gravity over a significant solid angle even if the mean radiation force averaged over 4π sr is stronger than gravity. This beaming could be accomplished by a disk (Nakano 1989; Nakano et al. 1995; Jijina and Adams 1996) or an outflow cavity (Krumholz et al. 2005), or by any other non-spherical feature that might be found in the flow.

3 The Formation of Very Massive Stars

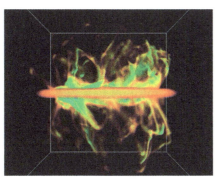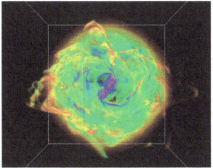

Fig. 3.2 Volume renderings of the density field in the central 4,000 AU of a simulation of the formation of a massive binary system including radiation pressure feedback (Krumholz et al. 2009). The *left image* shows the edge-on view of the disk, while the *right image* shows the face-on view. At the time shown in these images, the simulation contains a $41.5 + 29.2\,M_\odot$ binary, each with its own disk, and with the two disks embedded in a circumbinary disk. The filamentary structure above and below the disk is created by radiation Rayleigh-Taylor instability, and consists of dense filaments carrying mass onto the disk

The first radiation-hydrodynamic simulations in two dimensions found that beaming by the disk was indeed effective at channeling radiation away from an accreting star (Yorke and Kaisig 1995; Yorke and Bodenheimer 1999; Yorke and Sonnhalter 2002), but that nevertheless the radiation field was able to reverse the accretion flow and prevent formation of stars larger than $\sim 40\,M_\odot$. The first three-dimensional radiation-hydrodynamic simulation, on the other hand, found that there was no flow reversal, and that mass was able to accrete essentially without limit (Krumholz et al. 2009). Figure 3.2 shows a snapshot from this simulation. The key physical process uncovered in these simulations was radiation Rayleigh-Taylor instability (RRTI): the configuration of a radiation field attempting to hold up a dense, accreting gas is unstable to the development of fingers of high optical depth material that channel matter down toward the star, while radiation preferentially escapes through low optical depth chimneys that contain little matter. While the instability was first discovered numerically, subsequently Jacquet and Krumholz (2011) and Jiang et al. (2013) performed analytic stability calculations that allowed them to derive the linear stability condition and linear growth rate for RRTI.

This picture was somewhat complicated by the work of Kuiper et al. (2010, 2011, 2012), who pointed out that the numerical method used in the Krumholz et al. (2009), while it provided a correct treatment of the dust-reprocessed radiation field, did not properly include the radiation force produced by the direct stellar radiation field. When Kuiper et al. include this effect, they find that the extra acceleration provided to the circumstellar matter is such that gas tends to be ejected from a protostellar core before the RRTI has time to become non-linear. While there is no reason to doubt that the result is correct in the case of an initially-laminar protostellar core as considered by Kuiper et al., it is unclear how general this result is, since any pre-existing density structure in the core will "jump-start" the growth of the instability and allow it to become non-linear in far less time. Such pre-existing density structures are inevitable given the regions of massive star formation are highly turbulent (e.g. Shirley et al. 2003), and even in the absence of turbulence, gravitational instabilities in the accretion disk will tend to produce large density contrasts and possibly binary systems (Kratter et al. 2010).[2]

While there is debate about the role of RRTI, there is no debate about whether radiation pressure can actually halt accretion. Kuiper et al. (2011) and Kuiper and Yorke (2013) find that, even though radiation pressure is able to eject matter in their simulations, it is unable to eject the accretion disk, and thus that accretion can continue onto stars up to essentially arbitrary masses. Similarly, Cunningham et al. (2011), confirming the hypothesis of Krumholz et al. (2005), show that a protostellar outflow cavity produced by a massive star provides an efficient mechanism for radiation to escape, allowing accretion to continue essentially without any limit due to radiation pressure. Indeed, the presence of an outflow cavity removes the need for RRTI to occur, as it provides a pre-existing low-optical depth chimney through which radiation escapes. Thus, the consensus of modern radiation-hydrodynamic simulations of massive star formation that radiation pressure does not represent a serious barrier to the formation of stars up to essentially arbitrary masses.

3.2.3 Ionization Feedback

The third potential problem in forming very massive stars is photoionization: galactic molecular clouds generally have escape speeds below $10\,\mathrm{km\,s^{-1}}$ (e.g., Heyer et al. 2009), the sound speed in $\sim 10^4$ K photoionized gas. As a result, if the gas in a star-forming region becomes ionized, the gas pressure may drive a thermal wind that will choke off accretion. This process is thought to be a major factor in limiting the star formation efficiency of giant molecular clouds (e.g., Whitworth 1979; Matzner 2002; Krumholz et al. 2006). However, it is much less clear whether photoionization can limit the formation of individual massive stars. The key argument on this point was first made by Walmsley (1995), who noted that an accretion flow onto a massive star can sharply limit the radial extent of an H II region. This is simply a matter of the ionizing photon budget: the higher the mass inflow rate, the higher the density of matter around the star, and thus the higher the recombination rate and the smaller the Strömgren radius. If the radius of the ionized region is small enough that the escape speed from its outer edge is $>10\,\mathrm{km\,s^{-1}}$, then photoionized gas will not be able to flow away in a wind or escape. Walmsley (1995) considered an accretion flow in free-fall onto a star, and showed that the

---

[2]Although Kuiper et al.'s simulations are three-dimensional, they cannot model either turbulence of disk fragmentation, because the numerical method they use for radiation transport can only handle a single star whose location is fixed at the origin of their spherical grid.

escape speed from the edge of the ionized region will exceed the ionized gas sound speed c_i if the accretion rate satisfies

$$\dot{M}_* > \left[\frac{8\pi\mu_H^2 GM_*S}{2.2\alpha_B \ln(v_{\mathrm{esc},*}/c_i)}\right] = 4\times 10^{-5} \left(\frac{M_*}{100\, M_\odot}\right)^{1/2} \left(\frac{S}{10^{49}\, \mathrm{s}^{-1}}\right)^{1/2} M_\odot\, \mathrm{year}^{-1}, \tag{3.7}$$

where M_* is the stellar mass, S is the star's ionizing luminosity (photons per unit time), $\mu_H = 2.34\times 10^{-24}$ is the mean mass per H nucleus, $\alpha_B \approx 2.6\times 10^{-13}\,\mathrm{cm}^3\,\mathrm{s}^{-1}$ is the case B recombination coefficient, and $v_{\mathrm{esc},*}$ is the escape speed from the stellar surface. The factor of 2.2 in the denominator arises from the assumption that He is singly ionized. The numerical evaluation uses $v_{\mathrm{esc},*} = 1{,}000\,\mathrm{km}\,\mathrm{s}^{-1}$ and $c_i = 10\,\mathrm{km}\,\mathrm{s}^{-1}$, but the numerical result is only logarithmically-sensitive to these parameters. Thus an accretion rate of $\sim 10^{-4}\,M_\odot\,\mathrm{year}^{-1}$ is sufficient to allow continuing accretion onto even an early O star. Given the dense, compact environments in which massive stars appear to form, such high accretion rates are entirely expected (McKee and Tan 2003). Keto (2003) extended Walmsley's result by deriving a full solution for a spherical inflow plus ionization front in spherical symmetry, and reached the same qualitative conclusion. Keto and Wood (2006), Klaassen and Wilson (2007), and Keto and Klaassen (2008) provide direct observational evidence for accretion in photoionized regions.

This argument makes clear that whether photoionization can limit accretion onto massive stars depends critically on the interplay between the initial conditions, which determine the accretion rate, and stellar evolution, which determines the ionizing luminosity. If the accretion rate drops low enough, and the ionizing flux is high enough, then the ionized region will extend out to the point where photoionized gas can escape and accretion will be choked off. The geometry of the flow matters as well. Keto (2007) considered rotating infall, and showed that this may result in a configuration where the ionized region blows out in the polar direction, but accretion continues uninhibited through a denser equatorial disk that self-shields against the ionizing photons. In three dimensions turbulent structure may plan an analogous role. Dale et al. (2005) and Peters et al. (2010a, 2011) have simulated the formation of massive stars and star clusters including photoionization feedback, and they find that photoionization is generally unable to disrupt accretion flows. In the simulations, accretion tends to be highly aspherical, proceeding through disks and filaments, as illustrated in Fig. 3.3. Because these structures are dense, they have very high recombination rates and thus are resistant to being photoionized. The structure that tends to result in these simulations is that there are low-density ionized regions where material is escaping, but that the majority of the mass is contained in dense filaments where it continues to accrete. As a result, in Dale et al.'s simulations accretion is able to continue to masses of several hundred $M_\odot$, while in Peters et al.'s simulations reach a mass of $\sim 70\, M_\odot$ without photoionization halting accretion.

There has been considerably more work on whether ionization can halt accretion in the context of the formation of the first stars. McKee and Tan (2008) developed an analytic model for several forms of feedback, and argued that photoionizing

Fig. 3.3 Column density from a simulation of the formation of a massive star cluster including photoionization feedback (Dale et al. 2005). The central star begins the simulation with a mass of ≈30 $M_\odot$, but continues to grow over the course of the simulation, reaching >100 $M_\odot$ by the end. *White dots* are stars

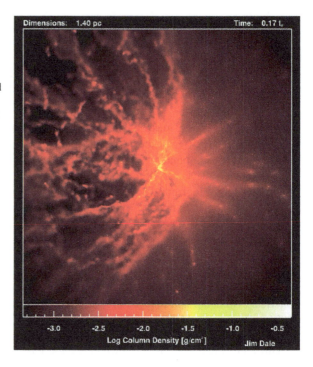

radiation will blow out the polar regions of a rotating accretion flow around an accreting star once its mass reaches ∼50–100 $M_\odot$, will thereafter go on to photoevaporate the disk. This process will halt accretion at a mass of ∼150 $M_\odot$. Hosokawa et al. (2011) conducted 2D simulations and obtained results qualitatively consistent with McKee and Tan's model, but with an even lower limiting mass of ∼40–50 $M_\odot$. Similar limiting masses were obtained from three-dimensional simulations by Stacy et al. (2012) and Susa (2013), and in a 2D simulation of metal free-star formation in gas that was externally ionized before collapsing (so-called population III.2 star formation), Hosokawa et al. (2012) found an even lower limiting mass of 20 $M_\odot$.

The fairly low limiting masses found in the simulations of primordial star formation appear to be in some tension with the results of the numerical simulations of present-day star formation. At first one might think that the presence or absence of dust opacity provides an obvious explanation for the difference, but it is not clear if this is the case. Even at Solar metallicity, most ionizing photons are absorbed by hydrogen atoms and not dust grains (see the Appendix of Krumholz and Matzner 2009 for a discussion of why this is), so dust is responsible for removing only a small fraction of ionizing photons. Similarly, primordial H II regions have somewhat higher temperatures (due to lack of metal line cooling) and metal-free stars have somewhat higher ionizing luminosities (due to the lack of metal opacity in the stellar atmosphere).

A more promising explanation has to do with the initial conditions, which determine the accretion rate and geometry of the inflow. In an isolated star-forming core, which is the initial condition employed in the primordial calculations, once the photoionized region escapes from the vicinity of the star it can choke off further accretion onto the disk, leaving the disk subject to photoevaporation. However this does not appear to happen in a flow that is continuously fed by large amounts of mass supplied from larger, ~ 1 pc scales, as occurs in the present-day star formation simulations. This mass supply into the filaments and disks appears to shield them against photoevaporation. If the initial conditions are the key difference, then for the case of present-day star formation this suggests that the mass limit imposed by photoionization is likely to depend on the large-scale environment, though exactly which environmental properties are important remains uncertain.

Finally, as a caveat, it is important to note that the treatments of ionizing radiative transfer used in the codes for the simulation of both the present-day and primordial star formation are based on a simple ray-trace using the "on-the-spot" approximation. In this approximation, one treats ionizing photons produced by recombinations in the ionized gas as having a mean free path of zero, so that photons produced by a recombination to the ground state are re-absorbed on the spot rather than propagating a finite distance. Thus the diffuse radiation field produced by recombinations is ignored, and shadowing is too perfect. This is potentially problematic for the treatment of accretion disks, as the photoevaporation of disks is probably dominated by the diffuse photons produced in the photoionized atmosphere above the disk, rather than by direct stellar radiation (Hollenbach et al. 1994; McKee and Tan 2008). Thus it is unclear if the numerical results are reliable. The question of whether photoionization might limit stellar masses thus remains an only partially-solved problem.

3.2.4 Stellar Winds

A final potential challenge for the formation of massive stars by accretion has received far less theoretical attention: stellar winds. Once the surface temperatures of stars exceed $\sim 2.5 \times 10^4$ K, they begin to accelerate fast, radiatively-driven winds (Leitherer et al. 1992; Vink et al. 2000, 2001). Zero-age main sequence stars reach this temperature at a mass of $\sim 40\,M_\odot$, and stars of this mass have such short Kelvin-Helmholtz timescales that, even if they are rapidly accreting, their radii and surface temperatures during formation are likely to be close to their ZAMS values (Hosokawa and Omukai 2009). The momentum carried by these winds is about half that of the stellar radiation field (Kudritzki et al. 1999; Richer et al. 2000; Repolust et al. 2004), and so if the direct stellar radiation field cannot stop accretion then the momentum carried by stellar winds will not either.

However, winds might yet be important, because the wind launch velocity is quite large, $\sim 1{,}000$ km s^{-1}. As a result, when the winds shock against the dense accretion flow, their post-shock temperature can be $>10^7$ K, well past the peak of the cooling curve (Castor et al. 1975; Weaver et al. 1977), and as a result the gas will stay hot rather than cooling radiatively. Should it become trapped, this hot gas could exert a pressure that is far greater than what would be suggested by its launch momentum – in effect, it could convert from a momentum-driven flow to an energy-driven one (cf. Dekel and Krumholz 2013). If this were to happen, it is possible that the stellar wind gas might be able to interfere with accretion.

There has been a great deal of work on the interaction of post-shock stellar wind gas with the ISM on the scale of star clusters (e.g., Tenorio-Tagle et al. 2007; Dale and Bonnell 2008; Rogers and Pittard 2013). This work suggests that the wind gas tends, much like radiation, to leak out through openings in the surrounding dense gas rather than becoming trapped and building up a large pressure. Indeed, on the cluster scale observations appear to confirm that the pressure exerted by the hot gas is subdominant (Lopez et al. 2011). However, there is no comparable work on the scale of individual stars, and it is therefore possible that the situation there could be different. Moreover, even when the wind gas does escape on cluster scales, as it flows past the colder, denser material it tends to entrain and carry of some of it. Again, the question of whether this might happen to the accretion flows around individual protostars has yet to be addressed. Given the lack of theoretical or observational attention, the best that can be said for now is that, if the interaction between stellar winds on small scales resembles those seen on larger scales, stellar winds are unlikely to set significant limits on the masses to which stars can grow by accretion.

3.3 The Formation of Very Massive Stars by Collision

The discussion in the previous section indicates that there is no strong argument against the idea that very massive stars form via the same accretion mechanisms that give rise to stars of lower masses. However, it is also possible for very massive stars to form through an entirely different channel: collisions between lower mass stars. The central challenge for forming massive stars via collisions is the very small cross-sectional area of stars compared to typical interstellar separations, and the relatively short times allowed for collisions by the lifetimes of massive stars. Very massive stars are found routinely in clusters with central densities $\sim 10^4 \, \text{pc}^{-3}$ (e.g. Hillenbrand and Hartmann 1998), and the highest observed central densities in young clusters are $\sim 10^5 \, \text{pc}^{-3}$ (Portegies Zwart et al. 2010, their Figure 9), with the possible exception of R136 (Selman and Melnick 2013). If gravitational focusing is significant in enhancing collision rates (usually the case for young clusters), the mean time between collisions in a cluster consisting of stars of number density n and velocity dispersion σ, each with mass m and radius r, is (Binney and Tremaine 1987)

$$t_{\text{coll}} = 7.1 n_4^{-2} \sigma_1 r_0^{-1} m_0^{-1} \text{ Myr}, \tag{3.8}$$

where $n_4 = n/10^4$ stars pc^{-3}, $\sigma_1 = \sigma/10$ km s^{-1}, $r_0 = R/R_\odot$, and $m_0 = m/M_\odot$. Thus under observed cluster conditions, we expect <1 collision between 1 $M_\odot$ stars to occur within the ~4 Myr lifetime of a massive star. Collision rates for stars more massive than the mean require a bit more care to calculate, but even under the most optimistic assumptions, production of very massive stars via collisions requires that clusters reach stellar densities much higher than the ~10^4 pc^{-3} seen in young clusters. This dense phase must be short-lived, since it is not observed. Models for the production of massive stars via collision therefore consist largely of proposals for how to produce such a short-lived, very dense phase. In this section I examine the collisional formation model. In Sects. 3.3.1 and 3.3.2 I describe two possible scenarios by which collisions could occur, and I discuss the mechanics of the collisions themselves, and the role of stellar evolution in mediating collisions, in Sect. 3.3.3.

3.3.1 Gas Accretion-Driven Collision Models

The first class of proposed mechanisms to raise the density high enough to allow collisional growth consists of processes that occur during the formation of a star cluster when it is still gas-rich. In a gas-rich cluster, stars can accrete gas, and this process is dissipative: it reduces the total gas plus stellar kinetic energy of the system, with the lost energy going into radiation from accretion shocks on the surfaces of protostars, and from Mach cones created by supersonic motion of stars through the gas. To see why this should lead to an increase in density, it is helpful to invoke the virial theorem. For a system where gravity, thermal pressure, and ram pressure are the only significant forces, the Lagrangian virial theorem states that (Chandrasekhar and Fermi 1953; Mestel and Spitzer 1956)

$$\frac{1}{2} \ddot{I} = 2\mathcal{T} - \mathcal{W}, \tag{3.9}$$

where

$$I = \int r^2 \, dm \tag{3.10}$$

$$\mathcal{T} = \int \left(\frac{3}{2} P + \frac{1}{2} \rho v^2 \right) dV \tag{3.11}$$

$$\mathcal{W} = -\int \rho \mathbf{r} \cdot \nabla \Phi \, dV \tag{3.12}$$

are the moment of inertia, the total kinetic plus thermal energy, and the gravitational binding energy, respectively.[3] The quantity Φ is the gravitational potential. If there are significant forces on the surface of the region, or significant magnetic forces, additional terms will be present, but for the moment we will ignore them.

The process of shock dissipation reduces $\mathscr{T}$ while leaving $\mathscr{W}$ unchanged, so the right-hand side becomes negative, and, on average, the system will tend to accelerate inward to smaller radii. This infall converts gravitational binding energy to kinetic energy, so both $\mathscr{T}$ and $-\mathscr{W}$ rise by the same amount. Because of the factor of 2 in front of $\mathscr{T}$ in Eq. (3.9), this tends to push the right-hand side back toward zero: the system is re-virializing, but at a smaller radius, higher density, and larger kinetic and (in absolute value) binding energy. However, this new equilibrium will last only as long as shocks do not keep decreasing $\mathscr{T}$. If shocks continue to happen, this will drive a continuous decrease in radius, and a continuous rise in density of both gas and stars. Bonnell et al. (1998) proposed the first version of this model, and argued that it could drive stellar densities to $\sim 10^8 \, \mathrm{pc}^{-3}$, at which point collisions would become common and massive stars could build up in this manner. The required density can be lowered significantly if all massive stars are in primordial hard binaries (Bonnell and Bate 2005), but even for such a configuration a significant rise in stellar density compared to observed conditions is required.

Bonnell et al. suggested that contraction would halt only when gas was removed by feedback from the forming massive stars. This halts contraction because, once the gas is removed, there is no longer a dissipation mechanism available to reduce $\mathscr{T}$. However, Clarke and Bonnell (2008) subsequently realized that at sufficiently high density two-body relaxation would become faster than dissipation, and this would halt further shrinkage. In terms of the virial theorem, the increase in $-\mathscr{W}$ required to increase $\mathscr{T}$ and balance the dissipation starts to come from stars forming tight binaries rather than from overall shrinkage of the system. The maximum density that can be reached therefore depends on the total cluster mass, in such a manner as to prevent collisions from becoming significant in clusters substantially smaller than $\sim 10^4 \, M_\odot$. It is important to note that this excludes the Orion Nebula Cluster, which contains a star of $\approx 38 \, M_\odot$ (Kraus et al. 2009), suggesting that stars of at least this mass at least can form via non-collisional processes.

These conclusions were based on analytic models, but more recently Moeckel and Clarke (2011) and Baumgardt and Klessen (2011) conducted N-body simulations including analytic prescriptions for the effects of gas accretion.[4] In these models, the gas is treated as a fixed potential that is reduced as the stars gain mass, eventually disappearing entirely when a prescribed amount has been accreted;

---

[3]The functional form of $\mathscr{W}$ is independent of whether or not there is an external gravitational field, but one can only identify the quantity $\mathscr{W}$ as the gravitational self-energy if the field is entirely due to self-gravity, with no external contribution.

[4]These authors did not include the effect of gas drag due to Mach cones, which for Bondi-Hoyle accretion flows is actually a factor of several larger than the change in stellar momentum due to accretion (Ruffert and Arnett 1994), but this is probably not the most serious limitation of the simulations.

3 The Formation of Very Massive Stars

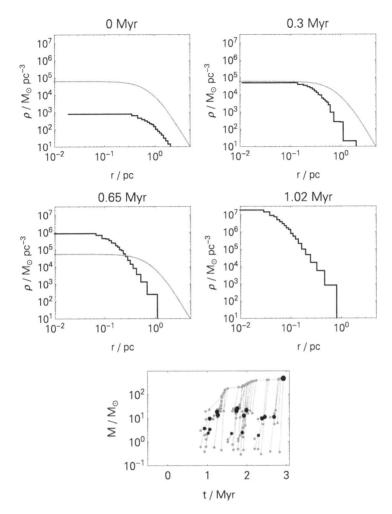

Fig. 3.4 Example results from the N-body plus accretion simulations of Moeckel and Clarke (2011). The *top set of four panels* shows the radially-averaged stellar density profile as a function of time in the simulations (*black lines*), together with the mass profile corresponding to the imposed gas potential (*gray lines*). The *bottom panel* shows the growth history of some of the most massive stars in the simulations. Points indicate stellar masses and the times when those stars first appear, and the convergence of two lines indicates a merger between two stars that yields a more massive star. *Black points* indicate stars that survive to the end of the simulation, while *gray points* indicate stars that merge before the end of the simulation

this sets the limit on the duration of the gas-dominated phase. Figure 3.4 shows an example output from one of these simulations. As anticipated by Clarke and Bonnell (2008), in these models the stars sink to the center until they form a stellar-dominated region in which two-body relaxation inhibits further contraction, though these regions can also undergo core collapse (see the next section). They both find that, as a result of this limitation, stellar collisions during the gas-dominated phase are negligible unless the initial conditions are already very compact or massive, with half-mass radii of ~ 0.1 pc or less and/or masses of $\sim 10^4$ $M_\odot$ or more.

The requirement for very high initial densities creates significant tension with observations. Moeckel and Clarke (2011) find that even the Arches cluster is insufficiently dense to have produced stellar collisions up to this point, despite the fact that it contains numerous very massive stars. Similarly, Baumgardt and Klessen (2011) find that, once the gas potential is removed and clusters re-virialize, those that began their evolution at densities high enough to induce significant numbers of collisions end up far too compact in comparison to observed open clusters, including those containing very massive stars. As a result of these findings, both sets of authors tentatively conclude that stellar collisions during the gas-dominated phase cannot be the dominant route to the formation of very high mass stars, though they cannot rule out the possibility that such collisions occur in rare circumstances.

Before relying on these conclusions too heavily, it is important to understand the limitations of these calculations. Undoubtedly the largest one is the simple prescription used to treat the gas. In these models, the shape of the gas potential (though not its depth) is fixed, the accretion rates onto stars are fixed and independent of stellar mass or position, and the final star formation efficiency is also fixed. Obviously none of these assumptions are fully realistic. In particular, depending on the effectiveness of stellar feedback, the gas potential might either shrink and promote increases in stellar density, or the gas potential might vary violently in time as gas is pushed around by stellar feedback, pumping energy into the stars and preventing contraction – indeed, the latter is seen to occur in at least some simulations that do treat the gas (Li and Nakamura 2006; Nakamura and Li 2007; Wang et al. 2010). It is unclear how the conclusions might change if these phenomena were treated more realistically.

3.3.2 Gas-Free Collision Models

The second class of models for inducing growth of very massive stars via collisions takes place in the context of gas-free clusters. These mechanisms, and their potential role in young massive clusters, were recently reviewed by Portegies Zwart et al. (2010), and I refer readers there for further details. The advantage of this approach compared to the gas-driven one is that the time available for collisions is longer, but the disadvantage is the lack of gas drag as a mechanism for raising the density.

Clusters of equal-mass stars are unstable to spontaneous segregation into a contracting core and an expanding envelope, in which the negative heat capacity of the system drives a continuous transfer of energy out of the core and thus ever-higher densities (Lynden-Bell and Wood 1968). In a cluster containing a spectrum of masses, contraction of the core is further enhanced by the tendency of the stars to mass-segregate, with the core consisting of more massive, dynamically-cool stars, and the envelope consisting of low-mass, dynamically-hot ones (Spitzer 1969; Gürkan et al. 2004). While there is no doubt that these processes operate, it is uncertain whether they are fast enough to produce collisions within the $\sim$4 Myr lifetime of the most massive stars. Portegies Zwart et al. (1999) conclude based on N-body simulations that collisions will occur before massive stars die only if the central density starts at $\sim 10^7$ stars pc^{-3}. In this case, the mergers themselves are dissipative and will trigger further core contraction, leading to runaway formation of a single very massive object. As in the case of gas-driven collisions, the existence of a large population of primordial hard binaries can somewhat reduce the density required to initiate this cascade (Portegies Zwart and McMillan 2002). Even so, the initial densities required in the simplest gas-free collision models would preclude the possibility of very massive stars forming by collisions except perhaps in R136. Models in which a significant fraction of very massive stars form via collision therefore generally posit a set of initial conditions that significantly increases the collision rate.

One way to shorten the time required for core collapse and the onset of collisions is to consider a cluster with primordial mass segregation, meaning that the cluster is mass-segregated even before the gas-free evolution begins (e.g. Ardi et al. 2008; Goswami et al. 2012; Banerjee et al. 2012a,b). Such a starting configuration reduces the time required for core collapse to begin because it provides both a higher density and a higher mean stellar mass (and thus a lower relaxation time) in the cluster center. Depending on the degree of mass segregation, the reduction in time to the onset of core collapse and collisions can be $\sim 1-2$ Myr, a non-trivial fraction of the lifetime of a very massive star, and simulations using sufficiently mass-segregated initial conditions generally find that collisions become common before massive stars end their lives.

The extent to which star clusters actually are primordially mass-segregated is unclear. Observations generally show at least some degree of mass segregation in present-day clusters, but the amount varies widely. At the low-mass end of clusters containing massive stars, in the Orion Nebula Cluster the Trapezium stars are all at the cluster center, but there is no observed mass segregation for any stars except these (Hillenbrand and Hartmann 1998; Huff and Stahler 2006). In NGC 3603 (Pang et al. 2013) and R136 (Andersen et al. 2009), the cluster is segregated throughout so that the mass function is flatter at small radii, but more massive stars are more segregated than less massive ones. However, all of these clusters are $\sim 1-3$ Myr old, so it is entirely possible that the segregation we see now is a product of dynamical evolution during this time, not primordial segregation – indeed, Pang et al. (2013) argue that the segregation they observe in NGC 3603 is more consistent with dynamical evolution from a weakly-segregated or unsegregated initial state than with primordial segregation. Unfortunately answering this question fully would require that observations probe the gas-enshrouded phase, which is possible only in the infrared, where low resolution creates severe difficulties with confusion. Indeed, confusion is a serious worry for measurements of mass segregation even in optically-revealed clusters (Ascenso et al. 2009).

Another way to accelerate the dynamical evolution of star clusters is to begin from unrelaxed initial conditions. Both observations (Fűrész et al. 2008; Tobin et al. 2009) and simulations (Offner et al. 2009a) of star clusters that are still gas-embedded show that the stars are subvirial with respect to the gas, and such cold conditions lead to more rapid dynamical evolution than virialized initial conditions (Allison et al. 2009). Larger star clusters may also be assembled via the mergers of several smaller clusters within the potential provided by a massive gas cloud (Bonnell et al. 2003; McMillan et al. 2007; Fujii et al. 2012). These smaller clusters, since they have smaller numbers of stars, also have smaller time-scales for core collapse. Allison et al. (2009) find that substructured initial conditions accelerate mass segregation, but it is unclear whether they do so enough to accelerate collisions. Fujii and Portegies Zwart (2013) find that the extent to which the formation of a large cluster out of multiple star clusters influences collisions depends on the ratio of the assembly time to the core collapse time of the initial subclusters. If the subclusters undergo core collapse before merging, then they may have a few internal collisions, but there are no collisions in the merged cluster, and collisional growth of stars is negligible overall. On the other hand, if core collapse does not occur in the subclusters before they merge, the evolution is similar to that of a cluster that formed as a single entity.

In summary, collisions during the gas-free phase are unlikely to contribute significantly to the growth of very massive stars if star clusters are born virialized and non-segregated, but in reality neither of these assumptions is likely to be exactly true. The viability of collisional formation models then turns sensitively on the extent to which these assumptions are violated, and this question is unfortunately poorly constrained by observations. Hydrodynamic simulations of the formation of massive star clusters that include the gas-dominated phase may be helpful in addressing this question, but to be credible these simulations will need to include feedback processes such as stellar winds, photoionization, and radiation pressure that are presently omitted from most models.

3.3.3 Stellar Evolution and Massive Star Mergers

One important subtlety for models of the growth of massive stars via mergers is that the outcome depends not just on N-body processes, but also on the physics of stellar collisions, and on the structure and subsequent evolution of stellar merger products. Both questions have been the subject of considerable study in the context of mergers between low-mass stars leading to the production of blue stragglers, but only a few authors have conducted similar simulations for mergers involving massive stars. Mergers involving massive stars (particularly very massive ones) may be substantially different than those involving low-mass stars because of the importance of radiation pressure for massive stars. As stellar mass increases, the increasing dominance of radiation pressure brings the structure close to that of an $n = 3$ polytrope, which is very weakly bound. Moreover, radiative forces may be non-negligible during the collision itself. For example, just as radiation pressure may be capable of inhibiting accretion, it may be capable of ejecting material that is flung off stellar surfaces during a collision, thereby increasing mass loss during the collision.

One quantity of interest from stellar merger simulations is the amount of prompt mass loss during the collision itself. Models for collisional growth generally assume that the mass loss is negligible, thus maximizing the collisional growth rate. Freitag and Benz (2005), Suzuki et al. (2007), and Glebbeek et al. (2013) all find in their simulations that losses are indeed small, with at most $\sim$10 % of the initial mass being ejected for realistic collision velocities. In three-body mergers produced when an intruder enters a tight binary system, the loss can be as large as $\sim$25 % (Gaburov et al. 2010). However, a very important limitation of these simulations is that they do not include any radiative transfer, and treat radiation pressure as simply an extra term in the equation of state, with the radiation pressure determined by the matter temperature, which in turn is determined by hydrodynamic considerations. This is likely to be a very poor approximation for the material that is flung outward from a collision, where illumination from the central merged object will dominate the thermodynamics, as it does in the case of accretion onto massive stars. In the approximation used by the existing merger simulations, it is impossible for radiation pressure to eject matter, and thus the $\sim$5–10 % mass loss found in the simulations simply represents the mass of material that is raised to escape velocities during the collision itself. This figure should therefore be thought of as representing a lower limit. There is a clear need to reinvestigate this problem using a true radiation-hydrodynamics code. If the mass loss has been underestimated, collisional growth will be harder than is currently supposed.

A second quantity of interest from merger simulations is the radius of the merger product. When stars merge, shocks during the collision raise the entropy of the stellar material, so that when hydrostatic equilibrium is re-established a few days after the merger, the resulting star will initially be very bloated compared to main sequence stars of the same mass, and will gradually shrink over a Kelvin-Helmholtz timescale (Dale and Davies 2006; Suzuki et al. 2007). Building up very massive stars via collisions likely requires multiple mergers, and at the very high densities required, the interval between mergers may be smaller than the KH timescale, so that the growing stars will have enlarged radii. Whether this will enhance or reduce the rate of collisional growth is unclear. Suzuki et al. (2007) point out that the enhanced radii of the merger products make them bigger targets that are more likely to collide with other stars. On the other hand, Dale and Davies (2006) note that the envelopes of the post-merger stars are even more weakly bound than those of massive main sequence stars, and as a result such collisions may actually erode the envelope rather than add to it, ultimately limiting collisional growth. Which effect dominates is unclear, as no merger simulations involving such swollen stars have been reported in the literature.

A final consideration for collisional growth models in the gas-free phase, where the timescales involved may be several Myr, is mass loss via stellar winds. At masses below $\sim$100 $M_\odot$, wind mass loss rates are considerable, but are unlikely to be able to counteract the effects of collisional growth if the density is high enough for runaway merging to begin. However, little is known about wind mass loss rates

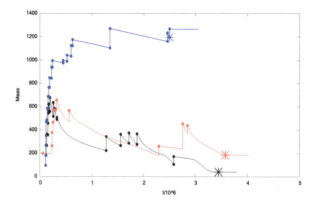

Fig. 3.5 Results from three simulations of massive stellar mergers driven by gas-free core collapse by Vanbeveren et al. (2009), reprinted by permission. The *blue line* shows a calculation using fairly modest wind mass losses, similar to those adopted by Portegies Zwart et al. (1999). The *black line* shows a calculation with identical initial conditions but using a wind prescription taken from Belkus et al. (2007) for Solar metallicity stars, and the *red line* is the same but using a metallicity of 5 % of Solar

at still higher masses, and there are good empirical arguments that they might be orders of magnitude larger (Belkus et al. 2007). N-body simulations using these enhanced winds find that they remove mass from stars faster than collisions can add it, yielding only very transient growth to large masses, followed by rapid shrinkage back to $\sim 100\,M_\odot$ (Yungelson et al. 2008; Vanbeveren et al. 2009; Glebbeek et al. 2009). Figure 3.5 provides an example. Moreover, the winds might remove mass so efficiently that they reduce the gravitational potential energy of the system fast enough to offset the loss of kinetic energy that occurs during mergers, halting the collisional cascade completely.

3.4 Observational Consequences and Tests

Having reviewed the various models for the origins of very massive stars, I now turn to the question of their predictions for observable quantities, and how these might be used to test the models. One can roughly divide these predictions into those that apply on the scale of star clusters, and those that apply on the scale of individual star systems.

3.4.1 The Shape of the Stellar Mass Function

On the cluster scale, one obvious difference between collisional and accretion-based models of massive star formation is their predictions for the form of the stellar mass function at the massive end – note that I refer here to the present-day mass function (PDMF) rather than the initial mass function (IMF), since in gas-free collision models very massive stars are absent in the IMF and only appear later due to collisions. As discussed above, in the case where massive stars form by the same accretion processes that produce low-mass stars, one naturally expects that very massive stars should simply be a smooth continuation of the Salpeter mass function seen at lower masses. The situation is very different for collisional models, where very massive stars form via a fundamentally different process than low mass ones. Not surprisingly, this different formation mechanism gives rise to a feature in the stellar mass function.

For the gas-driven collision gas, both Moeckel and Clarke (2011) and Baumgardt and Klessen (2011) find that the typical outcome of collisions is one or two objects whose masses are much greater than those of any other object, and a corresponding depletion of objects slightly less massive than the dominant one or two. Figure 3.6 shows an example result from Baumgardt and Klessen (2011). As the plot shows, collisions that yield very massive stars of several hundred $M_\odot$ tend to produce an overall mass function in which the range from ~ 10–$100\,M_\odot$ is actually significantly under-populated relative to the Salpeter slope found at lower masses, while the one or two most massive objects that are formed by collision represent a significant over-population relative to Salpeter. Unfortunately the authors of models in which collisional growth occurs during the gas-free phase have generally not reported the full mass functions produced in their simulations, but given that the mechanism for assembling the very massive stars is essentially the same as in the gas-driven models – runaway collisions that agglomerate many

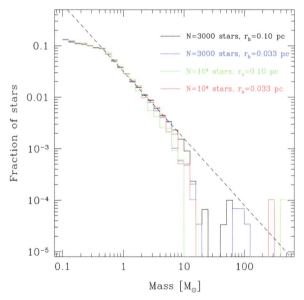

Fig. 3.6 Mass functions of stars in an N-body simulation of gas-driven stellar collisions by Baumgardt and Klessen (2011). The histograms are mass functions obtained 10 Myr after the beginning of the simulation, and the number of stars and initial half-mass radius used in each of the simulations are as indicated in the legend. The *straight dashed line* is the Salpeter mass function. Note that the simulations all predict a turndown in the mass function relative to Salpeter at masses from ~ 10 to $100\,M_\odot$

massive stars into one or two supermassive ones – it seems likely that these models would predict a similar functional form for the mass function.

At present there is no observational evidence for mass functions of this form. Massey and Hunter (1998) report that the mass function in R136 is well-approximated by a single powerlaw with a Salpeter-like slope from 3 to $120\,M_\odot$, and Andersen et al. (2009) report a continuous powerlaw slope over an even wider range of mass, with no evidence for a turn-down in the vicinity of $10\,M_\odot$. Similarly, Espinoza et al. (2009) examine the Arches cluster and report that the mass function above $10\,M_\odot$ is well-described by a powerlaw with a slope $\Gamma = -1.1 \pm 0.2$, consistent within the errors with the Salpeter value $\Gamma = -1.35$. There are significant systematic uncertainties on these values, arising mostly from the challenge of assigning masses to stars based on photometry, but it is important to note that a mass function of the form predicted by the collisional simulations should be apparent even from the *luminosity* function, independent of the mapping between luminosity and mass. Due to confusion, even luminosity functions can be difficult to measure in the cores of clusters dense enough to be candidates for collisions, but this data should improve significantly in the era of 30 m-class telescopes. Observations with these facilities should be able to settle the question of whether the mass and luminosity functions in cluster cores show the characteristic signature of a depletion from ∼10 to $100\,M_\odot$ coupled to a one or two stars at a few hundred $M_\odot$ that is predicted by collisional models.

3.4.2 Environmental-Dependence of the Stellar Mass Function

A second possible discriminant between accretion and collision as mechanisms for the formation of very massive stars is the way the stellar mass function depends on the large-scale properties of the cluster. As noted above, both gas-free and gas-driven collision models require very high stellar densities (even in the present-day state) and very high cluster masses; Baumgardt and Klessen (2011) argue that clusters where gas-driven collisions occur all end up too compact compared to observed ones, and Moeckel and Clarke (2011) argue that the Arches is not dense enough to be able to produce significant collisions. In contrast, accretion models either predict that the stellar IMF will be independent of environment, or that massive stars will be biased to regions of high surface density (Krumholz and McKee 2008; Krumholz et al. 2010). Accretion models do not predict that there should be an upper limit to stellar masses that is a function of either cluster mass or central stellar density.

This is a somewhat weaker test than the functional form of the stellar mass function, simply because the model predictions are somewhat less clear, but it may nonetheless provide a valuable complement. The challenge here will be obtaining a sample large enough to see if there is a statistically-significant correlation between the presence or absence of stars above some mass and properties of the environment like cluster mass or density. The major challenge is that one expects a correlation between maximum stellar mass and cluster size simply due to size of sample effects.

Observations must therefore remove the size of sample effect statistically, and search for a small correlation that might remain once the size of sample effect is removed. Some authors have claimed to detect such a correlation already in Galactic clusters (Weidner et al. 2010, 2013), while others have reported the absence of any such correlation in extra-Galactic environments (Calzetti et al. 2010; Fumagalli et al. 2011; Andrews et al. 2013). Given the poorly-understood selection issues associated with the Galactic sample (which is culled from the literature, rather than produced by a single survey), it seems likely that the extragalactic results based on uniform samples are more reliable, but the issue remains disputed.

An observation of a very massive star formed in relative isolation would also be strong proof that collisions are not required to make such stars, though it would not rule out the possibility that some stars form that way. There are several candidates for isolated stars with masses above $\sim 20\,M_\odot$ (and in some cases as much as $100\,M_\odot$; Bressert et al. 2012; Oey et al. 2013), and which appear unlikely to be runaways because they have small radial velocities and no bow shocks indicating large transverse motions. However, there remains the possibility that these are "slow runaways" with that were ejected very early and thus managed to reach fairly large distances from the cluster despite their fairly small space velocities (Banerjee et al. 2012a).

3.4.3 Companions to Massive Stars

The properties of massive star companions provide a final potential test for distinguishing accretion-based and collisional formation models. It is well known that massive stars are much more likely than stars of lower mass to be members of multiple systems. Mason et al. (2009) report a companion fraction of 75 % for O star primaries in Milky Way star clusters,[5] while Sana et al. (2013) find a companion fraction of 50 % for O stars in the Tarantula Nebula in the Large Magellanic Cloud. Sana et al. (2012) estimate that 70 % of O stars have a companion close enough that they will exchange mass with it at some point during their main sequence or post-main sequence evolution, and that 1/3 of O stars have a companion so close that they will merge.[6] From the standpoint of formation theories, a high binary fraction is expected regardless of whether massive stars are formed via accretion (e.g. Kratter et al. 2008, 2010; Krumholz et al. 2012) or collisions (e.g. Portegies Zwart et al. 1999; Bonnell and Bate 2005). However, much less is known about the prevalence of low-mass companions to massive stars, or to tight massive binaries,

---

[5] O stars outside clusters are likely to have been dynamically ejected from the cluster where they were born, and in the process stripped of companions.

[6] Mergers and mass transfer may also be significant during pre-main sequence evolution – see Krumholz and Thompson (2007).

and the statistics of low-mass companions to massive stars provide another potential test of formation models.

Accretion-based models predict that, in addition to their massive companions, massive stars are also very likely to have low-mass companions at separations of ∼100–1,000 AU (Kratter and Matzner 2006; Kratter et al. 2008, 2010; Krumholz et al. 2012). The authors of collisional models have not thus far published detailed predictions for massive binary properties, but these should be trivial to obtain from the simulations already run, and it seems likely that the dense dynamical environment required for collisions would strip any low-mass, distant companions from massive stars. Thus observations capable of probing large mass ratios at intermediate separations might be a valuable test of massive star formation models.

This range is unfortunately relatively hard to probe via observations, as the expected radial velocity shifts are too small to be measured against the broad lines of a massive star, and the large contrast ratio makes direct imaging difficult. Observations using high contrast techniques like speckle imaging (Mason et al. 2009), adaptive optics (Sana et al. 2010), and lucky imaging (Maíz Apellániz 2010) are starting to push into this range, but still have some distance to go. Consider a primary massive star of mass M_p with a companion of mass qM_p (with $q < 1$) in a circular orbit with semi-major axis a. The system is a distance d from the Sun. Spectroscopic surveys are generally limited in their ability to detect companions by the velocity semi-amplitude $v_{\lim}$ to which they are sensitive, which is generally ∼5–10 km s^{-1} depending on the linewidths of the primary star (e.g. Kiminki et al. 2007; Kobulnicky and Fryer 2007). The companion is detectable only if

$$a < \left(\frac{q^2}{q+1}\right)\frac{GM_p}{v_{\lim}^2} \approx 5.3 \left(\frac{q}{0.1}\right)^2 \left(\frac{M_p}{60\,M_\odot}\right) \left(\frac{v_{\lim}}{10\text{ km s}^{-1}}\right)^{-2} \text{ AU}. \quad (3.13)$$

Imaging surveys are limited by the contrast they can achieve. For example, Sana et al. (2010) estimate that their detection threshold is a contrast of $\Delta K_s \approx \Delta K_{s,0}(r - 0\farcs24)^{1/3}$, where $\Delta K_{s,0} = 6$ mag and r is the angular separation in arcsec and ΔK_s is the contrast in the K_s band. Given a mass-magnitude relationship $K_s(M)$, a companion will be detectable if

$$|K_s(M_p) - K_s(qM_p)| < \Delta K_{s,0} \left[2.06 \times 10^5 \left(\frac{\text{arcsec}}{\text{rad}}\right)\frac{a}{d} - 0\farcs24\right]^{1/3}. \quad (3.14)$$

Figure 3.7 shows the ranges of q and a over which companions to massive stars are detectable given these sensitivities.

The next generation of high-contrast systems designed for planet imaging, such as the Gemini Planet Imager (GPI) and Spectro-Polarimetric High-contrast Exoplanet Research instrument (SPHERE) should push much farther and be able to detect even very low mass companions to massive stars. Indeed, the contrast ratios these instruments can achieve is such that, outside of their occulting stops, they should be sensitive to companions to O stars down to the hydrogen burning limit. Figure 3.7 shows an estimate of the sensitivity region for GPI. Observations using

3 The Formation of Very Massive Stars

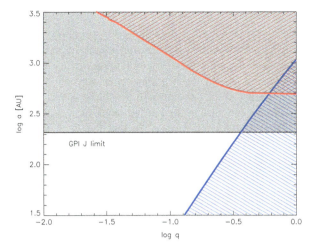

Fig. 3.7 Estimated detectability of companions to massive stars as a function of mass ratio q and semi-major axis a using spectroscopic surveys (*blue hashed region*), adaptive optics imaging surveys (*red dashed region*), and using a next-generation instrument like GPI (*gray region*). These sensitivities are calculated for a hypothetical primary of mass $M_p = 60\,M_\odot$ at a distance $d = 2$ kpc. The limit for spectroscopy is computed using Eq. (3.13) assuming a velocity semi-amplitude limit $v_{\mathrm{lim}} = 5\,\mathrm{km\,s}^{-1}$. The limit for adaptive optics is computed from Eq. (3.14) using the Padova mass-magnitude relations for ZAMS stars (Marigo et al. 2008; Girardi et al. 2010). The GPI limit shown is the physical size corresponding to the $0''.11$ size of the GPI occulting stop in J band

these instruments provide a definitive census of massive star companions at high mass ratio and intermediate separation. This is likely to prove a powerful constraint on formation models.

> **Conclusions and Summary: Does Star Formation Have an Upper Mass Limit?**
> Having discussed the two main formation scenarios, I now return to the question of whether star formation has a mass limit. To review, there is at present no really convincing evidence that any mechanism is capable of halting the growth of stars by accretion. The classical mechanism for limiting stellar masses is radiation pressure, but non-spherical accretion, produced by some combination of disks, outflow cavities, and instabilities appears to defeat this limit. Similarly, the problem of gas fragmenting too strongly to form massive stars appears to be solved by a combination of radiative heating and magnetic fields, though the possibility that disk fragmentation might at some point limit stellar masses remains. Photoionization and stellar winds are somewhat more promising as mechanisms that might limit stars' growth, but these remain at best possibilities. There are no real analytic
>
> (continued)

models applicable to present-day (as opposed to primordial) star formation that suggest what limits these mechanisms might impose, and there are no simulations demonstrating that either of these processes are capable of terminating accretion. A fair description of the state of the field a decade ago might have been that the presumption was in favor of feedback limiting massive star formation, and that the burden of proof was on those trying to show that feedback could be overcome. The last decade of work has reversed that situation, with all tests thus-far performed showing that accretion is very difficult to reverse. This does not prove that no mechanism can limit stellar masses, but does mean that such a limit would need to be demonstrated.

For collisions, the question is not whether but where they can create very massive stars. There is no doubt that collisions and collisional growth will occur if the conditions are dense enough, and the only question is the frequency with which such dense conditions are created in nature, which in turn will determine the contribution of the collisional formation channel to the overall population of very massive stars. No presently-observed star cluster has a density high enough for collisions to be likely, but it is possible that these clusters experienced a very dense phase during which collisions occurred. This could have been either an early gas-dominated phase or a later phase of core collapse aided by primordial mass segregation and high levels of primordial substructure. However, the threshold density required to achieve significant collisional growth depends on details of massive star mergers and winds that are poorly understood. Even for favorable assumptions about these uncertain parameters, it is not clear that the observed present-day properties of massive star clusters can be reconciled with an evolutionary history in which they were once dense enough to have produced collisions.

There are a number of observational tests that may be able to settle the question of which of these mechanisms is the dominant route to the formation of the most massive stars. Accretion models predict that massive stars are simply the tip of the iceberg of normal star formation, so that the high end of the stellar mass function is continuous and does not depend radically on the environment, and that massive stars are likely to have low-mass as well as high-mass companions. Although the observable consequences of the collisional formation models have received somewhat less attention, such models appear to predict quite different results: there should be a large gap in stellar mass functions separating the bulk of the accretion-formed stellar population from the few collisionally-formed stars, and this feature should appear only in the most massive and densest clusters. It seems likely that these collisionally-formed stars will lack low-mass companions. It should be possible to perform most or all of these observational tests with the coming generation of telescopes and instruments, which will provide higher angular resolution and contrast sensitivity than have previously been possible.

Acknowledgements I thank all the authors who provided figures for this review: H. Baumgardt, J. Dale, N. Moeckel, A. C. Myers, and D. Vanbeveren. During the writing of this review I was supported by NSF CAREER grant AST-0955300, NASA ATP grant NNX13AB84G, and NASA TCAN grant NNX14AB52G. I also thank the Aspen Center for Physics, which is supported by NSF Grant PHY-1066293, for hospitality during the writing of this review.

References

Allison, R. J., Goodwin, S. P., Parker, R. J., et al. (2009). *Astrophysical Journal Letters, 700*, L99.
Andersen, M., Zinnecker, H., Moneti, A., et al. (2009). *Astrophysical Journal, 707*, 1347.
Andrews, J. E., Calzetti, D., Chandar, R., et al. (2013). *Astrophysical Journal, 767*, 51.
Ardi, E., Baumgardt, H., & Mineshige, S. (2008). *Astrophysical Journal, 682*, 1195.
Ascenso, J., Alves, J., & Lago, M. T. V. T. (2009). *Astronomy and Astrophysics, 495*, 147.
Banerjee, S., Kroupa, P., & Oh, S. (2012a). *Astrophysical Journal, 746*, 15.
Banerjee, S., Kroupa, P., & Oh, S. (2012b). *Monthly Notices of the Royal Astronomical Society, 426*, 1416.
Bastian, N., Covey, K. R., & Meyer, M. R. (2010). *Annual Review of Astronomy and Astrophysics, 48*, 339.
Bate, M. R. (2009). *Monthly Notices of the Royal Astronomical Society, 392*, 1363.
Bate, M. R. (2012). *Monthly Notices of the Royal Astronomical Society, 419*, 3115.
Baumgardt, H., & Klessen, R. S. (2011). *Monthly Notices of the Royal Astronomical Society, 413*, 1810.
Belkus, H., Van Bever, J., & Vanbeveren, D. (2007). *Astrophysical Journal, 659*, 1576.
Beuther, H., & Schilke, P. (2004). *Science, 303*, 1167.
Beuther, H., Sridharan, T. K., & Saito, M. (2005). *Astrophysical Journal Letters, 634*, L185.
Binney, J., & Tremaine, S. (1987). *Galactic dynamics*. Princeton: Princeton University Press.
Bonnell, I. A., & Bate, M. R. (2005). *Monthly Notices of the Royal Astronomical Society, 362*, 915.
Bonnell, I. A., Bate, M. R., & Vine, S. G. (2003). *Monthly Notices of the Royal Astronomical Society, 343*, 413.
Bonnell, I. A., Bate, M. R., & Zinnecker, H. (1998). *Monthly Notices of the Royal Astronomical Society, 298*, 93.
Bonnell, I. A., Clarke, C. J., Bate, M. R., & Pringle, J. E. (2001). *Monthly Notices of the Royal Astronomical Society, 324*, 573.
Bontemps, S., Motte, F., Csengeri, T., & Schneider, N. (2010). *Astronomy and Astrophysics, 524*, A18+.
Bressert, E., Bastian, N., Evans, C. J., et al. (2012). *Astronomy and Astrophysics, 542*, A49.
Calzetti, D., Chandar, R., Lee, J. C., et al. (2010). *Astrophysical Journal Letters, 719*, L158.
Castor, J., McCray, R., & Weaver, R. (1975). *Astrophysical Journal Letters, 200*, L107.
Chandrasekhar, S., & Fermi, E. (1953). *Astrophysical Journal, 118*, 116.
Clarke, C. J., & Bonnell, I. A. (2008). *Monthly Notices of the Royal Astronomical Society, 388*, 1171.
Commerçon, B., Hennebelle, P., & Henning, T. (2011). *Astrophysical Journal Letters, 742*, L9.
Crutcher, R. M. (2012). *Annual Review of Astronomy and Astrophysics, 50*, 29.
Cunningham, A. J., Klein, R. I., Krumholz, M. R., & McKee, C. F. (2011). *Astrophysical Journal, 740*, 107.
Dale, J. E., & Bonnell, I. A. (2008). *Monthly Notices of the Royal Astronomical Society, 391*, 2.
Dale, J. E., Bonnell, I. A., Clarke, C. J., & Bate, M. R. (2005). *Monthly Notices of the Royal Astronomical Society, 358*, 291.
Dale, J. E., & Davies, M. B. (2006). *Monthly Notices of the Royal Astronomical Society, 366*, 1424.
Dekel, A., & Krumholz, M. R. (2013). *Monthly Notices of the Royal Astronomical Society, 432*, 455.

de Mink, S. E., Langer, N., Izzard, R. G., Sana, H., & de Koter, A. (2013). *Astrophysical Journal, 764*, 166.
Dobbs, C. L., Bonnell, I. A., & Clark, P. C. (2005). *Monthly Notices of the Royal Astronomical Society, 360*, 2.
Espinoza, P., Selman, F. J., & Melnick, J. (2009). *Astronomy and Astrophysics, 501*, 563.
Freitag, M., & Benz, W. (2005). *Monthly Notices of the Royal Astronomical Society, 358*, 1133.
Fujii, M. S., & Portegies Zwart, S. (2013). *Monthly Notices of the Royal Astronomical Society, 430*, 1018.
Fujii, M. S., Saitoh, T. R., & Portegies Zwart, S. F. (2012). *Astrophysical Journal, 753*, 85.
Fumagalli, M., da Silva, R. L., & Krumholz, M. R. (2011). *Astrophysical Journal Letters, 741*, L26.
Fűrész, G., Hartmann, L. W., Megeath, S. T., Szentgyorgyi, A. H., & Hamden, E. T. (2008). *Astrophysical Journal, 676*, 1109.
Gaburov, E., Lombardi, Jr., J. C., & Portegies Zwart, S. (2010). *Monthly Notices of the Royal Astronomical Society, 402*, 105.
Girardi, L., Williams, B. F., Gilbert, K. M., et al. (2010). *Astrophysical Journal, 724*, 1030.
Girichidis, P., Federrath, C., Banerjee, R., & Klessen, R. S. (2012). *Monthly Notices of the Royal Astronomical Society, 420*, 613.
Glebbeek, E., Gaburov, E., de Mink, S. E., Pols, O. R., & Portegies Zwart, S. F. (2009). *Astronomy and Astrophysics, 497*, 255.
Glebbeek, E., Gaburov, E., Portegies Zwart, S., & Pols, O. R. (2013). *Monthly Notices of the Royal Astronomical Society, 434*, 3497.
Goswami, S., Umbreit, S., Bierbaum, M., & Rasio, F. A. (2012). *Astrophysical Journal, 752*, 43.
Gürkan, M. A., Freitag, M., & Rasio, F. A. (2004). *Astrophysical Journal, 604*, 632.
Hennebelle, P., & Chabrier, G. (2008). *Astrophysical Journal, 684*, 395.
Hennebelle, P., & Chabrier, G. (2009). *Astrophysical Journal, 702*, 1428.
Hennebelle, P., & Chabrier, G. (2013). *Astrophysical Journal, 770*, 150.
Hennebelle, P., Commerçon, B., Joos, M., et al. (2011). *Astronomy and Astrophysics, 528*, A72+.
Heyer, M., Krawczyk, C., Duval, J., & Jackson, J. M. (2009). *Astrophysical Journal, 699*, 1092.
Hillenbrand, L. A., & Hartmann, L. W. (1998). *Astrophysical Journal, 492*, 540.
Hollenbach, D., Johnstone, D., Lizano, S., & Shu, F. (1994). *Astrophysical Journal, 428*, 654.
Hopkins, P. F. (2012). *Monthly Notices of the Royal Astronomical Society, 423*, 2037.
Hosokawa, T., Offner, S. S. R., & Krumholz, M. R. (2011). *Astrophysical Journal, 738*, 140.
Hosokawa, T., & Omukai, K. (2009). *Astrophysical Journal, 691*, 823.
Hosokawa, T., Yoshida, N., Omukai, K., & Yorke, H. W. (2012). *Astrophysical Journal Letters, 760*, L37.
Huff, E. M., & Stahler, S. W. (2006). *Astrophysical Journal, 644*, 355.
Jacquet, E., & Krumholz, M. R. (2011). *Astrophysical Journal, 730*, 116.
Jiang, Y.-F., Davis, S. W., & Stone, J. M. (2013). *Astrophysical Journal, 763*, 102.
Jijina, J., & Adams, F. C. (1996). *Astrophysical Journal, 462*, 874.
Kahn, F. D. (1974). *Astronomy and Astrophysics, 37*, 149.
Keto, E. (2003). *Astrophysical Journal, 599*, 1196.
Keto, E. (2007). *Astrophysical Journal, 666*, 976.
Keto, E., & Klaassen, P. (2008). *Astrophysical Journal Letters, 678*, L109.
Keto, E., & Wood, K. (2006). *Astrophysical Journal, 637*, 850.
Kim, S. S., Figer, D. F., Kudritzki, R. P., & Najarro, F. (2006). *Astrophysical Journal Letters, 653*, L113.
Kiminki, D. C., Kobulnicky, H. A., Kinemuchi, K., et al. (2007). *Astrophysical Journal, 664*, 1102.
Klaassen, P. D., & Wilson, C. D. (2007). *Astrophysical Journal, 663*, 1092.
Kobulnicky, H. A., & Fryer, C. L. (2007). *Astrophysical Journal, 670*, 747.
Kratter, K. M., & Matzner, C. D. (2006). *Monthly Notices of the Royal Astronomical Society, 373*, 1563.
Kratter, K. M., Matzner, C. D., & Krumholz, M. R. (2008). *Astrophysical Journal, 681*, 375.

Kratter, K. M., Matzner, C. D., Krumholz, M. R., & Klein, R. I. (2010). *Astrophysical Journal, 708*, 1585.
Kraus, S., Weigelt, G., Balega, Y. Y., et al. (2009). *Astronomy and Astrophysics, 497*, 195.
Krumholz, M. R. (2006). *Astrophysical Journal Letters, 641*, L45.
Krumholz, M. R., Cunningham, A. J., Klein, R. I., & McKee, C. F. (2010). *Astrophysical Journal, 713*, 1120.
Krumholz, M. R., Klein, R. I., & McKee, C. F. (2007). *Astrophysical Journal, 656*, 959.
Krumholz, M. R., Klein, R. I., & McKee, C. F. (2011). *Astrophysical Journal, 740*, 74.
Krumholz, M. R., Klein, R. I., & McKee, C. F. (2012). *Astrophysical Journal, 754*, 71.
Krumholz, M. R., Klein, R. I., McKee, C. F., Offner, S. S. R., & Cunningham, A. J. (2009). *Science, 323*, 754.
Krumholz, M. R., & Matzner, C. D. (2009). *Astrophysical Journal, 703*, 1352.
Krumholz, M. R., Matzner, C. D., & McKee, C. F. (2006). *Astrophysical Journal, 653*, 361.
Krumholz, M. R., & McKee, C. F. (2008). *Nature, 451*, 1082.
Krumholz, M. R., McKee, C. F., & Klein, R. I. (2005). *Astrophysical Journal Letters, 618*, L33.
Krumholz, M. R., & Thompson, T. A. (2007). *Astrophysical Journal, 661*, 1034.
Kudritzki, R. P., Puls, J., Lennon, D. J., et al. (1999). *Astronomy and Astrophysics, 350*, 970.
Kuiper, R., Klahr, H., Beuther, H., & Henning, T. (2010). *Astrophysical Journal, 722*, 1556.
Kuiper, R., Klahr, H., Beuther, H., & Henning, T. (2011). *Astrophysical Journal, 732*, 20.
Kuiper, R., Klahr, H., Beuther, H., & Henning, T. (2012). *Astronomy and Astrophysics, 537*, A122.
Kuiper, R., & Yorke, H. W. (2013). *Astrophysical Journal, 763*, 104.
Larson, R. B., & Starrfield, S. (1971). *Astronomy and Astrophysics, 13*, 190.
Leitherer, C., Robert, C., & Drissen, L. (1992). *Astrophysical Journal, 401*, 596.
Li, Z.-Y., & Nakamura, F. (2006). *Astrophysical Journal Letters, 640*, L187.
Lopez, L. A., Krumholz, M. R., Bolatto, A. D., Prochaska, J. X., & Ramirez-Ruiz, E. (2011). *Astrophysical Journal, 731*, 91.
Lynden-Bell, D., & Wood, R. (1968). *Monthly Notices of the Royal Astronomical Society, 138*, 495.
Maíz Apellániz, J. (2010). *Astronomy and Astrophysics, 518*, A1+.
Marigo, P., Girardi, L., Bressan, A., et al. (2008). *Astronomy and Astrophysics, 482*, 883.
Maschberger, T., Clarke, C. J., Bonnell, I. A., & Kroupa, P. (2010). *Monthly Notices of the Royal Astronomical Society, 404*, 1061.
Mason, B. D., Hartkopf, W. I., Gies, D. R., Henry, T. J., & Helsel, J. W. (2009). *Astronomical Journal, 137*, 3358.
Massey, P., & Hunter, D. A. (1998). *Astrophysical Journal, 493*, 180.
Matzner, C. D. (2002). *Astrophysical Journal, 566*, 302.
McKee, C. F., & Ostriker, E. C. (2007). *Annual Review of Astronomy and Astrophysics, 45*, 565.
McKee, C. F., & Tan, J. C. (2003). *Astrophysical Journal, 585*, 850.
McKee, C. F., & Tan, J. C. (2008). *Astrophysical Journal, 681*, 771.
McMillan, S. L. W., Vesperini, E., & Portegies Zwart, S. F. (2007). *Astrophysical Journal Letters, 655*, L45.
Mestel, L., & Spitzer, L., Jr. (1956). *Monthly Notices of the Royal Astronomical Society, 116*, 503.
Moeckel, N., & Bonnell, I. A. (2013). *Monthly Notices of the Royal Astronomical Society*. Submitted, arXiv:1301.6959
Moeckel, N., & Clarke, C. J. (2011). *Monthly Notices of the Royal Astronomical Society, 410*, 2799.
Myers, A. T., McKee, C. F., Cunningham, A. J., Klein, R. I., & Krumholz, M. R. (2013). *Astrophysical Journal, 766*, 97.
Nakamura, F., & Li, Z.-Y. (2007). *Astrophysical Journal, 662*, 395.
Nakano, T. (1989). *Astrophysical Journal, 345*, 464.
Nakano, T., Hasegawa, T., & Norman, C. (1995). *Astrophysical Journal, 450*, 183.
Oey, M. S., Lamb, J. B., Kushner, C. T., Pellegrini, E. W., & Graus, A. S. (2013). *Astrophysical Journal, 768*, 66.

Offner, S. S. R., Hansen, C. E., & Krumholz, M. R. (2009a). *Astrophysical Journal Letters, 704*, L124
Offner, S. S. R., Klein, R. I., McKee, C. F., & Krumholz, M. R. (2009b). *Astrophysical Journal, 703*, 131
Padoan, P., & Nordlund, Å. (2002). *Astrophysical Journal, 576*, 870
Padoan, P., Nordlund, Å., Kritsuk, A. G., Norman, M. L., & Li, P. S. (2007). *Astrophysical Journal, 661*, 972.
Pang, X., Grebel, E. K., Allison, R. J., et al. (2013). *Astrophysical Journal, 764*, 73.
Peters, T., Banerjee, R., Klessen, R. S., & Mac Low, M. (2011). *Astrophysical Journal, 729*, 72.
Peters, T., Banerjee, R., Klessen, R. S., et al. (2010a). *Astrophysical Journal, 711*, 1017.
Peters, T., Klessen, R. S., Mac Low, M., & Banerjee, R. (2010b). *Astrophysical Journal, 725*, 134.
Portegies Zwart, S. F., Makino, J., McMillan, S. L. W., & Hut, P. (1999). *Astronomy and Astrophysics, 348*, 117.
Portegies Zwart, S. F., & McMillan, S. L. W. (2002). *Astrophysical Journal, 576*, 899.
Portegies Zwart, S. F., McMillan, S. L. W., & Gieles, M. (2010). *Annual Review of Astronomy and Astrophysics, 48*, 431.
Preibisch, T., Sonnhalter, C., & Yorke, H. W. (1995). *Astronomy and Astrophysics, 299*, 144.
Repolust, T., Puls, J., & Herrero, A. (2004). *Astronomy and Astrophysics, 415*, 349.
Richer, J. S., Shepherd, D. S., Cabrit, S., Bachiller, R., & Churchwell, E. (2000). In *Protostars and planets IV* (p. 867). Tucson: University of Arizona Press.
Rogers, H., & Pittard, J. M. (2013). *Monthly Notices of the Royal Astronomical Society, 431*, 1337.
Ruffert, M., & Arnett, D. (1994). *Astrophysical Journal, 427*, 351.
Sana, H., de Koter, A., de Mink, S. E., et al. (2013). *Astronomy and Astrophysics, 550*, A107.
Sana, H., de Mink, S. E., de Koter, A., et al. (2012). *Science, 337*, 444.
Sana, H., Momany, Y., Gieles, M., et al. (2010). *Astronomy and Astrophysics, 515*, A26+.
Selman, F. J., & Melnick, J. (2013). *Astronomy and Astrophysics, 552*, A94.
Shakura, N. I., & Sunyaev, R. A. (1973). *Astronomy and Astrophysics, 24*, 337
Shirley, Y. L., Evans, II, N. J., Young, K. E., Knez, C., & Jaffe, D. T. (2003). *Astrophysical Journal Supplement, 149*, 375.
Spitzer, L, Jr. (1969). *Astrophysical Journal Letters, 158*, L139.
Stacy, A., Greif, T. H., & Bromm, V. (2012). *Monthly Notices of the Royal Astronomical Society, 422*, 290.
Stahler, S. W., & Palla, F. (2005). *The formation of stars*. Berlin: Wiley-VCH.
Susa, H. (2013). *Astrophysical Journal, 773*, 185.
Suttner, G., Yorke, H. W., & Lin, D. N. C. (1999). *Astrophysical Journal, 524*, 857.
Suzuki, T. K., Nakasato, N., Baumgardt, H., et al. (2007). *Astrophysical Journal, 668*, 435.
Tenorio-Tagle, G., Wünsch, R., Silich, S., & Palouš, J. (2007). *Astrophysical Journal, 658*, 1196.
Tobin, J. J., Hartmann, L., Furesz, G., Mateo, M., & Megeath, S. T. (2009). *Astrophysical Journal, 697*, 1103.
Toomre, A. (1964). *Astrophysical Journal, 139*, 1217.
Vanbeveren, D., Belkus, H., van Bever, J., & Mennekens, N. (2009). *Astrophysics and Space Science, 324*, 271.
Vink, J. S., de Koter, A., & Lamers, H. J. G. L. M. (2000). *Astronomy and Astrophysics, 362*, 295.
Vink, J. S., de Koter, A., & Lamers, H. J. G. L. M. (2001). *Astronomy and Astrophysics, 369*, 574
Walmsley, M. (1995). In *Revista Mexicana de Astronomia y Astrofisica Conference Series* (p. 137), Cozumel, Mexico.
Wang, P., Li, Z., Abel, T., & Nakamura, F. (2010). *Astrophysical Journal, 709*, 27.
Weaver, R., McCray, R., Castor, J., Shapiro, P., & Moore, R. (1977). *Astrophysical Journal, 218*, 377.
Weidner, C., Kroupa, P., & Bonnell, I. A. D. (2010). *Monthly Notices of the Royal Astronomical Society, 401*, 275.
Weidner, C., Kroupa, P., & Pflamm-Altenburg, J. (2013). *Monthly Notices of the Royal Astronomical Society, 434*, 84.
Whitworth, A. (1979). *Monthly Notices of the Royal Astronomical Society, 186*, 59.

Wolfire, M. G., & Cassinelli, J. P. (1986). *Astrophysical Journal, 310*, 207
Wolfire, M. G., & Cassinelli, J. P. (1987). *Astrophysical Journal, 319*, 850.
Yorke, H. W., & Bodenheimer, P. (1999). *Astrophysical Journal, 525*, 330.
Yorke, H. W., & Kaisig, M. (1995). *Computer Physics Communications, 89*, 29.
Yorke, H. W., & Sonnhalter, C. (2002). *Astrophysical Journal, 569*, 846.
Yungelson, L. R., van den Heuvel, E. P. J., Vink, J. S., Portegies Zwart, S. F., & de Koter, A. (2008). *Astronomy and Astrophysics, 477*, 223.

Chapter 4
Mass-Loss Rates of Very Massive Stars

Jorick S. Vink

Abstract We discuss the basic physics of hot-star winds and we provide mass-loss rates for (very) massive stars. Whilst the emphasis is on theoretical concepts and line-force modelling, we also discuss the current state of observations and empirical modelling, and we address the issue of wind clumping.

4.1 Introduction

Mass loss via stellar winds already plays a well-documented role in the evolution of canonical 20–60 $M_\odot$ O stars, because of the removal of *mass* from the outer layers, as well as the removal of *angular momentum*. However, nowhere is mass loss more dominant than for the most massive stars. As very massive stars (VMS) evolve structurally close to chemically homogeneously, the detailed mixing processes due to rotation and magnetic fields are less relevant than for canonical massive stars. Instead, VMS evolution is determined by mass loss (Yungelson et al. 2008; Yusof et al. 2013; Köhler et al. 2015). However, there is uncertainty regarding the quantitative mass-loss rates, partly because of uncertain physics in close proximity to the Eddington (Γ) limit, and partly because O-star winds are inhomogeneous and clumpy, implying that empirical mass-loss rates are overestimated if one does not properly take clumping effects into account in the analysis.

In this mass-loss chapter, we start off in Sect. 4.2 with the mass-loss theory of canonical 20–60 $M_\odot$ O-star winds, which are optically thin, and where the traditional CAK theory due to Castor et al. (1975) is applicable. For VMS, the role of radiation pressure over gas pressure is even more important than for normal massive stars, and as VMS are in closer proximity to the Γ limit, at some point their winds are expected to become optically thick.

In Sect. 4.3, we discuss the optically thick wind theory for classical Wolf-Rayet (WR) stars with very strong emission lines and dense winds. Once we have reached

J.S. Vink (✉)
Armagh Observatory, Armagh, Ireland
e-mail: jsv@arm.ac.uk

a basic understanding of both optically thin and optically thick winds,[1] we discuss the transition from O to WR star winds in the context of VMS in Sect. 4.4. VMS are associated with WR stars of the WNh subtype. WNh implies the presence of both hydrogen (H) and nitrogen (N) at the surface. The latter is thought to have originated in the CNO-cycle, having reached the surface through mass loss and (rotational) mixing. WNh stars are thought to be core H-burning (see Martins' Chap. 2) and can thus be considered "O-stars on steroids". The reason they have a WR type spectrum is due to their strong winds, because of the proximity to the Eddington limit.

Another group of objects that may be relevant for VMS evolution are the Luminous Blue Variables (LBVs). Already in quiescence these objects reside in dangerous proximity to the Eddington limit, where they are subjected to outbursts and mass ejections. A discussion of both "quiet" and super-Eddington winds relevant to both the characteristic "moderate" S Dor variations and the "giant" outbursts, such as displayed by Eta Car in 1840, as well as the theory of super-Eddington winds are thus discussed in Sect. 4.6.

After this theoretical overview of homogeneous stellar winds, we consider clumped winds. To this purpose, we first discuss the diagnostics of smooth winds (Sect. 4.7) before turning to clumped winds in Sect. 4.8. We finish Sect. 4.8 with potential theories that may cause wind clumping, as well as some possibilities to quantify the number of clumps, before we summarize.

For the 2D effects of rotation on stellar winds, we refer to the review by Puls et al. (2008) and for more recent calculations to Müller and Vink (2014), which also includes a discussion of the diagnostics of axi-symmetric outflows.

4.2 O Stars with Optically Thin Winds

As each photon carries a momentum, $P = h\nu/c$, it was thought as early as the 1920s (e.g. Milne 1926) that radiative acceleration on spectral lines might selectively "eject" metal ions (such as iron, Fe) from stellar photospheres. However, it was not until the arrival of ultraviolet (UV) observations in the late 1960s that the theory of radiative line driving became the established theory describing the stationary outflows from massive OB stars. Lucy and Solomon (1970) and CAK showed that in case the momentum imparted on metal ions was shared through Coulomb interactions with the more abundant H and helium (He) species[2] in the atmospheric plasma, this would result in a substantial rate of mass loss $\dot{M}$, affecting the evolution of massive stars significantly (Conti 1976; Langer et al. 1994; Meynet and Maeder

---

[1] Note however that these winds are also driven by a myriad of lines, forming a "pseudo" continuum of lines.

[2] Note that for every Fe atom there are as many as 2,500 H atoms (for a solar abundance pattern; see Anders and Grevesse 1989).

2003; Eldridge and Vink 2006; Limongi and Chieffi 2006; Belkus et al. 2007; Brott et al. 2011; Hirschi's Chapter 6; and Woosley & Heger's Chapter 7).

4.2.1 Stellar Wind Equations

The basic idea of momentum transfer by line-scattering is that absorbed photons originate from a preferred direction, whereas the subsequent re-emission is averaged to be (more or less) isotropic. This change in direction angle θ leads to a *radial* transfer of momentum, $\Delta P = h/c(v_{in} \cos \theta_{in} - v_{out} \cos \theta_{out})$ – comprising the key to the momentum transfer with its associated line acceleration g_{rad}^{line}. The mass-loss rate through a spherical shell with radius r that surrounds the star is conserved, as may be noted from the equation of mass continuity

$$\dot{M} = 4\pi r^2 \rho(r) v(r). \qquad (4.1)$$

The equation of motion is given by:

$$v \frac{dv}{dr} = -\frac{GM}{r^2} - \frac{1}{\rho} \frac{dp}{dr} + g_{rad}, \qquad (4.2)$$

with inwards directed gravitational acceleration $g_{grav} = GM/r^2$ and an outwards directed gas pressure (p) term and total (continuum and line) radiative acceleration (g_{rad}). The wind initiation condition is that the total radiative acceleration, $g_{rad} = g_{rad}^{line} + g_{rad}^{cont}$ exceeds gravity beyond a certain point. With the equation of state, $p = a^2 \rho$, where a is the isothermal sound speed, the equation becomes:

$$\left(1 - \frac{a^2}{v^2}\right) v \frac{dv}{dr} = \frac{2a^2}{r} - \frac{da^2}{dr} - \frac{GM}{r^2} + g_{rad}. \qquad (4.3)$$

The prime challenge lies in accurately computing g_{rad}. For free electrons this concerns the Thomson opacity, $\sigma_e = s_e \rho$ (s_e proportional to cross section) and the flux:

$$g_{rad}^{Th} = \frac{1}{c\rho} \frac{\sigma_e L}{4\pi r^2} = g_{grav} \Gamma, \qquad (4.4)$$

with the Eddington parameter Γ representing the radiative acceleration over gravity, given by:

$$\Gamma = \frac{\kappa L}{4\pi cGM}. \qquad (4.5)$$

Spectral lines provide the dominant contribution to the overall radiative acceleration. The reason is that line scattering is intrinsically much stronger than electron scattering because of the resonant nature of bound-bound transitions (Gayley 1995), and although photons and matter are only allowed to interact at specific frequencies, they can be made to resonate over a wide range of stellar wind radii via the Doppler effect (see Owocki's Chap. 5).

For a single line at frequency v, with line optical depth τ, the line acceleration can be approximated by local quantities (Sobolev 1960). This approximation is valid as long as opacity, source function, and the velocity gradient (dv/dr) do not change significantly over a velocity range $\Delta v = v_{\text{th}}$, corresponding to a *spatial* region $\Delta r \approx v_{\text{th}}/(dv/dr)$, i.e. the Sobolev length. In the Sobolev approximation, the line acceleration becomes:

$$g_{\text{rad},i}^{\text{line}} = \frac{L_\nu \nu}{4\pi r^2 c^2} \left(\frac{dv}{dr}\right) \frac{1}{\rho} (1 - e^{-\tau}), \quad (4.6)$$

with L_ν the luminosity at the line frequency, and with

$$\tau = \bar{\kappa}\lambda/(dv/dr), \quad (4.7)$$

where $\bar{\kappa}$ represents the frequency integrated line-opacity and λ is the wavelength of the transition. For optically thin lines ($\tau < 1$) the line acceleration has the same $1/r^2$ dependence as electron scattering (Eq. 4.4), whereas for optically thick lines ($\tau > 1$) it depends on the velocity gradient (dv/dr), which is the root cause for the peculiar nature of line driving.

4.2.2 CAK Solution

The next step is to sum the line acceleration over all lines. In the CAK theory this is achieved through the line-strength distribution function that describes the statistical dependence of the number of lines on frequency position and line-strength (e.g. Puls et al. 2000). Combining the radiative line acceleration (Eq. 4.6) with the distribution of lines, the *total* line acceleration can be calculated by integration. It can be expressed in terms of the Thomson acceleration (Eq. 4.4) multiplied by the famous force-multiplier $M(t)$,

$$M(t) = \frac{g_{\text{rad}}^{\text{line}}}{g_{\text{rad}}^{\text{TH}}} = k\, t^{-\alpha} \propto \left(\frac{dv/dr}{\rho}\right)^\alpha, \quad (4.8)$$

where k and α are the so-called force multiplier parameters.

For the complete distribution of lines, the radiative acceleration depends on (dv/dr) through the power of α. CAK postulated that this term has a similar meaning as the velocity gradient entering the inertial term on the left hand side of Eq. (4.3). Assuming this is the case, the equation of motion becomes non-linear, and can be solved through a critical point that sets the mass-loss rate $\dot{M}$:

$$\dot{M} \propto (kL)^{1/\alpha} (M(1-\Gamma))^{1-1/\alpha}. \qquad (4.9)$$

And with velocity:

$$v(r) = v_\infty (1 - R/r)^\beta \qquad (4.10)$$

$$v_\infty = C_\infty \Big(\frac{2GM(1-\Gamma)}{R_*}\Big)^{\frac{1}{2}} = C_\infty v_{\rm esc}, \qquad (4.11)$$

where $C_\infty \approx 2.6$ for O stars, and $v_{\rm esc}$ is the photospheric escape velocity corrected for Thomson electron acceleration. β is exactly 0.5 for a point source, and in the range $\beta \approx 0.8 - 1$ for more realistic (finite sized) objects (Pauldrach et al. 1986; Müller and Vink 2008). For O stars, $\alpha \simeq 0.6$ and k is of the order of 0.1.

Using these relations, one can construct the modified wind momentum rate, $D_{\rm mom} = \dot{M} v_\infty (R_*/R_\odot)^{1/2}$. Given that v_∞ scales with the escape velocity (Eq. 4.11), $D_{\rm mom}$ scales with luminosity and effective line number only, and as long as $\alpha \simeq 2/3$, the effective mass $M(1-\Gamma)$ conveniently cancels from the product $\dot{M} v_\infty$, resulting in:

$$\log D_{\rm mom} \approx x \, \log(L/L_\odot) + D, \qquad (4.12)$$

(with slope x and offset D, depending on the flux-weighted number of driving lines), the "wind momentum luminosity relationship (WLR)" (Kudritzki et al. 1995; Puls et al. 1996; Vink et al. 2000). The relationship played an instrumental role in determining the empirical mass-loss metallicity (Z) dependence for O stars in the Local Group (Mokiem et al. 2007), and observed and predicted WLRs can be compared to test the validity of the theory, and to highlight potential shortcomings, e.g. concerning wind clumping. One should also realise that $\dot{M}$ is not only a function of L but also parameters like $T_{\rm eff}$. One should properly account for this multivariate behaviour of $\dot{M}$ when one attempts to compare observations to theory, and when one wishes to properly assess the effects of stellar wind mass loss in stellar evolution modelling.

We note that all CAK-type relations are only valid for spatially constant force multiplier parameters, k and α, which is not the case in more realistic models (Vink et al. 2000; Kudritzki 2002; Muijres et al. 2012a). Other assumptions involve the adoption of a core-halo structure, and the neglect of multi-line effects.

4.2.3 Predictions Using a Monte Carlo Radiative Transfer Approach

An alternative approach to CAK involves the Monte Carlo method developed by Abbott and Lucy (1985). Here photon-scattering histories are tracked on their journey outwards. At each interaction, momentum and energy are transferred from

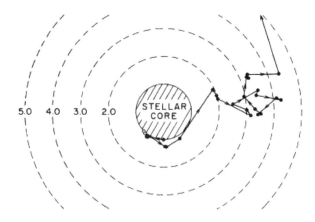

Fig. 4.1 Cartoon explaining the Monte Carlo method: photon path histories are tracked on their outwards journey (From Abbott and Lucy 1985)

the photons to the ions (see Fig. 4.1). One of the major advantages of the Monte Carlo method is that it easily allows for multi-line scattering, which becomes important in denser winds. Prior to the year 2000, theoretical mass-loss rates fell short of the observed rates for dense O star and WR winds, whilst for weak winds the oft-used single line approach overestimated mass-loss rates. The crucial point is that multiply scattered photons add radially outward momentum to the wind, and the momentum may exceed the single-scattering limit, i.e., $\eta = \dot{M} v_\infty / (L/c)$ can become larger than unity. The overall $\dot{M}$ can be obtained from global energy conservation:

$$\frac{1}{2} \dot{M} (v_\infty^2 + v_{\text{esc}}^2) = \Delta L, \qquad (4.13)$$

where ΔL is the total energy transferred per second from the radiation to the outflowing particles.

Vink et al. (2000) and Vink et al. (2001) used the Monte Carlo method to derive a mass-loss recipe, where for objects hotter than the so-called bi-stability jump at $\simeq 25{,}000$ K, the rates roughly scale as:

$$\dot{M} \propto L^{2.2} M^{-1.3} T_{\text{eff}} (v_\infty/v_{\text{esc}})^{-1.3}. \qquad (4.14)$$

The success of the Monte Carlo method is highlighted through the comparison of observed and predicted mass-loss rates in Vink (2006). Figures 4.1 and 4.4 of that review display the level of agreement between modified CAK models and observations on the one hand, and the Vink et al. (2000) predictions on the other hand. Despite remaining uncertainties due to an unknown amount of wind clumping, by properly including multiple scatterings, the results were shown to be equally successful for relatively weak (with $\dot{M} \sim 10^{-7}\ M_\odot\ \text{year}^{-1}$) as dense O-star winds (with $\dot{M} \sim 10^{-5}\ M_\odot\ \text{year}^{-1}$). The predictions can also be expressed via the WLR. For O-stars hotter than 27,500 K, the relation is shown in Fig. 4.2 and given by Eq. (4.12) with a slope $x = 1.83$.

Fig. 4.2 Predicted WLR for O stars hotter than 27 kK for a range of (L, M)-combinations in the upper HR diagram (From Vink et al. 2000)

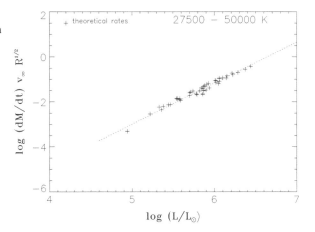

Traditionally, the prime drawback of the Monte Carlo approach was the usage of a pre-determined v_∞ (guided by accurate empirical values) but this assumption can be dropped, as discussed in the following.

4.2.4 Line Acceleration Formalism $g(r)$ for Monte Carlo Use

In solving the equation of motion self-consistently without relying on any free parameters, Müller and Vink (2008) determined the velocity field through the use of a parameterised description of the line acceleration that only depends on radius (rather than explicitly on the velocity gradient dv/dr as in CAK theory.) The line acceleration was obtained from Monte Carlo radiative transfer calculations. As this acceleration is determined in a statistical way, it shows scatter, and given the delicate nature of the equation of motion it should be represented by an appropriate analytic fit function. Müller and Vink (2008) motivated:

$$g_{\text{rad}}^{\text{line}} = \begin{cases} 0 & \text{if } r < r_\circ \\ g_\circ (1 - r_\circ/r)^\gamma / r^2 & \text{if } r \geq r_\circ, \end{cases} \quad (4.15)$$

where $g_\circ$, $r_\circ$, and γ are fit parameters to the Monte Carlo line acceleration. Müller and Vink (2008) derived an analytic solution of the velocity law in the outer wind, which was compared to the standard CAK β-law and subsequently used to derive v_∞ and the most representative β value.

Equation 4.3 is a critical point equation, where the left- and right-hand side vanish at the point $v(r_s) = a_\circ$, i.e. where r_s is the sonic-point radius. Müller and Vink (2008) showed that for the isothermal case and a line acceleration as described in Eq. (4.15), analytic expressions for all types of solutions of Eq. (4.3) can be constructed by means of the Lambert W function. A useful approximate wind solution for the velocity law can be constructed if the gas pressure related terms $2a^2/r$ and a/v are neglected. After some manipulation one obtains the approximate velocity law:

$$v(r) = \sqrt{\frac{R_* v_{esc}^2}{r} + \frac{2}{r_o} \frac{g_o}{(1+\gamma)} \left(1 - \frac{r_o}{r}\right)^{\gamma+1} + C}, \qquad (4.16)$$

where C is an integration constant. From this equation the terminal wind velocity can be derived if the integration constant C can be determined, which can be done by assuming that at radius r_o the velocity approaches zero, resulting in:

$$C = -\frac{R_* v_{esc}^2}{r_o}. \qquad (4.17)$$

In the limit $r \to \infty$:

$$v_\infty = \sqrt{\frac{2}{r_o} \frac{g_o}{(1+\gamma)} - \frac{R_* v_{esc}^2}{2}}. \qquad (4.18)$$

The terminal velocity v_∞ can also be determined from the equation of motion. At the critical point, the left-hand and right-hand side of Eq. (4.3) both equal zero. Introducing v_∞ in relation to g_o as expressed in Eq. (4.18), one obtains

$$v_{\infty,\text{new}} = \sqrt{\frac{2}{r_o}\left[\left(\frac{r_s}{r_s - r_o}\right)^\gamma \frac{r_s}{(1+\gamma)}\left(\frac{v_{esc}}{2} - 2r_s\right) - v_{esc}^2\right]}. \qquad (4.19)$$

A direct comparison to the β-law can be made for the supersonic regime of the wind, resulting in

$$\beta = \frac{1+\gamma}{2}. \qquad (4.20)$$

The procedure to obtain the best-β solution is that in each iteration step of the Monte Carlo simulation the values of g_o, r_o, and γ are determined by fitting the output line acceleration. Using these values and the radius of the sonic point, Eqs. (4.18)–(4.20) are used to determine v_∞ and β. v_∞ derived from Eq. (4.19), the predicted mass-loss rate, and the expression derived for β serve as input for the next model, with iterations continuing until convergence is achieved.

Muijres et al. (2012a) tested the Müller and Vink (2008) wind solutions through explicit numerical integrations of the fluid equation, also accounting for a temperature stratification, obtaining results that were in excellent agreement with the Müller & Vink solutions. These solutions were extended to 2D in Müller and Vink (2014).

4.3 Wolf-Rayet Stars with Optically Thick Winds

4.3.1 Wolf-Rayet (WR) Stars

WR stars can be divided into nitrogen-rich WN stars and carbon/oxygen rich WC/WO stars. The principal difference between the two subtypes is believed to be that the N-enrichment in WN stars is a by-product of H-burning, whereas the C/O in WC/WO stars is due to the arrival of He-burning products at the surface, showing strong emission lines of He, C and O.

The WR classification is purely spectroscopic, signalling the presence of strong and broad emission lines. Such spectra can originate in evolved stars, or alternatively from objects that formed with high initial masses and luminosities, the VMS. This latter group of WR stars may thus include objects still in their core H-burning phase of evolution: WNh stars.

Stellar radii determined from sophisticated non-LTE models are a factor of several ($\sim$3) larger than those predicted for the He-main sequence by stellar evolution modelling. In other words, there is a radius problem, and a potential solution might involve the inflation of a clumped outer envelope (Gräfener et al. 2012; See Chap. 5 for more details).

4.3.2 WR Wind Theory

WR stars have strong winds with large mass-loss rates, typically a factor of 10 larger than O-star winds with the same luminosity (see Fig. 4.3), and they are not easily explained by the optically thin line-driven wind theory by CAK. The observed wind efficiency η values are typically in the range of 1–5, i.e. well above the single-scattering limit. So, *if* WR-type winds are driven by radiation, photons must be

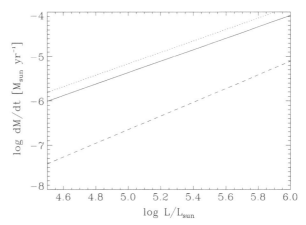

Fig. 4.3 Comparison of mass-loss rates from WR and Galactic O supergiants (From Puls et al. (2008)). *Solid* and *dotted lines* represent mean relations for H-poor WN (*solid*) and WC stars (*dotted*) (From Nugis & Lamers, 2000). The *dashed line* corresponds to Galactic O supergiants – taken (From the Mokiem et al. (2007) WLR)

scattered more than once. As the ionisation equilibrium decreases outwards, photons can interact with lines from a variety of different ions on their way out, whilst gaps between lines become "filled in" (see Lucy and Abbott 1993; Schaerer and Schmutz 1994; Springmann 1994; Gayley et al. 1995).

The initiation of the mass loss relies on the condition that the winds are already optically *thick* at the sonic point and that the photospheric line acceleration due to the high opacity "iron peak" may overcome gravity, thus driving a wind (Nugis and Lamers 2002).

The crucial point in such a critical-point analysis for optically thick winds is that due to their large mass-loss rates, the atmospheres become so extended and the sonic point of the wind is already reached at large flux-mean optical depth τ_s, which implies that the radiation can be treated in the diffusion approximation. The equation for the radiative acceleration can then be approximated to:

$$g_{\rm rad} = \frac{1}{c}\int \kappa_\nu F_\nu d\nu = \kappa_{\rm Ross}\frac{L_\star}{4\pi r^2 c}, \qquad (4.21)$$

where $\kappa_{\rm Ross}$ is the Rosseland mean opacity which can be taken from for instance the OPAL opacity tables (Iglesias and Rogers 1996). As $g_{\rm rad}$ does not depend on $(\frac{dv}{dr})$ Eq. (4.3) has a critical point at the sonic point r_s where $v = a$.

A finite value of $(\frac{dv}{dr})$ can only be obtained if the right hand side of Eq. (4.3) is zero at this point.

$$0 = -\frac{GM}{r_s^2} + \frac{2a^2}{r_s} - \frac{da^2}{dr_s} + g_{\rm rad}. \qquad (4.22)$$

For reasonable wind parameters the second and third term on the right-hand side of Eq. (4.22) become zero such that

$$\frac{GM}{r_s^2} \simeq g_{\rm rad}(r_s) \equiv \kappa_{\rm crit}\frac{L_\star}{4\pi r_s^2 c}. \qquad (4.23)$$

The Eddington limit with respect to the Rosseland mean opacity is thus crossed at the sonic point, and $\kappa_{\rm crit}$ for the Rosseland mean opacity can be computed for stellar parameters in terms of the (L/M) ratio.

In Fig. 4.4 the solution of Eq. (4.23) is plotted. This figure shows the relation between density and temperature with $\kappa_{\rm Ross}(\rho, T) = \kappa_{\rm crit}$, for a typical WC star. Below the sonic point, r_s, the radiative acceleration must be sub-Eddington, and $\kappa_{\rm Ross}$ thus needs to increase outward with decreasing density. Figure 4.4 shows how this condition is fulfilled at the hot edges of two Fe opacity peaks, one "cool" one at $\sim$70 kK and a "hot" one above 160 kK. The resulting mass-loss rates on these parts of the curve are given by $\dot{M} = 4\pi R_\star^2 \rho a$. To determine the actual density and temperature at the sonic point, Nugis and Lamers (2002) utilised the approximate relation between temperature and optical depth due to Lucy (1971) (see also Gräfener and Vink 2013):

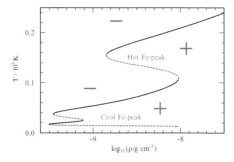

Fig. 4.4 Solution of Eq. (4.23) in the ρ-T plane. The sonic-point conditions for an optically thick wind, i.e. $\kappa_{\rm Ross} = \kappa_{\rm crit}$ with *outward increasing* $\kappa_{\rm Ross}$, are fulfilled at the *solid parts* of the curve around 70 kK, and above 160 kK. The Rosseland opacities are taken from the OPAL opacity tables (From Gräfener & Hamann)

$$T_S^4(r) = \frac{3}{4}T_{\rm eff}^4\left(\tau_S(r) + \frac{4}{3}W(r)\right), \qquad (4.24)$$

with the modified optical depth τ_S and the dilution factor W, which is close to unity. τ_S is obtained from the assumption that the outer wind is driven by radiation, and by combining Eqs. (4.34) and (4.35) of Nugis and Lamers (2002) for the optical depth and the temperature stratification, the resulting mass-loss rate for optically-thick winds is:

$$\dot{M} = C\frac{aT_s^4 R_S^3}{M}.$$

Nugis and Lamers (2002) found that the observed WR mass-loss rates are in agreement with this optically-thick wind assumption, and with a *bifurcation* of two sonic-point temperature regimes: a "cool" regime corresponding to late-type WN (WNL) stars, and a hot regime for early-type WR (WC and WN) stars.

4.3.3 Hydrodynamic Optically Thick Wind Models

Gräfener and Hamann (2005) included the OPAL Fe-peak opacities of the ions Fe IX–XVII in more sophisticated models that treat the full set of non-LTE population numbers in combination with the radiation field in the co-moving frame (CMF). Combining these models with the equations of hydrodynamics, Gräfener & Hamann obtained a self-consistent model for the WC5 star WR 111. The resulting wind acceleration and Fe-ionisation structure are depicted in Fig. 4.5. $g_{\rm rad}$ was obtained from an integration of the product of opacity and flux over frequency (see Eq. 4.21). Wind clumping was treated in the optically thin ("micro") clumping approach (see Sect. 4.8.1). With a mass-loss rate of $\dot{M} = 10^{-5.14}\,M_\odot$/year and

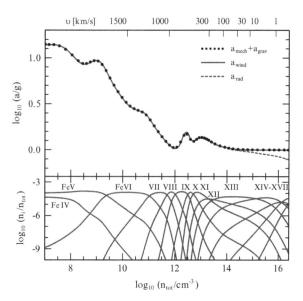

Fig. 4.5 *Top panel*: the radiative acceleration of the Gräfener and Hamann (2005) WC5 star model WR 111 (expressed in units of the local gravity). The wind acceleration g_{wind} due to radiation and gas pressure balances the mechanical and gravitational acceleration $g_{mech} + g_{grav}$. *Bottom panel*: the Fe-ionisation structure

terminal wind velocity of $v_\infty = 2{,}010\,\mathrm{km/s}$, the observed spectrum was also reproduced, although the electron scattering wings highlighted that the assumed clumping factor of $D = 50$ was rather (too) large given that WC stars generally seem to have clumping factors of the order of $D = 10$, as determined from electron scattering wings (Hillier 1991; Hamann and Koesterke 1998). The models might therefore underestimate the mass loss rate by a factor of $\sqrt{5}$. This is likely due to the omission of opacities of intermediate-mass elements, such as Cl, Ne, Ar, S, and P, which according to Monte Carlo models may account for up to half of the total line acceleration in the outer wind (Vink et al. 1999).

4.4 VMS and the Transition Between Optically Thin and Thick Winds

There are many uncertainties in the quantitative mass-loss rates of both VMS as well as canonical 20–60 $M_\odot$ massive stars. One reason is related to the role of wind clumping, which will be discussed later, but there are also uncertainties related to modelling techniques. Nevertheless, arguably the most pressing uncertainty is actually still qualitative! Do VMS winds become optically thick in Nature?[3] And if so, would this lead to an accelerated increase of $\dot{M}$? And if so, at what point does the transition occur?

---

[3]Note that Pauldrach et al. (2012) argue that VMS winds remain optically thin.

4.4.1 Analytic Derivation of Transition Mass-Loss Rate

As hydrostatic equilibrium is a good approximation for the subsonic part of the wind the terms on the right-hand side of Eq. (4.2) cancel each other. In the supersonic portion of the wind the gas pressure gradient becomes small, and through multiplying Eq. (4.2) by $4\pi r^2$, it reads:

$$4\pi \rho r^2 v dv = 4\pi r^2 (g_{\text{rad}} - g) dr. \tag{4.25}$$

Employing the mass-continuity equation, one obtains

$$\dot{M} dv = 4\pi GM(\Gamma(r) - 1)\rho dr. \tag{4.26}$$

Where $\Gamma(r)$ the Eddington factor with respect to the total flux-mean opacity κ_F: $\Gamma(r) = \frac{\kappa_F L}{4\pi cGM}$. Using the wind optical depth $\tau = \int_{r_s}^{\infty} \kappa_F \rho \, dr$, one obtains

$$\frac{\dot{M}}{L/c} dv = \kappa_F \rho \frac{\Gamma - 1}{\Gamma} dr = \frac{\Gamma - 1}{\Gamma} d\tau. \tag{4.27}$$

Assuming hydrostatic equilibrium below the sonic point, in integral form this becomes:

$$\int_0^{v_\infty} \frac{\dot{M}}{L/c} dv = \frac{\dot{M} v_\infty}{L/c} = \int_{r_s}^{\infty} \frac{\Gamma - 1}{\Gamma} d\tau \simeq \tau_S. \tag{4.28}$$

Where it is assumed that Γ is significantly larger than one in the supersonic region, such that the factor $\frac{\Gamma-1}{\Gamma}$ becomes close to unity, and

$$\dot{M} v_\infty = \frac{L}{c} \tau. \tag{4.29}$$

Vink and Gräfener (2012) derived a condition for the wind efficiency number η:

$$\eta = \frac{\dot{M} v_\infty}{L/c} = \tau = 1. \tag{4.30}$$

The key point is that one can employ the unique condition $\eta = \tau = 1$ right at the transition from optically thin O-star winds to optically-thick WR winds. In other words, if one were to have a data-set containing luminosities for O and WR stars, the transition mass-loss rate $\dot{M}_{\text{trans}}$ is obtained by simply considering the transition luminosity L_{trans} and the terminal velocity v_∞ representing the transition point from O to WR stars:

$$\dot{M}_{\text{trans}} = \frac{L_{\text{trans}}}{v_\infty c} \tag{4.31}$$

This transition point can be obtained by purely spectroscopic means, *independent of any assumptions regarding wind clumping*.

As $\Gamma = g_{\rm rad}/g$ is expected to be connected to the ratio $(v_\infty + v_{\rm esc})/v_{\rm esc} = v_\infty/v_{\rm esc} + 1$, and $f \simeq \frac{\Gamma-1}{\Gamma}$, Vink & Gräfener followed a model-independent approach, adopting β-type velocity laws, as well as full hydrodynamic wind models, computing the integral $\tau = \int_{r_s}^{\infty} \kappa \rho\, dr$ numerically using the flux-mean opacity $\kappa_F(r)$. The mean opacity κ_F follows from the resulting radiative acceleration $g_{\rm rad}$

$$g_{\rm rad}(r) = \kappa_F(r) \frac{L}{4\pi c r^2}. \tag{4.32}$$

Whilst $g_{\rm rad}$ follows from the prescribed density $\rho(r)$ and velocity structures $v(r)$ – via the equation of motion:

$$v \frac{dv}{dr} = g_{\rm rad} - \frac{1}{\rho}\frac{dp}{dr} - \frac{GM}{r^2}, \tag{4.33}$$

where a grey temperature structure can be assumed to compute the gas pressure p. The sole assumption entering this analysis is that the winds are radiatively driven. The resulting mean opacity κ_F thus captures all physical effects that could affect the radiative driving, including clumping and porosity. The obtained values for the correction factor is 0.6 ± 0.2. The transition between O and WR spectral types should in reality occur at:

$$\dot{M} = f \frac{L_{\rm trans}}{v_\infty c} \simeq 0.6 \dot{M}_{\rm trans}. \tag{4.34}$$

There is a transition between O and WR spectral types. The *spectroscopic* transition for spectral subtypes O4-6If+ occurs at $\log(L) = 6.05$ and $\log(\dot{M}_{\eta=1}/M_\odot\,{\rm year}^{-1}) = -4.95$. This is the transition mass-loss rate for the Arches cluster. The only remaining uncertainties are due to uncertainties in the terminal velocity and the stellar luminosity L, with potential errors of at most $\sim 40\%$, and several factors lower than the order-of-magnitude uncertainties in mass-loss rates resulting form clumping and porosity.

4.4.2 Models Close to the Eddington Limit

The predictions of the O star recipe of Vink et al. (2000) and Eq. (4.14) are only valid for objects at a sufficient distance from the Eddington limit, with $\Gamma \leq 0.5$. There are two regimes where this is no longer the case: (i) stars that have formed with large initial masses and luminosities, i.e. very massive stars (VMS) with $M > 100\,M_\odot$, and (ii) less extremely luminous "normal" stars that approach the Eddington limit

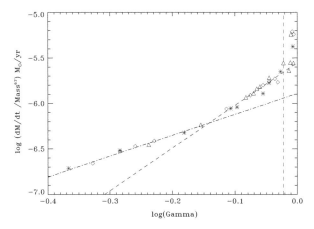

Fig. 4.6 Mass-loss predictions versus the Eddington parameter Γ – divided by $M^{0.7}$. Symbols correspond to models of different mass ranges (Vink et al. 2011)

when they have evolved significantly. Examples of the latter category are LBVs and classical WR stars.

For LBVs, Vink and de Koter (2002) and Smith et al. (2004) showed with Monte Carlo computations that the mass-loss rate increases more rapidly than Eq. (4.14) indicates. This implies that not only does the mass-loss rate increase when the Eddington limit is approached, but the mass-loss rate increases *more strongly*, which leads to a positive feedback effect on the total mass lost over time. For VMS, Vink et al. (2011) discovered a kink in the slope of the mass-loss vs. Γ relation at the transition from optically thin O-type to optically thick WR-type winds. Bestenlehner et al. (2014) performed a homogeneous spectral analysis of >60 Of-Of/WN-WNh stars in 30 Doradus, and confirmed the kink empirically.

Figure 4.6 depicts mass-loss predictions for VMS as a function of the Eddington parameter Γ from Monte Carlo modelling. For ordinary O stars with "low" Γ the $\dot{M} \propto \Gamma^x$ relationship is shallow, with $x \simeq 2$. There is a steepening at higher Γ, where x becomes $\simeq 5$. Here the optical depths and wind efficiencies exceed unity.

4.5 Predictions for Low Metallicity Z and Pop III Stars

For objects in a Z-range representative for the observable Universe with $Z/Z_\odot > 1/100$, Monte Carlo mass-loss predictions were provided by Vink et al. (2001). Extending the predictions to extremely low $Z/Z_\odot < 10^{-2}$, $\dot{M}$ is still expected to drop until the winds reach a point where they become susceptible to ion-decoupling and multi-component effects (Krticka et al. 2003). In order to maintain a one-fluid wind model is by increasing the Eddington factor – by pumping up the stellar mass and luminosity.

For the case of Pop III stars with truly "zero" metallicity, i.e. only H and He present, it seems unlikely that these objects develop stellar winds of significant strength (Kudritzki 2002; Muijres et al. 2012b). However, other physical effects may contribute to the driving. Interesting possibilities include stellar rotation and pulsations, although pure vibration models for Pop III stars also indicate little mass loss via pulsations alone (Baraffe et al. 2001). Perhaps a combination of several effects could result in large mass loss close to the Eddington limit. Moreover, we know that even in the present-day Universe a significant amount of mass is lost in LBV type eruptions, potentially driven by *continuum* radiation pressure, which might be also relevant for the First Stars (Vink and de Koter 2005; Smith and Owocki 2006).

Despite the fact that the first generations of massive stars start their evolutionary clocks with fewer metals, as the First Stars may be highly luminous and/or rapidly rotating, it is not inconceivable that they enrich their atmospheres with nitrogen and carbon (Meynet and Maeder 2006), thereby inducing a stellar wind (Vink 2006).

In a first attempt to investigate the effects of self-enrichment on the total wind strength, Vink and de Koter (2005) performed a pilot study of WR mass loss versus Z. The prime interest in WR stars here is that these objects, especially those of WC subtype, show the products of core burning in their outer atmospheres.

The reasoning behind the assertion that WR winds may not be Z-dependent was that WR stars enrich themselves by burning He into C, and it could be the large C-abundance that is the most relevant ion for the WC wind driving, rather than the sheer number of Fe lines. Figure 4.7 shows that despite the fact that the C ions overwhelm the amount of Fe, both late-type WN (dark line) and WC (light line)

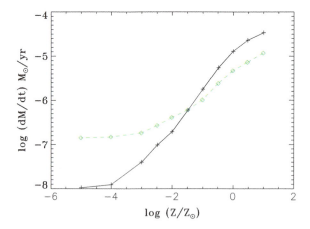

Fig. 4.7 Monte Carlo WR mass-loss predictions as a function of Z. The *dark line* represents the late-type WN stars, whilst the *lighter dashed line* shows the results for late-type WC stars. The slope for the WN models is similar to the predictions for OB-supergiants, whereas the slope is shallower for WC stars. At low Z, the slope becomes smaller, flattening off entirely at $Z/Z_\odot = 10^{-3}$ (The computations are from Vink and de Koter 2005)

show a strong $\dot{M}$-Z dependence, basically because Fe has such a complex electronic structure.

The implications of Fig. 4.7 are two-fold. First, WR mass-loss rates decrease steeply with Z. This may be of key relevance for black hole formation and the progenitor evolution of long duration GRBs. The collapsar model of MacFadyen and Woosley (1999) requires a rapidly rotating stellar core prior to collapse, but at solar metallicity stellar winds are expected to remove the bulk of the core angular momentum (Zahn 1992). The WR $\dot{M}$-Z dependence from Fig. 4.7 provides a route to maintain rapid rotation, as the winds are weaker at lower Z prior to final collapse.

The second point is that mass loss is no longer expected to decrease when $Z/Z_\odot$ falls below $\sim 10^{-3}$ (due to the dominance of driving by carbon lines). This suggests that once massive stars enrich their outer atmospheres, radiation-driven winds might still exist, even if stars started their lives with extremely small amounts of metals.

Whether the mass-loss rates are sufficiently high to alter the evolutionary tracks of the First Stars remains to be seen, but it is important to keep in mind that the mass-loss physics does not only quasi-linearly depend on Z, but that other factors, such as the proximity to the Γ limit, should also be considered.

4.6 Luminous Blue Variables

4.6.1 What Is an LBV?

Luminous Blue Variables represent a short-lived ($\sim 10^4$–10^5 years) phase of massive star evolution during which the objects are subjected to humongous changes in their stellar radii by about an order of magnitude. They come in two flavors. The largest population of ~ 30 LBVs in the Galaxy and the Magellanic Clouds is that of the S Doradus variables with magnitude changes of 1–2 magnitudes on timescales of years to decades (Humphreys and Davidson 1994). These are the characteristic S Dor variations, represented by the dotted horizontal lines in Fig. 4.8. The general understanding is that the S Dor cycles occur at approximately constant bolometric luminosity (which has yet to be proven) – principally representing temperature variations. The second type of LBV instability involves objects that show truly giant eruptions with magnitude changes of order 3–5 during which the bolometric luminosity most certainly increases. In the Milky Way it is only the cases of P Cygni and Eta Carina which have been noted to exhibit such giant outbursts.

Whether these types of variability occur in similar or distinct objects is not yet clear, but in view of the "unifying" properties of the object P Cygni it is rather probable that the S Dor variables and giant eruptors are subject to the same type of instabilities near the Eddington limit (see Vink (2012)).

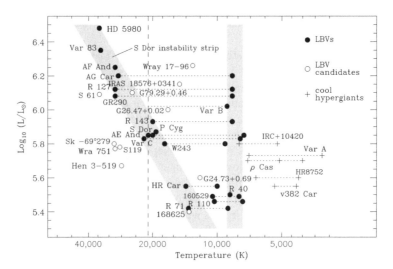

Fig. 4.8 The LBVs in the Hertzsprung-Russell diagram. The slanted band running from 30 kK at high $L/L_\odot$ to 15 kK at lower luminosity is the S Dor instability strip. The vertical band at a temperature of ∼8,000 K represents the position of the LBVs "in outburst". The vertical line at 21,000 K is the position of the observed bi-stability jump (Lamers et al. 1995) (Adapted from Vink 2012 and Smith et al. 2004)

4.6.2 Do LBVs Form Pseudo-photospheres?

Although it *appears* that the photospheric temperatures of the objects in Fig. 4.8 change during HRD transits, there is an alternative possibility that the underlying star does *not* change its actual temperature but that the star undergoes changes in mass-loss properties instead. The second case is normally referred to as the formation of a "pseudo-photosphere" resulting from the formation of an optically thick wind.

Eta Car's outstanding wind density with $\dot{M} \sim 10^{-3} M_\odot$ year^{-1} Hillier et al. (2001) places $R(\tau_{\text{Ross}} = 2/3)$ at 80% of the terminal velocity, impeding any derivation of the hydrostatic radius, but it is not yet clear whether the general LBV population of S Dor variables have $\dot{M}$ values high enough to produce pseudo photospheres.

As a result of enhanced mass loss during maximum it is hypothetically possible to form a pseudo-photosphere. Until the late 1980s this was the leading idea to explain the colour changes of S Dor variables. Using more advanced non-LTE model atmosphere codes, Leitherer et al. (1989) and de Koter et al. (1996) predicted colours based on *empirical* LBV mass-loss rates that are not red enough to make an LBV appear cooler than the temperature of its underlying surface. Despite the proximity of LBVs to the Eddington limit, current consensus is that LBV winds are generally not sufficiently optically thick.

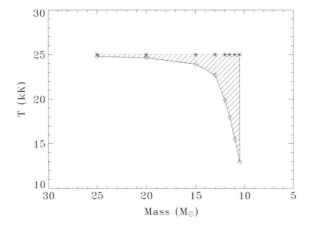

Fig. 4.9 Pseudo-photosphere formation in a relatively low mass LBV with a high L/M ratio. The difference in inner (*dashed*) and apparent temperature (representative for the size of the computed pseudo-photosphere) is plotted against the stellar mass. These computations have been performed for a constant luminosity of log $L/L_\odot = 5.7$. The mass is gradually decreased whilst the LBV approaches the Eddington limit: the apparent temperature drops as a result of the lower effective gravity, and the higher mass loss results in the formation of a pseudo-photosphere Smith et al. 2004

Figure 4.9 shows the potential formation of an optically thick wind for a relatively low-mass (with high L/M) LBV in close proximity to the bi-stability jump (Pauldrach and Puls 1990; Vink and de Koter 2002; Groh and Vink 2011). The size of the temperature difference (dashed vs. solid) is a proxy for the extent of the pseudo-photosphere. The figure demonstrates that for masses in the range 15–25 $M_\odot$ and fixed luminosity, the winds remain optically thin, but when the stellar mass approaches values as low as 10 $M_\odot$, and the star enters the mass-loss regime near the Eddington limit, the photospheric scale-height blows up, which results in the formation of a pseudo-photosphere.

4.6.3 Winds During S Doradus Variations

Although most S Dor variables have been subject to photometric monitoring, only a few have been analysed in sufficient detail to understand the driving mechanism of their winds. Mass-loss rates are of the order of $10^{-3} - 10^{-5} M_\odot$ year^{-1}, whilst terminal wind velocities are in the range $\sim$100–500 km s^{-1}. Obviously, these values vary with L and M, but there are indications that the mass loss varies as a function of $T_{\rm eff}$ when the S Dor variables transit the upper HRD on timescales of years, providing an ideal laboratory for testing the theory of radiation-driven winds.

The Galactic LBV AG Car is one of the best monitored and analysed S Dor variables. Vink and de Koter (2002) predicted $\dot{M}$ rises in line with radiation-driven wind models for which the $\dot{M}$ variations are attributable to ionisation shifts of Fe. Sophisticated non-LTE spectral analysis have since confirmed these predictions (Groh et al. 2011; Groh and Vink 2011).

It is relevant to mention here that this variable wind concept (wind bi-stability; see also Pauldrach and Puls 1990) has been suggested to be responsible for circumstellar density variations inferred from modulations in radio light-curves and Hα spectra of supernovae (Kotak and Vink 2006; Trundle et al. 2008). However most stellar evolution models would have predicted massive stars with $M \geq 25\, M_\odot$ to explode at the end of the WR phase, rather than after the LBV phase. The implications could be gigantic, impacting our most basic understanding of massive star death in the Universe (see Smith's Chap. 8).

4.6.4 Super-Eddington Winds

Whilst during "quiet" phases, LBVs may lose mass via ordinary line-driving, some objects, like Eta Car also seem to be subject to phases of more extreme mass loss. For instance, the giant eruption of η Car with a cumulative loss of $\sim 10\, M_\odot$ between 1840 and 1860 (Smith et al. 2003) which resulted in the Homunculus nebula corresponds to $\dot{M} \approx 0.1 - 0.5\, M_\odot$ year^{-1}, which is a factor of 1,000 larger than that expected from line-driven wind models for an object of that luminosity.

Shaviv (1998) and Owocki et al. (2004) studied the theory of porosity-moderated continuum driving in objects that formally exceed the Eddington limit. It is possible that continuum-driven winds in super-Eddington stars reach mass-loss rates close to the *photon tiring limit*, $\dot{M}_{\mathrm{tir}} = L_*/(GM_*/R_*)$, which could result in a stagnating flow that may lead to spatial structure (van Marle et al. 2008). However, it should be noted that alternatively, wind clumping may be the result of other instabilities, possibly related to the presence of the Fe opacity peak (Cantiello et al. 2009; Gräfener et al. 2012; Gräfener and Vink 2013; Glatzel and Kiriakidis 1993), especially for objects approaching the Γ-limit.

The *general* equation of motion for a stellar wind (ignoring gas pressure) is given by:

$$v\left(1 - \frac{a^2}{v^2}\right)\frac{dv}{dr} \simeq g_{\mathrm{grav}}(r) + g_{\mathrm{rad}}(r) = -\frac{GM}{r^2}(1 - \Gamma(r)). \quad (4.35)$$

At the sonic point, r_s: $v = a$, and thus $g_{\mathrm{rad}} = -g_{\mathrm{grav}}$ implying $\Gamma(r_s) = 1$. Thus, $\Gamma(r)$ must be <1 below the sonic point and $\Gamma(r)$ must be >1 above the sonic point. An accelerating wind solution thus implies an increasing opacity $\frac{d\bar{\kappa}}{dr}|_s > 0$ (given that $\Gamma(r) = \frac{\bar{\kappa}(r) L_*}{4\pi GMc}$).

If, on the other hand, the entire atmosphere is super-Eddington, i.e. $\Gamma(r) > 1$ throughout the atmosphere, continuum driving might nonetheless become possible. The reason is that when atmospheres exceed the Eddington limit, instabilities may arise which could make them clumpy: outward travelling photons may avoid regions of enhanced density, which means that the medium may behave in a porous manner, leading to a lower g_{rad}. This means that the effective Eddington parameter can drop below unity. However, further out in the wind, the clumps become optically thinner as a result of expansion, and the porosity effect decreases. $\Gamma_{\text{eff}}^{\text{cont}}$ can now become larger than unity. In other words, a wind solution with $\Gamma_{\text{eff}}^{\text{cont}}$ crossing unity is feasible, even when the stars are formally above the Eddington limit.

Owocki et al. (2004) expressed the effective opacity in terms of the so-called porosity length (see Sect. 4.8). They showed that $\dot{M}$ might become substantial when the porosity length is of the order of the pressure scale height H. Owocki et al. developed the concept of a power-law distributed porosity length (in analogy to CAK-type the line-strength distribution function), and showed that even the gigantic mass-loss rate during Eta Car's giant eruption might be explained by some form of radiative driving.

4.7 Observed Wind Parameters

Radiation-driven wind models can provide predictions for two global wind parameters: the mass-loss rate, $\dot{M}$, and the terminal velocity, v_∞. Most studies rely on the assumption of a smooth wind. The mass-loss rate then follows from the continuity equation (Eq. 4.1), and most diagnostics are based on a wind model with a prescribed β-type velocity field.

A useful concept involves the optical-depth invariant Q parameter (Puls et al. 1996), where Q_{res} can be utilised for resonance lines with line opacity $\propto \rho$. Alternatively, recombination is a 2-body process and Q_{rec} is useful for recombination based line processes such as H$_\alpha$ which thus have opacities $\propto \rho^2$,

$$Q_{\text{res}} = \frac{\dot{M}}{R_* v_\infty^2}, \qquad Q_{\text{rec}} = \frac{\dot{M}}{(R_* v_\infty)^{1.5}}. \qquad (4.36)$$

Most diagnostics rely on the use of non-LTE model atmospheres. Stellar and wind parameters, such as $\dot{M}$ can be determined by fitting resonance and recombination lines simultaneously. Smooth wind models constitute the ideal case, but the optical depth invariant Q_{rec} as defined in Eq. (4.36) can easily be modified for the case that the winds are clumped (Sect. 4.8, Eq. 4.46).

A more detailed discussion of the various methods to derive wind parameters is given in Puls et al. (2008). The most common line profiles in a stellar wind are (i) UV P Cygni profiles with a blue absorption trough and a red emission peak, and (ii) optical emission lines (such as Hα). These line shapes are caused by different population mechanisms of the upper energy level of the transition. In a P Cygni scattering line, the upper level is populated by the balancing act between absorption from and spontaneous decay to the lower level. An emission line is formed if the upper level is populated by recombinations from above (see however Puls et al.

(1998) and Petrov et al. (2014) for the formation of P Cygni Hα lines in the cooler BA supergiants).

4.7.1 Ultraviolet P Cygni Resonance Lines

P Cygni lines may be used to determine the velocity field in stellar winds, and in particular v_∞. Hα is generally utilised to derive $\dot{M}$ (or Q_{rec}). UV P-Cygni lines from hot stars (e.g. C IV and P V) are usually analysed by means of the Sobolev optical depth:

$$\tau_{\text{Sob}}(r) = \frac{\frac{\pi e}{m_e c} f n_1(r) \lambda}{dv/dr} \frac{R_*}{v_\infty}, \qquad (4.37)$$

where f is the oscillator-strength and n_1 the lower occupation number of the transition. Relating the occupation number, n_1, to the density:

$$\tau_{\text{Sob}}(r) = \frac{1}{r^2 v dv/dr} E(r) q(r) \frac{\dot{M}}{R_* v_\infty^2} \frac{(\pi e^2)/(m_e c)}{4\pi m_H} \frac{A_k}{1 + 4Y} f \lambda, \qquad (4.38)$$

where E is the excitation factor of the lower level, q the ionisation fraction, A_k the abundance of the element, and Y the He abundance. This quantity is invariant with respect to $Q_{\text{res}} = \dot{M}/(R_* v_\infty^2)$ (see Eq. 4.36) as long as the ground-state population is proportional to the density ρ. Thus, $\dot{M}$ can be derived from resonance line P Cygni profiles when the ionisation fraction is known. Most P Cygni lines however are saturated and mass-loss rate derivations become unfeasible, such that only lower limits on $\dot{M}$ can be determined.

UV resonance lines have been considered relatively clean from clumping effects, but this might not be the case if porosity effects become important.

4.7.2 The Hα Recombination Emission Line

The most oft-used diagnostics to derive $\dot{M}$ for O-star winds involves Hα, for which there is hardly any uncertainty due to ionisation. The Hα opacity scales with ρ^2, and

$$\tau_{\text{Sob}}(r) \propto \frac{\dot{M}^2}{(R_* v_\infty)^3} \frac{b_2(r)}{r^4 v^2 dv/dr}, \qquad (4.39)$$

i.e., the scaling invariant quantity is now Q_{rec}^2 (Eq. 4.36), and b_2 is the non-LTE departure coefficient of n_2.

The challenge with H$_\alpha$ concerns its ρ^2 dependence. Any notable inhomogeneity will necessarily result in an $\dot{M}$ overestimate if clumping is neglected in the analysis. An advantage is the fact that H$_\alpha$ remains optically thin in the main part of the emitting wind, such that porosity effects can be neglected (which is not the case for UV resonance lines).

4.7.3 Radio and (Sub)millimetre Continuum Emission

A somewhat different approach to measure mass-loss rates is to utilize long wavelength radio and (sub)millimetre continua. In fact this approach may lead to the most accurate results, as they are model-independent. The basic concept is to measure the excess wind flux over that from the stellar photosphere. This excess flux is emitted by free-free and bound-free processes. The reason the excess flux becomes more important at longer (sub)-mm/radio wavelengths is due to the λ^2 dependence of the opacities.

Following Wright and Barlow (1975), Panagia and Felli (1975), and Lamers and Cassinelli (1999), the dominant free-free opacity (in units of cm^{-1}) at frequency ν can be written as:

$$\kappa_\nu \propto n_i n_e g_\nu \left(\frac{1}{\nu^2}\right) \propto \frac{\dot{M}}{v_\infty}^2 \left(\frac{1}{r^4}\right) g_\nu \left(\frac{1}{\nu^2}\right), \qquad (4.40)$$

in cm^{-3}, and g_ν is the Gaunt factor for free-free emission. For an isothermal wind and frozen-in ionisation, $\bar{z}$ (the mean value of the atomic charge) and μ_e and μ_i remain constant, and:

$$\kappa_\nu \propto g_\nu \lambda^2 \rho^2, \qquad (4.41)$$

which increases with λ and ρ. As the continuum becomes optically thick in the wind in free-free opacity the emitting wind volume increases as a function of λ, leading to the formation of a radio photosphere where the the radio emission dominates the stellar photospheric emission. For a typical O supergiant this occurs at about 100 stellar radii. At such large distances the outflow reaches its terminal wind velocity and an analytic solution of the radiative transfer problem becomes possible:

$$F_\nu \propto \left(\frac{\dot{M}}{v_\infty}\right)^{4/3} \frac{(\nu g_\nu)^{2/3}}{d^2}, \qquad (4.42)$$

where F_ν is the observed radio flux measured in Jansky, $\dot{M}$ in units of $M_\odot$ year^{-1}, v_∞ in km s^{-1}, distance d to the star in kpc and frequency ν in Hz. Thus, the spectral index of *thermal* wind emission is close to 0.6.

4.8 Wind Clumping

Hα and long-wavelength continuum diagnostics depend on the density squared, and are thus sensitive to clumping, whereas UV P Cygni lines such as Pv are insensitive to clumping, as they depend linearly on density. In the canonical optically thin (micro-clumping) approach the wind is divided into a portion of the wind that contains all the material with a volume filling factor f (the reciprocal of the clumping factor D), whilst the remainder of the wind is assumed to be void. In reality however, clumped winds are porous with a range of clump sizes, masses, and optical depths.

Wind clumping has been extensively discussed for canonical 20–60 $M_\odot$ O-type stars and WR stars in a dedicated clumping workshop (Hamann et al. 2008). Here one may also find studies of X-ray observations (see also Cohen et al. (2014) and references therein for more recent work).

4.8.1 Optically Thin Clumping ("Micro-clumping")

The general concept of optically-thin micro clumping is simply based on the assumption that the wind is made up of large numbers of small-scale density clumps. Largely motivated by the results from hydrodynamic simulations including the line-deshadowing instability (LDI; see Owocki's Chap. 5), the inter-clump gas is usually assumed to be void. The average density $\langle \rho \rangle = \dot{M}/(4\pi r^2 v)$ is given by:

$$\langle \rho \rangle = f\rho_C, \quad \langle \rho^2 \rangle = f(\rho_C)^2 \qquad (4.43)$$

where ρ_C is the density inside the over-dense clumps, and $\langle \rho^2 \rangle$ is the mean of the squared density. Thus, the clumping factor:

$$D = \langle \rho^2 \rangle/\langle \rho \rangle^2 \quad \Rightarrow \quad D = f^{-1} \quad \text{and} \quad \rho_C = D\langle \rho \rangle, \qquad (4.44)$$

measures the clump over-density. As the inter-clump space is assumed to be void, matter is only present inside the clumps, with density ρ_C, and with its opacity given by $\kappa = \kappa_C(D\langle \rho \rangle)$, where C represents the quantities inside the clump. Optical depths may be calculated via $\tau = \int \kappa_C(D\langle \rho \rangle) f \, dr$ with a reduced path length ($f dr$) as to correct for the volume where clumps are actually present.

The formulation is only correct as long as the clumps are optically thin, and optical depths may be expressed by a mean opacity $\bar{\kappa}$:

$$\bar{\kappa} = \kappa_C(D\langle \rho \rangle) f = \frac{1}{D}\kappa_C(D\langle \rho \rangle). \qquad (4.45)$$

Thus, for processes that are linearly dependent on density, the mean opacity of a clumped medium is exactly the same as for a smooth wind, whilst for processes that scale with the density squared, mean opacities are enhanced by the clumping factor D.

It should be noted that processes described by the optically thin micro-clumping approach do not depend on clump size nor geometry, but only the sheer clumping *factor*. The enhanced opacity for ρ^2 dependent processes implies that $\dot{M}$ derived by such diagnostics are a factor of $\sqrt{D}$ lower than older mass-loss rates derived with the assumption of smooth winds. As a result, the optical depth invariant, Q_{rec} (see Eq. 4.36) transforms into:

$$Q_{\text{rec}} = \frac{\dot{M}\sqrt{D}}{(R_* v_\infty)^{1.5}}. \qquad (4.46)$$

Note that also for the case of thermal radio and (sub)-mm continuum emission the scaling invariant is proportional to $\dot{M}/R_*^{1.5}$, i.e. very similar to Q_{rec} for optical emission lines, such as Hα. Abbott et al. (1981) studied the effects of clumping on the wind radio emission as a function of the volume filling factor and the density ratio between clumped and inter-clump material. For the standard assumption of vanishing inter-clump density, Abbott et al. showed that the radio flux may be a factor $f^{-2/3}$ larger than that from a smooth wind with the same $\dot{M}$. In other words, using Eq. (4.42), it can be noted that radio mass-loss rates derived from clumped winds must also be lower than those derived from smooth winds.

4.8.2 The P v Problem

Due to the very low cosmic abundance of phosphorus (P), the P v doublet remains unsaturated, even when P^{+4} is dominant. This allows for a direct estimate of the product $\dot{M}\langle q \rangle$, where $\langle q \rangle$ is a spatial average of the ion fraction. Unfortunately, $\langle q \rangle$ estimates for a given resonance line are uncertain due to shocks and associated X-ray ionisation. Empirical determination of ionisation fractions is normally not feasible, as resonance lines from consecutive ionisation stages are not generally available. Nevertheless, for P v, insight is gained from FUSE data: for those O-stars in a certain $\langle q \rangle \simeq 1$ region, the P v line should provide an accurate estimate of $\dot{M}$, as the pure linear character with ρ makes it clumping independent.

Fullerton et al. (2006) selected a large sample of O-stars, which also had ρ^2 (from Hα/radio) estimates available, and compared both ρ-linear UV and ρ-quadratic dependent methods. They found enormous discrepancies, with a median $\dot{M}(\rho^2)/(\dot{M}(\text{P v})\langle q \rangle) = 20$ in mid-O supergiants, implying an extreme clumping factor D $\simeq$ 400 if the winds could indeed be treated in an optically thin (micro-clumping) approach (see also Bouret et al. (2003)).

4.8.3 Optically Thick Clumping ("Macro"-clumping)

With studies yielding clumping factors ranging from D up to 400, one may wonder whether a pure micro-clumping analysis is physically sound. Most of the atmospheric codes only consider density variations, but hydrodynamic simulations also reveal strong velocity changes inside the clumps. Most worrisome is probably the assumption that all clumps are assumed to be optically thin.

Within the optically thin approach, a clump has a size smaller than the photon mean free path. However, in an optically thick clump, photons may interact with the gas several times before they escape through the inter-clump gas. Whether a clump is optically thin or thick depends on the abundance, ionisation fraction, and cross-section of the transition.

For optically thick clumps, photons care about the distribution, the size and the geometry of the clumps (see Fig. 4.10). The conventional description of macro-clumping is based on a clump size, $l(r)$, and an average spacing of a statistical distribution of clumps, $L(r)$, which are related to f:

$$f = \left(\frac{l}{L}\right)^3 = \frac{1}{D}. \tag{4.47}$$

Following Eq. (4.45), the optical depth across a clump of size l and opacity κ_C becomes:

$$\tau_C = \kappa_C l = \bar{\kappa} D l = \bar{\kappa} \frac{L^3}{l^2} = \bar{\kappa} h, \tag{4.48}$$

with mean opacity $\bar{\kappa}$ (Eq. 4.45) and porosity length $h = L^3/l^2$. The porosity length h involves the key parameter to define a clumped medium, as h corresponds to the photon mean free path in a medium consisting of optically thick clumps.

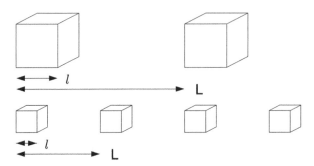

Fig. 4.10 Schematic explanation of porosity, involving a notable difference between the volume filling fraction f (and its reciprocal clumping factor $D = 1/f$), which is the same for the top and bottom case, and the separation of the clumps L, which is larger in the top case than the bottom case (From Muijres et al. 2011)

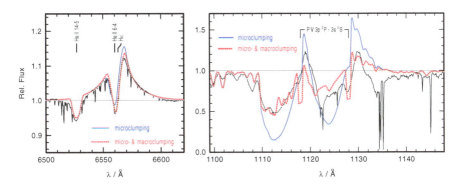

Fig. 4.11 Porosity as a possible solution for the PV problem (Adapted from Oskinova et al. 2007)

The effective clump cross section, i.e., the *spatial* cross section now corrected for the fraction of transmitted radiation, becomes:

$$\sigma_C = l^2 \left(1 - e^{-\tau_C}\right), \tag{4.49}$$

and the effective opacity becomes:

$$\kappa_{\text{eff}} = n_C \sigma_C = \frac{l^2 \left(1 - e^{-\tau_C}\right)}{L^3} = \bar{\kappa} \frac{\left(1 - e^{-\tau_C}\right)}{\tau_C}, \tag{4.50}$$

where n_C is the clump number density. The key point is that this very equation holds for clumps of any optical thickness! For instance, in the optically thin limit, the micro-clumping approximation is recovered: $\kappa_{\text{eff}} = \bar{\kappa}$, which depends on f and not on clump size or distribution. In the optically thick case, the effective opacity is indeed reduced appropriately, $\kappa_{\text{eff}} = \bar{\kappa}/\tau_C = h^{-1}$ now only depending on h.

Oskinova et al. (2007) employed the effective opacity concept in the formal integral for the line profile modelling of the O supergiant ζ Pup. Figure 4.11 shows that the most pronounced effect involves strong resonance lines, such as P v which can be reproduced by this macro-clumping approach – without the need for extremely low $\dot{M}$ – resulting from an effective opacity reduction when clumps become optically thick. Given that H$_\alpha$ remains optically thin for O stars it is not affected by porosity,[4] and it can be reproduced simultaneously with P v. This enables a solution to the P v problem (see also Surlan et al. 2013).

However, this porosity concept was developed for continuum processes, whilst line processes may also be affected by velocity-field changes. Owocki (2008) performed LDI simulations where the line strength was described through a velocity-clumping factor. These simulations resulted in a reduced wind absorption

---

[4]This might be different for B supergiants below the bi-stability jump (see Petrov et al. 2014).

due to porosity in velocity space, which has been termed "vorosity". The issue with explaining a reduced P v line-strength through vorosity is that one needs to have a relatively large number of substantial velocity gaps, which does not easily arise from the LDI simulations. In any case, there is still a need to study scenarios including both porosity and vorosity, as well as how they interrelate (Sunqvist et al. 2012).

4.8.4 Quantifying the Number of Clumps

In the traditional view of line-driven winds of O-type stars via the CAK theory and the associated LDI, clumping would be expected to develop in the wind when the wind velocities are large enough to produce shocked structures. For typical O star winds, this is thought to occur at about half the terminal wind velocity at about 1.5 stellar radii.

Various observational indications, involving the existence of linear polarisation (e.g. Davies et al. 2005) as well as radial dependent spectral diagnostics (Puls et al. 2006) however show that clumping must already exist at very low wind velocities, and more likely arise in the stellar photosphere. Cantiello et al. (2009) suggested that waves produced by the subsurface convection zone associated with the Fe opacity peak could lead to velocity fluctuations, and possibly density fluctuations, and thus be the root cause for the observed wind clumping at the stellar surface (see Fig. 4.12).

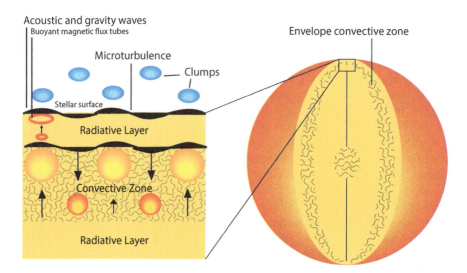

Fig. 4.12 Cartoon of the physical processes involved in sub-surface convection. Acoustic and gravity waves are emitted in the convective zone, and travel through the radiative layers, reaching the stellar surface, thereby inducing density and velocity fluctuations. In this picture, clumping starts at the wind base (From Cantiello et al. 2009)

Assuming the horizontal extent of the clumps to be comparable to the sub-photospheric pressure scale height H_p, one may estimate the number of convective cells by dividing the stellar surface area by the surface area of a convective cell finding that it scales as $(R/H_P)^2$. For main-sequence O stars in the canonical mass range 20–60 $M_\odot$, pressure scale heights are within the range 0.04–0.24 $R_\odot$, corresponding to a total number of clumps $6 \times 10^3 - 6 \times 10^4$. These estimates may in principle be tested through linear polarisation variability, which probes wind asphericity at the wind base.

In an investigation of WR linear polarisation variability Robert et al. (1989) uncovered an anti-correlation between the wind terminal velocity and the scatter in polarisation. They interpreted this as the result of blobs that grow or survive more effectively in slow winds than fast winds. Davies et al. (2005) found this trend to continue into the regime of LBVs, with even lower v_∞. LBVs are are thus an ideal test-bed for constraining clump properties, due to the larger wind-flow times. Davies et al. showed that over 50 % of LBVs are intrinsically polarised. As the polarisation angle was found to vary irregularly with time, the polarisation line effects were attributed to wind clumping. Monte Carlo models for scattering off wind clumps have been developed by Code and Whitney (1995), Rodrigues and Magalhaes (2000), and Harries (2000), whilst analytic models to produce the variability of the linear polarisation may be found in Davies et al. (2007), Li et al. (2009), and Townsend and Mast (2011).

An example of an analytic model that predicts the time-averaged polarisation for the LBV P Cygni is presented in Fig. 4.13. The clump ejection rate per wind flow-time $\mathcal{N}$ is defined as $\mathcal{N} = \dot{N} t_{fl} = \dot{N} R_\star/v_\infty$, where the clump ejection rate, $\dot{N}$, is related to $\dot{M}$ as $\dot{M} = \dot{N} N_e \mu_e m_H$, where N_e is the number of electrons in each clump, and μ_e is the mean mass per electron. There are two regimes where the observed polarisation level can be achieved. One is where the ejection rate is low and a few very optically thick clumps are expelled; the other one is that of a very large number of clumps. These two cases can be distinguished via

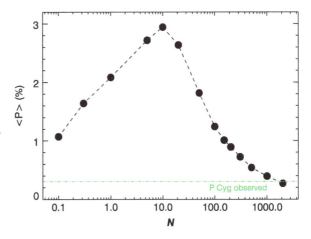

Fig. 4.13 Time-averaged polarisation over a range of ejection rates per wind flow-time. At $\mathcal{N} \sim 20$, the optical depth per clump exceeds unity and the overall polarisation falls off (see Davies et al. 2007 for details). The observed polarisation level for the LBV P Cygni is given by the dash-dotted line. There are two ejection-rate regimes where the required polarisation level can be achieved

time resolved polarimetry. Given the relatively short timescale of the observed polarisation variability, Davies et al. argued that LBV winds consist of order thousands of clumps near the photosphere.

Nevertheless, for main-sequence O stars the derivation of wind-clump sizes from polarimetry has not yet been feasible as very high signal-to-noise data are required. LBVs however provide an excellent group of test-objects owing to the combination of higher mass-loss rates, and lower terminal wind velocities. Davies et al. (2007) showed that in order to produce the observed polarisation variability of P Cygni, the wind should consist of $\sim$1,000 clumps per wind flow-time. In order to check whether this is compatible with the sub-surface convection scenario ultimately being the root cause for wind clumping, one would need to consider the sub-surface convective regions of an object with global properties similar to those of P Cygni. Due to the lower LBV gravity, the pressure scale height is about $4R_\odot$, i.e. significantly larger than for O-type stars. As a result, the same estimate for the number of clumps drops to about 500 clumps per wind-flow time, which appears to be consistent with that derived for P Cygni from observations (see Fig. 4.13).

4.8.5 Effects on Mass-Loss Predictions

Muijres et al. (2011) studied the possible effects of both optically thin and thick wind clumping (porosity) on mass-loss predictions for O-type stars.

Because of the non-linear character of the equation of motion, the CAK solution is complex, with the physics involving instabilities due to the LDI (e.g. Owocki et al. 1988). One of the key implications of the LDI is that in hydro-dynamical simulations the time-averaged $\dot{M}$ is *not* anticipated to be affected by wind clumping, as it has the same average $\dot{M}$ as the smooth CAK solution. However, the shocked velocity structure and its associated density structure are expected to result in effects on the mass-loss diagnostics.

In contrast to the LDI simulations, Muijres et al. (2011) studied the effects of clumping on g_{rad} due to changes of the ionisation structure, as well as the effects of wind porosity, using Monte Carlo simulations. When only accounting for optically thin (micro) clumping g_{rad} was found to *increase* for certain clumping stratifications $D(r)$, but only for an extremely high clumping factor of $D \sim 100$ (see Fig. 4.14 for a range of clumping factors and stratifications). The reason g_{rad} may increase is the result of recombination yielding more flux-weighted opacity from lower Fe ionisation stages (similar to the bi-stability physics). For $D = 10$ the effects were however found to be relatively minor.

When simultaneously also accounting for optically thick (macro) clumping, the effects were partially reversed, as photons could now escape in between the clumps without interaction, and the predicted g_{rad} goes down, as well as up (see Fig. 4.15 for a range of clumping stratifications). Nevertheless, again, for $D = 10$ the effects were found to be rather modest.

A fully consistent study of the impact of wind-clumping on predicted wind properties has yet to be performed.

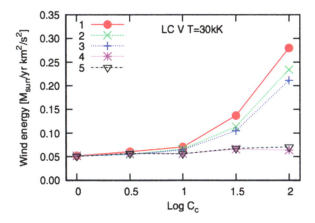

Fig. 4.14 The effect of optically thin micro-clumping on the wind kinetic energy in Monte Carlo simulations for different clumping stratifications for 30,000 K OV-type stars. The smooth wind models have $D = 1$. The numbers 1–5 refer to different clumping stratifications (see Muijres et al. 2011 for details), but clumping in the outer winds (stratifications 1 through 3) results in an increase of the kinetic wind energy due to a larger number of effective driving lines

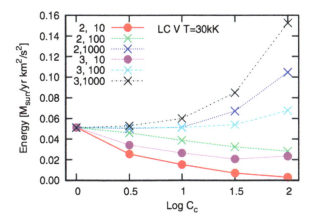

Fig. 4.15 The effects of optically thick macro-clumping on the wind kinetic energy in Monte Carlo simulations for different clumping *and porosity* stratifications for 30,000 K OV-type stars. The smooth wind models have $D = 1$. The numbers 1–5 refer to different clumping stratifications (see Muijres et al. 2011 for details)

Summary and Conclusion
As we mentioned in Sect. 4.1 (see also Chaps. 6 and 7) the evolution and fate of VMS are predominantly determined by $\dot{M}$. Current stellar evolution models for VMS (e.g. Yusof et al. 2013; Köhler et al. 2015) utilise the smooth Monte Carlo *theoretical* predictions of Vink et al. (2000).

(continued)

However, it has become clear that *empirical* $\dot{M}$ rates have been overestimated when determined from ρ^2 diagnostics such as H_α. According to Repolust et al. (2004) and Mokiem et al. (2007) the non-clumping corrected empirical rates are a factor 2–3 *higher* than the Vink et al. (2000) rates, meaning that moderate clumping effects (with $D = 4$–10) are indirectly accounted for in stellar evolution models, noting that all recently reviewed stellar models employ Vink et al. (2000) rates according to Martins and Palacios (2013).

However, there has been a breakthrough in our understanding of $\dot{M}$ for VMS in close proximity to the Eddington Γ limit. Vink et al. (2011) discovered a "kink" in the $\dot{M}$ vs. Γ relation at the transition from optically thin O-type to optically thick winds. For ordinary O stars with "low" Γ the $\dot{M} \propto \Gamma^x$ relationship is shallow, with $x \simeq 2$. There is a steepening at higher Γ, where x becomes $\simeq 5$. This mass-loss enhancement due to VMS in proximity to the Γ-limit has not yet been included in evolutionary models of VMS, and is likely to be crucial for their ultimate fate.

We also discussed a methodology that involves a model-*independent* $\dot{M}$ indicator: the transition mass-loss rate $\dot{M}_{trans}$ – located right at the transition from optically thin to optically thick stellar winds (Vink and Gräfener 2012). As $\dot{M}_{trans}$ is model independent, *all* that is required is to postulate the *spectroscopic* transition point in a given data-set and to determine the far more accurate L parameter. In other words $\dot{M}_{trans}$ is extremely useful for calibrating wind mass loss, and assessing its role in mass loss during stellar evolution. As was mentioned, current stellar models use Vink et al. mass-loss rates that have been reduced by factors of 2–3 compared to previous unclumped empirical rates, and there is thus no immediate reason to reduce them further, unless clumping factors would be higher than ~ 10.

Furthermore, we have also seen in Sect. 4.8 that clumping can affect the Monte Carlo mass-loss predictions in various ways, involving both reductions and increases in $\dot{M}$. We have also highlighted that both the origin and onset of wind clumping remain unclear. Polarisation measurements call for clumping to be already present in the stellar photosphere, but how this would interact with the hydro-dynamical LDI simulations further out, and how this would need to be consistently incorporated into radiative transfer calculations and mass-loss predictions is as yet unclear. For these reasons, the search for the nature and implications of wind clumping should continue!

References

Abbott D. C., & Lucy L. B. (1985). *Astrophysics Journal, 288*, 679.
Abbott, D. C., Bieging, J. H., & Churchwell, E. (1981). *Astrophysics Journal, 250*, 645.
Anders, E., & Grevesse N. (1989). *Geochimica et Cosmochimica Acta, 53*, 197.

Baraffe I., Heger A., & Woosley S. E. (2001). *Astrophysics Journal, 550*, 890.
Belkus H., Van Bever J., & Vanbeveren D. (2007). *Astrophysics Journal, 659*, 1576.
Bestenlehner, J.M., Grafener, G., & Vink, J.S. (2014). arXiv1407.1837
Bouret S.-C., Lanz T., & Hillier D. J. (2003). *Astrophysics Journal, 595*, 1182.
Brott, I., de Mink, S. E., & Cantiello, M., et al. (2011). *Astronomy and Astrophysics, 530*, 115.
Cantiello M., Langer N., & Brott I., et al. (2009). *Astronomy and Astrophysics, 499*, 279.
Castor J., Abbott D. C., & Klein R. I. (1975). *Astrophysics Journal, 195*, 157.
Code, A. D., & Whitney, B. A. (1995). *Astrophysics Journal, 441*, 400.
Cohen, D. H., Wollman, E. E., & Leutenegger, M., et al. (2014). *Monthly Notices of the Royal Astronomical Society, 439*, 908.
Conti P. S. (1976). *MSRSL 9*, 193.
Davies B., Oudmaijer R. D., & Vink J. S. (2005). *Astronomy and Astrophysics, 439*, 1107.
Davies B., Vink J. S., & Oudmaijer R. D. (2007). *Astronomy and Astrophysics 469*, 1045.
de Koter, A., Lamers, H. J. G. L. M., & Schmutz, W. (1996). *Astronomy and Astrophysics, 306*, 501.
Eldridge, J. J., & Vink, J. S. (2006). *Astronomy and Astrophysics, 452*, 295.
Fullerton, A. W., Massa, D. L., & Prinja, R. K. (2006). *Astrophysics Journal, 637*, 1025.
Gayley, K. G. (1995). *Astrophysics Journal, 454*, 410.
Gayley, K. G., Owocki S. P., Cranmer S. R. (1995). *Astrophysics Journal, 442*, 296.
Glatzel, W., Kiriakidis, M. (1993). *Monthly Notices of the Royal Astronomical Society, 263*, 375.
Gräfener G., & Hamann W.-R. (2005). *Astronomy and Astrophysics, 432*, 633.
Gräfener, G., & Vink, J. S. (2013). *Astronomy and Astrophysics 560*, 6.
Gräfener, G., Owocki, S. P., & Vink, J. S. (2012). *Astronomy and Astrophysics, 538*, 40.
Groh, J. H., & Vink, J. S. (2011). *Astronomy and Astrophysics 531L*, 10.
Groh, J. H., Hillier, D. J., & Damineli, A. (2011). *Astrophysics Journal, 736*, 46.
Hamann W.-R., & Koesterke L. (1998). *Astronomy and Astrophysics, 335*, 1003.
Hamann, W.-R., Feldmeier, A., & Oskinova, L. M. (2008). *CIHW conference, 30*.
Harries, T. J. (2000). *Monthly Notices of the Royal Astronomical Society, 315*, 722.
Hillier, D. J. (1991). *Astronomy and Astrophysics, 247*, 455.
Hillier, D. J., Davidson, K., Ishibashi, K., & Gull, T. (2001) *Astrophysics Journal, 553*, 837.
Humphreys, R. M., & Davidson, K. (1994). *Publications of the Astronomical Society of the Pacific, 106*, 1025.
Iglesias, C. A., & Rogers, F. J. (1996). *Astrophysics Journal, 464*, 943.
Kotak, R., Vink, J. S. (2006). *Astronomy and Astrophysics, 460L*, 5.
Krticka, J., Owocki, S. P., Kubat, J., Galloway, R. K., & Brown, J. C. (2003). *Astronomy and Astrophysics, 402*, 713.
Kudritzki, R.-P., Lennon, D.J., & Puls, J. (1995) *SVLT conference, 246.*
Kudritzki, R.-P. (2002). *Astrophysics Journal, 577*, 389.
Lamers, H. J. G. L. M., & Cassinelli, J. P. (1999). ISW Book, CUP.
Lamers, H. J. G. L. M., Snow, T. P., & Lindholm, D. M. (1995). *Astrophysics Journal, 455*, 269.
Langer, N., Hamann, W.-R., & Lennon, M., et al. (1994). *Astronomy and Astrophysics, 290*, 819.
Leitherer, C., Schmutz, W., & Abbott, D. C., et al. (1989). *Astrophysics Journal, 346*, 919.
Li, Q.-K., Cassinelli, J. P., Brown, J. C., & Ignace, R. (2009). *Research in Astronomy and Astrophysics, 9*, 558.
Limongi, M., & Chieffi, A. (2006). *Astrophysics Journal, 647*, 483.
Lucy, L. B. (1971). *Astrophysics Journal, 163*, 95.
Lucy L. B., & Solomon P. M. (1970. *Astrophysics Journal, 159*, 879.
Lucy L. B., & Abbott D. C. (1993). *Astrophysics Journal, 405*, 738.
MacFadyen, A. I., & Woosley, S. E. (1999). *Astrophysics Journal, 524*, 262.
Martins, F., & Palacios, A. (2013). *Astronomy and Astrophysics, 560*, 16.
Meynet G., & Maeder A. (2003). *Astronomy and Astrophysics, 404*, 975.
Meynet, G., Ekstrom, S., & Maeder, A. (2006). *Astronomy and Astrophysics, 447*, 623.
Milne, E. A. (1926). *Monthly Notices of the Royal Astronomical Society, 86*, 459.
Mokiem M. R., de Koter A., & Vink J. S., et al. (2007). *Astronomy and Astrophysics, 473*, 603.
Muijres L., de Koter A., & Vink J. S., et al. (2011). *Astronomy and Astrophysics, 526*, 32.

Muijres L., Vink J. S., & de Koter A., et al. (2012a). *Astronomy and Astrophysics, 537*, 37.
Muijres, L., Vink, J. S., & de Koter, A., et al. (2012b). *Astronomy and Astrophysics, 546*, 42.
Müller P. E., & Vink J. S. (2008). *Astronomy and Astrophysics, 492*, 493.
Müller P. E., & Vink J. S. (2014). *Astronomy and Astrophysics, 564*, 57.
Nugis, T., & Lamers, H. J. G. L. M. (2002). *Astronomy and Astrophysics, 389*, 162.
Oskinova, L. M., Hamann, W.-R., & Feldmeier, A. (2007). *Astronomy and Astrophysics, 476*, 1331.
Owocki, S. P. (2008). *CIHW conference, 121*.
Owocki, S. P., Castor, J. I., & Rybicki, G. B. (1988). *Astrophysics Journal, 335*, 914.
Owocki S. P., Gayley K. G., & Shaviv N. J. (2004). *Astrophysics Journal, 616*, 525.
Panagia, N., & Felli, M. (1975). *Astronomy and Astrophysics, 39*, 1.
Pauldrach, A. W. A., & Puls, J. (1990). *Astronomy and Astrophysics, 237*, 409.
Pauldrach A. W. A., Puls J., & Kudritzki R. P. (1986). *Astronomy and Astrophysics, 164*, 86.
Pauldrach, A.W.A., Vanbeveren, D., Hoffmann, T. L. (2012) . *Astronomy and Astrophysics, 538*, 75.
Petrov, B., Vink, J. S., & Gräfener, G. (2014, in press). *Astronomy and Astrophysics*, (arXiv1403.4097).
Puls, J., Kudritzki R. P., & Herrero A., et al. (1996). *Astronomy and Astrophysics, 305*, 171.
Puls, J., Springmann, U., & Owocki, S. P. (1998). *CVSW conference, 389*.
Puls, J., Springmann, U., & Lennon, M. (2000). *Astronomy and Astrophysics Supplement Series, 141*, 23.
Puls, J., Markova, N., & Scuderi, S., et al. (2006). *Astronomy and Astrophysics, 454*, 625.
Puls J., Vink J. S., & Najarro F. (2008). *Astronomy and Astrophysics Review, 16*, 209.
Repolust, T., Puls, J., & Herrero, A. (2004). *Astronomy and Astrophysics, 415*, 349.
Robert, C., Moffat, A. F. J., Bastien, P., Drissen, L., & St.-Louis, N. (1989). *Astrophysics Journal, 347*, 1034.
Rodrigues, C. V., & Magalhaes, A. M. (2000). *Astrophysics Journal, 540*, 412.
Schaerer, D., & Schmutz, W. (1994). *Astronomy and Astrophysics, 288*, 231.
Shaviv N. J. (1998). *Astrophysics Journal, 494*, L193.
Smith, N., & Owocki, S. P. (2006). *Astrophysics Journal, 645L*, 45.
Smith, N., Gehrz, R. D., & Hinz, P. M., et al. (2003). *Astronomical Journal, 125*, 1458.
Smith N., Vink J. S., & de Koter A. (2004). *Astrophysics Journal, 615*, 475.
Sobolev, V.V. (1960). Harvard University Press, Circumstellar Envelopes.
Springmann, U. (1994). *Astronomy and Astrophysics, 289*, 505.
Sundqvist, J. O., Owocki, S. P., & Cohen, D. H., et al. (2012). *Monthly Notices of the Royal Astronomical Society, 420*, 1553.
Surlan, B., Hamann, W.-R., & Aret, A., et al. (2013). *Astronomy and Astrophysics, 559*, 130.
Townsend, R. H. D., & Mast, N. (2011). *Institute of Architecture and Urban & Spatial Planning of Serbia, 272*, 216.
Trundle, C., Kotak, R., Vink, J. S., & Meikle, W. P. S. (2008). *Astronomy and Astrophysics, 483*, 47.
van Marle A. J., Owocki S. P., & Shaviv N. J. (2008). *Monthly Notices of the Royal Astronomical Society, 389*, 1353.
Vink, J. S. (2006). *American Shetland Pony Club, 353*, 113 (astro-ph/0511048).
Vink J. S. (2012). *Astrophysics and Space Science Library, 384*, 221. Eta Carinae, Springer (astro-ph/0905.3338).
Vink J. S., & de Koter, A. (2002). *Astronomy and Astrophysics, 393*, 543.
Vink J. S., & de Koter, A. (2005). *Astronomy and Astrophysics, 442*, 587.
Vink, J. S., & Gräfener, G. (2012). *Astrophysics Journal, 751*, 34.
Vink J. S., de Koter, A., & Lamers, H. J. G. L. M. (1999). *Astronomy and Astrophysics, 345*, 109.
Vink J. S., de Koter, A., & Lamers, H. J. G. L. M. (2000). *Astronomy and Astrophysics, 362*, 295.
Vink J. S., de Koter, A., & Lamers, H. J. G. L. M. (2001). *Astronomy and Astrophysics, 369*, 574.
Vink, J. S., Muijres, L. E., & Anthonisse, B., et al. (2011). *Astronomy and Astrophysics, 531*, 132.
Wright, A. E., & Barlow, M. J. (1975). *Monthly Notices of the Royal Astronomical Society, 170*, 41.

Yungelson, L. R., van den Heuvel, E. P. J., & Vink, J. S., et al. (2008). *Astronomy and Astrophysics, 477*, 223.
Yusof, N., Hirschi, R., & Meynet, G., et al. (2013). *Monthly Notices of the Royal Astronomical Society, 433*, 1114.
Zahn, J.-P. (1992). *Astronomy and Astrophysics, 265*, 115.

Chapter 5
Instabilities in the Envelopes and Winds of Very Massive Stars

Stanley P. Owocki

Abstract The high luminosity of Very Massive Stars (VMS) means that radiative forces play an important, dynamical role both in the structure and stability of their stellar envelope, and in driving strong stellar-wind mass loss. Focusing on the interplay of radiative flux and opacity, with emphasis on key distinctions between continuum vs. line opacity, this chapter reviews instabilities in the envelopes and winds of VMS. Specifically, we discuss how: (1) the iron opacity bump can induce an extensive inflation of the stellar envelope; (2) the density dependence of mean opacity leads to strange mode instabilities in the outer envelope; (3) desaturation of line-opacity by acceleration of near-surface layers initiates and sustains a line-driven stellar wind outflow; (4) an associated line-deshadowing instability leads to extensive small-scale structure in the outer regions of such line-driven winds; (5) a star with super-Eddington luminosity can develop extensive atmospheric structure from photon bubble instabilities, or from stagnation of flow that exceeds the "photon tiring" limit; (6) the associated porosity leads to a reduction in opacity that can regulate the extreme mass loss of such continuum-driven winds. Two overall themes are the potential links of such instabilities to Luminous Blue Variable (LBV) stars, and the potential role of radiation forces in establishing the upper mass limit of VMS.

5.1 Background: VMS M-L Relation and the Eddington Limit

A hallmark of very massive stars (VMS) is that they are very, very luminous. For example, a star of a hundred solar masses typically has a luminosity that is of order a *million* times the solar luminosity. This means that, from the realm of solar to very massive stars, the luminosity scales roughly with the *cube* of the stellar mass, $L \sim M^3$ (justifying perhaps adding even a third "very" to "luminous"). This is *not* (as sometimes inferred) a consequence of the core nuclear burning source of

---

S.P. Owocki (✉)
Department of Physics & Astronomy, University of Delaware, Newark, DE 19716, USA
e-mail: owocki@udel.edu

the stellar luminosity; instead, as worked out by Eddington (1926) and others even before nuclear burning was fully understood, this follows from the basic equations of stellar structure, namely the dual requirements of *hydrostatic* pressure support against stellar gravity, and *radiative transport* of energy from the interior to the surface (see Sect. 5.3). As was also recognized (most notably by Eddington) from these early studies of stellar structure, this cubic scaling of luminosity with mass can not be maintained to arbitrarily large masses, essentially because at high luminosity the associated *radiation pressure* becomes significant in the star's gravitational support.

Radiation pressure is a consequence of the fact that, in addition to their important general role as carriers of energy, photons also have an associated momentum, set by their energy divided by the speed of light c. The trapping of radiative energy within a star thus inevitably involves a trapping of its associated momentum, leading to an outward radiative force, or for a given mass, an outward *radiative acceleration* g_{rad}, that can compete with the star's gravitational acceleration g. For a local radiative energy flux F (energy/time/area), the associated momentum flux (force/area, or pressure) is just F/c. The material acceleration resulting from absorbing this radiation depends on the effective cross sectional area σ for absorption, divided by the associated material mass m,

$$g_{rad} = \frac{\sigma}{m}\frac{F}{c} \equiv \frac{\kappa F}{c}. \qquad (5.1)$$

The latter equality defines the *opacity* $\kappa = \sigma/m$, which is just a measure of the total effective absorption cross section per unit mass of absorbing material.

For a star of luminosity L, the radiative flux at some radial distance r is just $F = L/4\pi r^2$. This gives the radiative acceleration the same inverse-square radial decline as the stellar gravity, $g = GM/r^2$ (with G the gravitation constant); their ratio, generally referred to as the "Eddington parameter", thus tends to be relatively constant, set by the ratio of luminosity to mass,

$$\Gamma \equiv \frac{g_{rad}}{g} = \frac{\kappa F}{gc} = \frac{\kappa L}{4\pi GMc} = \Gamma_e \frac{\kappa}{\kappa_e} \approx 2.6 \times 10^{-5} \frac{\kappa}{\kappa_e} \frac{L/M}{L_\odot/M_\odot}. \qquad (5.2)$$

The last two equalities provide scalings in terms of the classical Eddington parameter, Γ_e, defined for the electron scattering opacity, $\kappa_e \equiv \sigma_{Th}/\mu_e$, where $\sigma_{Th} = 6.7 \times 10^{-25}\,\mathrm{cm}^2\,\mathrm{g}^{-1}$ is the Thompson cross section for free electron scattering, and μ_e is the mean mass per free electron. The latter scales with the Hydrogen mass m_H rather than the much smaller electron mass, because even for free electrons, maintaining overall charge neutrality requires an effective coupling between electrons and the ions that are the main contributors to the material mass. For a fully ionized gas with Hydrogen mass fraction X, $\mu_e = 2m_H/(1+X)$, giving $\kappa_e = 0.2(1+X) \approx 0.34\,\mathrm{cm}^2\,\mathrm{g}^{-1}$ for standard (solar) mass fraction $X \approx 0.7$. Applying this in (5.2), the last equality shows that, for a star with the solar luminosity to mass ratio $L_\odot/M_\odot$, the electron Eddington parameter $\Gamma_{e\odot}$ is very small, implying that for such solar-type stars the electron scattering acceleration $g_e = \Gamma_e g$ is entirely negligible compared to gravity.

But if one assumes an overall cubic scaling of luminosity with mass, then

$$\Gamma_e = \Gamma_{e\odot}(M/M_\odot)^2, \tag{5.3}$$

which would reach the classical Eddington limit $\Gamma_e = 1$ for a mass $M_{Edd} = M_\odot/\sqrt{\Gamma_{e\odot}} \approx 195 M_\odot$. Rather remarkably, this agrees quite well with modern empirical estimates for the most massive observed stars, which are in the range 150–300 $M_\odot$ (Figer 2005; Oey and Clarke 2005; Crowther et al. 2010; Crowther 2012). (See Chaps. 1 and 2.)

More complete analyses that account for the effect of radiation pressure in the hydrostatic support against gravity show that associated adjustments in the stellar structure formally allow gravitationally bound ($\Gamma < 1$) stars without *any* upper mass limit. For example, for a homogenous star with solar Hydrogen mass fraction $X \approx 0.7$ and radially constant Γ, the so-called "Eddington Standard Model" (ESM) gives the scaling (Eddington 1926),

$$\frac{\Gamma}{(1-\Gamma)^4} \approx \left(\frac{M}{48 M_\odot}\right)^2. \tag{5.4}$$

The $1 - \Gamma$ factor comes from the reduction in effective gravity from radiation pressure; its presence as a quartic in the denominator represents a strong repeller against the Eddington limit, $\Gamma \to 1$. As illustrated in Fig. 5.1, this forces the low-mass cubic scaling $L \sim M^3$ to reduce to a linear scaling $L \sim M$ at large mass, always keeping below the limit.

The ESM assumption of a radially fixed Γ also implies (see Sect. 5.3.1) a fixed ratio of the radiation pressure to gas pressure,

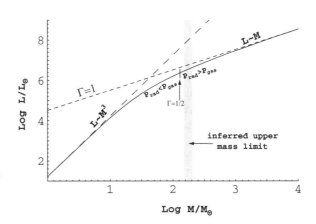

Fig. 5.1 Log-log plot of the stellar luminosity L vs. mass M for the Eddington Standard Model scaling of Eq. (5.4)

$$\frac{P_{\text{rad}}}{P_{\text{gas}}} \approx \frac{\Gamma}{1-\Gamma}. \tag{5.5}$$

The case $\Gamma = 1/2$, with ESM mass $M = 136\,M_\odot$, has $P_{\text{rad}} = P_{\text{gas}}$, and so marks the transition from gas to radiation as the dominant source of pressure support against gravity. In analogy to having a heavier fluid be supported by a lighter one, an envelope with gravitational support that is predominately from radiation pressure is expected to be intrinsically unstable, sometimes dubbed the "photon bubble instability" (Spiegel and Tao 1999; Shaviv 1998, 2000, 2001). If the nonlinear growth of this and related instabilities can sufficiently disrupt the stellar structure, for example inducing strong episodes of extensive mass loss, it could be a key factor in setting an effective upper limit for stellar mass.

Indeed, the above-mentioned observationally inferred mass limit (Crowther 2012) (Chaps. 1 and 2) is just above the ESM mass for transition to radiation pressure dominance. For electron scattering opacity, the corresponding ESM luminosity $\sim 5 \times 10^6 L_\odot$ ($M_{\text{bol}} \sim -12$) is near the luminosities of the intrinsically brightest observed stars, for example, η Carinae or the 'Pistol star', both of which show evidence for past episodes of strong mass loss. These stars are prototypes of a "giant eruption" subclass of "Luminous Blue Variable" (LBV) stars. On the Hertzsprung-Russell diagram they lie very near the Humphreys-Davidson (H-D) limit (Humphreys and Davidson 1979) that delineates the most luminous observed stars. A second subtype, the S-Doradus LBVs, can occur a factor ten or more below the H-D limit; in contrast to the strong brightenings of eruptive LBV's, they are characterized by year timescale variations in effective temperature, but with a roughly *constant* bolometric magnitude.

Focusing on the interplay of opacity, radiative flux and gravity, the remaining sections of this chapter review how the strong radiative acceleration in luminous VMS can lead to strong mass loss and induce instabilities in both their interior envelopes and stellar wind outflows. The goal is to provide a good physical basis for exploring the potential role of such radiative acceleration and the associated mass loss and instability for understanding both giant eruption and S-Doradus variability in LBVs, as well as for the inferred VMS upper mass limit.

Building on methods (Sect. 5.2) for estimating the flux-weighted mean opacity, Sect. 5.3 examines the effect of radiative forces on the structure and inflation of the hydrostatic, gravitationally bound stellar envelope. We next (Sect. 5.4) write the general time-dependent equations for conservation of mass, momentum, and energy, and apply these (Sect. 5.4.2) to a linear perturbation analysis of 'strange-mode' instabilities in the stellar envelope, and to write the basic equations for steady stellar wind outflow (Sect. 5.4.3). Applying the latter to the standard case of a line-driven wind (Sect. 5.5), we derive steady solutions for the mass loss rate and wind velocity (Sect. 5.5.1), and then discuss (Sects. 5.5.2 and 5.5.3) the extensive structure (clumping and porosity) that arises in time-dependent models that account for the strong *Line-Deshadowing Instability* (LDI) intrinsic to line-driving. Finally, for the giant eruption LBVs with a super-Eddington luminosity, we review (Sect. 5.6) how the much stronger mass loss – which can approach the "photon tiring" limit for the luminosity to lift material out of the star's gravitational potential – can be modeled in terms of a quasi-steady *continuum-driven* wind regulated by a porosity reduction in the effective opacity. Finally section "Concluding Summary" presents a concluding summary.

5.2 Mean Opacity Formulations

5.2.1 Flux-Weighted Mean Opacity

In addition to the free electron scattering that provides a nearly fixed, frequency-independent (gray), baseline opacity, there are additional contributions associated with electron interactions with ions, namely through free-free (f-f), bound-free (b-f) and bound-bound (b-b) processes. The resonant nature of bound-bound transitions makes the associated *line* opacities very strong, and these, especially from complex heavy ions of iron-group elements, turn out to be particularly important in the structure of near-surface layers and in driving stellar wind mass loss.

To account for the strong frequency dependence of the associated opacity, the simple expression (5.1) for the radiative acceleration must now be generalized to the integral form,

$$g_{rad}(r) = 2\pi \int_{-1}^{1} d\mu\, \mu \int_{0}^{\infty} d\nu\, \kappa_\nu\, I_\nu(\mu, r)/c = \int_{0}^{\infty} d\nu\, \kappa_\nu\, F_\nu(r)/c. \quad (5.6)$$

where the integrals are over frequency ν and radial direction cosine $\mu = \hat{\mathbf{n}} \cdot \hat{\mathbf{r}}$ for radiation in vector direction $\hat{\mathbf{n}}$ with specific intensity I_ν and associated opacity κ_ν. If this opacity is *isotropic*,[1] the evaluation reduces to just the latter frequency integral of κ_ν times the associated energy flux F_ν.

To compute the radiative acceleration in terms of the local *bolometric* flux $F \equiv \int_0^\infty F_\nu d\nu$, as done in Eq. (5.1), the appropriate opacity is now a *flux-weighted mean*,

$$\bar{\kappa}_F \equiv \int_{0}^{\infty} \frac{\kappa_\nu F_\nu}{F} d\nu. \quad (5.7)$$

The dependence on local gas and radiation conditions means this opacity, and thus the Eddington parameter from Eq. (5.2), both now generally vary with the local radius r.

While notationally convenient in connecting back to the simple gray opacity scalings, it is important to realize that computation of $\bar{\kappa}_F$ can be very difficult,

---

[1] As discussed in Sect. 5.5.1, even when the opacity is formally isotropic in the atom's frame, a spherical wind expansion can lead to an anisotropy for *line* opacity in the stellar frame, through the directional dependence of the local velocity gradient.

in principal requiring a *global* integral solution of the generally *nonlocal* radiative transport to obtain the frequency dependence of the local flux F_ν, accounting for the frequency dependence of the opacity κ_ν, as well as its dependence on ionization and excitation level of the absorbing ions.

5.2.2 Planck Mean and its Dominance by Line Opacity

To illustrate the potentially dominant importance of line opacity, let's first consider an *optically thin* limit in which absorbing ions are exposed fully to the local continuum radiation, unattenuated by any self-absorption within the lines. If we model this continuum flux spectrum as being given by the broad Planck blackbody function for the star's surface temperature, so that $F_\nu/F = B_\nu/B$, we see that the flux-mean opacity $\bar{\kappa}_F$ of Eq. (5.7) is just given in terms of a *Planck mean* opacity, defined by

$$\bar{\kappa}_P \equiv \int_0^\infty \frac{\kappa_\nu B_\nu}{B} d\nu . \tag{5.8}$$

As a direct mean, $\bar{\kappa}_P$ is dominated by the strongest opacity sources, namely by the cumulative contribution from individual spectral lines. Relative to electron scattering, the line opacity of an individual line with index i has the form,

$$\frac{\kappa_{\nu_i}}{\kappa_e} = \frac{n_i}{n_e} f_i \frac{\sigma_{cl}}{\sigma_e} \phi(\nu - \nu_i), \tag{5.9}$$

where n_i and n_e are the number densities of absorbing ions and electrons, and the line-profile function, $\phi(\nu - \nu_i)$, is narrowly peaked around the line-center frequency ν_i, with unit normalization $\int_0^\infty \phi(\nu - \nu_i) d\nu = 1$. The quantum mechanical oscillator strength, f_i, corrects the frequency-integrated cross-section σ_{cl} (with dimensions of area times frequency) obtained from the "classical oscillator" model of line absorption. In terms of the classical electron radius $r_e \equiv e^2/m_e c^2$, the frequency-integrated line-cross-section is enhanced by the dimensionless factor

$$Q_{\lambda_i} \equiv \frac{\sigma_{cl}}{\nu_i \sigma_e} = \frac{\pi r_e c}{\nu_i \, 8\pi r_e^2/3} = \frac{3}{8} \frac{\lambda_i}{r_e} = 1.5 \times 10^8 \frac{\lambda_i}{1{,}000\,\text{Å}}, \tag{5.10}$$

where $\lambda_i = c/\nu_i$ is the line-center wavelength, and the notation Q_{λ_i} is chosen because it is related (by just a factor π^2) to the resonance *quality* $Q = \nu_i/\gamma_i$, with γ_i the radiative damping rate (Gayley 1995). The very large numerical value for a sample UV wavelength stems from the resonance nature of line transitions, showing that, even when integrated over a broad frequency range that is much larger than the line width, a bound electron has an enormously larger cross section than a free electron. The effect is somewhat analogous to blowing into a whistle vs. just open

5 Instabilities in VMS Envelopes and Winds

air; the response is very strong, but concentrated in a narrow frequency range near the resonance.

Upon integration over the individual line-profiles, we can thus approximate the Planck opacity as a sum over the line index i, weighted by a factor $W_i = v_i B_{v_i}/B$ that reflects the blackbody strength at the line frequency v_i,

$$\frac{\kappa_P}{\kappa_e} = \sum_i \frac{n_i}{n_e} f_i Q_{\lambda_i} W_i \equiv \overline{Q} \approx 2{,}000 \left(\frac{Z}{0.02}\right), \tag{5.11}$$

where the notation and evaluation in the last two equalities are due to Gayley (1995). For allowed transitions near the peak of the Planck function, both f_i and W_i are order unity; but for hot stars with high ionization and near-solar metallicity, only a relatively small fraction $\sim 10^{-5}$ of electrons remain bound in metal ions, implying a similarly small cumulative abundance ratio $\sum_i n_i/n_e$ that counters the large resonance factor Q_{λ_i}, leaving a more moderately strong average resonance quality $\overline{Q} \approx 2{,}000$.

The upshot here is that, when unsaturated in this way, the radiative force from line-opacity can approach an upper limit that is substantially enhanced over that associated with electron scattering, with an associated Eddington parameter

$$\Gamma_{max} \approx \overline{Q}\Gamma_e. \tag{5.12}$$

This means that for lines the requirement $\Gamma > 1$ to overcome gravity and drive a wind outflow can occur in any stars with electron Eddington parameters $\Gamma_e > 1/\overline{Q} \approx 0.0005$ (see Sect. 5.5.1). For VMS with Γ_e only a factor few below unity, it indicates the potential for strong *line-deshadowing instability* (LDI), with any optically thin portions of a wind outflow having radiative acceleration approaching a thousand times the acceleration of gravity (see Sects. 5.5.2 and 5.5.3).

5.2.3 Rosseland Opacity and Radiative Diffusion in Stellar Interior

Of course this full brunt of line opacity does not apply in the dense, opaque stellar interior because, in diffusing outward, radiation preferentially leaks through the inter-line frequencies of lower continuum opacity, leaving only a significantly reduced flux within the lines. Within such a *diffusion approximation* for radiation transport, the local frequency-dependent flux now scales as

$$F_v(r) \approx -\left[\frac{4\pi}{3\kappa_v \rho} \frac{\partial B_v}{\partial T}\right] \frac{dT}{dr}, \tag{5.13}$$

where ρ is the local density and $B_\nu(T)$ is the frequency-dependent Planck function for the interior temperature $T(r)$ at local radius r. Application of (5.13) in (5.7) shows that in this diffusion limit, the flux-weighted opacity is now approximated by the so-called "*Rosseland mean*",

$$\kappa_R \equiv \frac{\int_0^\infty \frac{\partial B_\nu}{\partial T} d\nu}{\int_0^\infty \frac{1}{\kappa_\nu} \frac{\partial B_\nu}{\partial T} d\nu} . \tag{5.14}$$

As a *harmonic mean*, κ_R is dominated by the weaker components of the opacity, with generally little relative contribution from individual spectral lines.[2] The numerator can be readily evaluated by taking the temperature derivative outside the frequency integral,

$$\int_0^\infty \frac{\partial B_\nu}{\partial T} d\nu = \frac{\partial}{\partial T} \int_0^\infty B_\nu \, d\nu = \frac{\partial B}{\partial T} = \frac{\partial}{\partial T} \left(\frac{\sigma_B T^4}{\pi} \right) = \frac{4\sigma_B T^3}{\pi}, \tag{5.15}$$

where σ_B is the Stefan-Boltzmann constant. One can then write the frequency-integrated flux as a *radiative diffusion equation*,

$$F(r) = -\left[\frac{16\sigma_B}{3} \frac{T^3}{\kappa_R \rho} \right] \frac{dT}{dr} . \tag{5.16}$$

The next section examines the stellar structure scalings for such radiative envelopes, both in terms of the classical ESM M-L scaling, and for detailed opacity models based on the OPAL tables.

5.3 Effect of Radiation Pressure on Stellar Envelope

5.3.1 Mass-Luminosity Scaling for Radiative Envelope

This diffusion form for energy transport, along with the requirement for momentum balance through hydrostatic equilibrium, provide the basic stellar structure constraints that set the mass-luminosity scaling. To see this, let us rewrite Eq. (5.16) in terms of the radial gradient of the *radiation* pressure $P_{\text{rad}} \equiv 4\sigma_B T^4/3c$,

$$\frac{dP_{\text{rad}}}{dr} = -\rho \frac{\kappa_R F}{c} = -\rho g_{\text{rad}} = -\rho \Gamma \frac{GM}{r^2} . \tag{5.17}$$

---

[2] An exception is when the spectral density of lines become high enough to lead to an effective "line-blanketing" effect, as occurs in the iron opacity bump discussed in Sect. 5.3.3.

5 Instabilities in VMS Envelopes and Winds

When modified to account for the $1 - \Gamma$ radiative reduction in the effective gravity, the requirement for hydrostatic equilibrium sets the gradient of the *gas* pressure,

$$\frac{dP_{\text{gas}}}{dr} = -\rho \frac{GM}{r^2}(1 - \Gamma). \tag{5.18}$$

Together these imply that the relative variation of gas to radiation pressure depends only on the Eddington parameter,

$$\frac{dP_{\text{gas}}}{dP_{\text{rad}}} = \frac{1 - \Gamma}{\Gamma}. \tag{5.19}$$

As noted in Sect. 5.1, for the Eddington Standard Model with constant Γ, this gives $P_{\text{gas}}/P_{\text{rad}} \approx (1 - \Gamma)/\Gamma = \text{constant}$, which leads to the simple ESM scaling (5.4) for the mass dependence of the Eddington parameter. A related, commonly quoted quantity is the gas pressure fraction β of the total pressure, which in the interior of an ESM model is just set by the Eddington parameter,

$$\beta \equiv \frac{P_{\text{gas}}}{P_{\text{gas}} + P_{\text{rad}}} = 1 - \Gamma. \tag{5.20}$$

These ESM scalings can be understood from average gradients in (5.17) and (5.18) in terms of stellar mass M and radius R. Using the ideal gas law $P_{\text{gas}} \sim \rho T$ with $\rho \sim M/R^3$, (5.18) implies the characteristic interior temperature scales as

$$T \sim \frac{M(1 - \Gamma)}{R}. \tag{5.21}$$

With the further proportionalities $F \sim L/R^2$ and $P_{\text{rad}} \sim T^4$, the radiative diffusion (5.17) gives

$$L \sim \frac{R^4 T^4}{M}. \tag{5.22}$$

Combining (5.21) and (5.22), we can eliminate both R and T to find

$$L \sim (1 - \Gamma)^4 M^3 \quad \text{or} \quad \frac{\Gamma}{(1 - \Gamma)^4} \sim M^2, \tag{5.23}$$

which agrees with the ESM scaling (5.4). As noted, this scaling does not depend explicitly on the nature of energy generation in the stellar core, but is strictly a property of the envelope structure.[3]

---

[3]Of course, this simple scaling relation has to be modified to accommodate gradients in the molecular weight as a star evolves from the zero-age main sequence, and it breaks down altogether

5.3.2 Virial Theorem and Stellar Binding Energy

This hydrostatic balance of a stellar envelope can also be used to derive a relation – known as the *virial theorem* – between the internal thermal energy U and the gravitational binding energy Φ of the whole star (Kippenhahn et al. 2013),

$$\Phi = -3(\gamma - 1)U. \tag{5.24}$$

where γ is the ratio of specific heats. The total stellar energy is thus given by

$$E \equiv \Phi + U = \frac{3\gamma - 4}{3\gamma - 3}\Phi. \tag{5.25}$$

For the case of a monotonic ideal gas $\gamma = 5/3$, the total energy is just half the gravitational binding energy, $E = \Phi/2$.

However, in very massive stars near the Eddington limit $\Gamma \to 1$, the internal energy can become dominated by radiation instead of gas, since $P_{\text{rad}}/P_{\text{gas}} = \Gamma/(1-\Gamma) \to \infty$. In this limit of a radiation gas, $\gamma \to 4/3$, which by Eq. (5.25) implies a *vanishing* total energy $E \to 0$. This is another factor toward making VMS unstable.

5.3.3 OPAL Opacity

Let us next examine how the Rosseland opacity $\bar{\kappa}_R$, and its associated Eddington parameter Γ, can change through the stellar envelope due to changes in temperature and density. For this, we adopt the widely used tabulations from the OPAL[4] opacity project (Iglesias and Rogers 1996), using the specific OPAL tables given by Grevesse and Noels (1993), and taking the case with standard solar values $X = 0.7$ and $Z = 0.02$ for the Hydrogen and metal mass fractions. The OPAL tabulations are given in terms of temperature T and a parameter $\mathscr{R} \equiv \rho/(T/10^6 K)^3$, but to make a clear connection to the above discussion, let us here cast the latter in the equivalent terms of gas to radiation pressure, $P_{\text{gas}}/P_{\text{rad}}$.

The left panel of Fig. 5.2 plots contours of $\log(\kappa/\kappa_e)$ in the $\log(T)$ vs. $\log(P_{\text{rad}}/P_{\text{gas}})$ plane, oriented such that the high-temperature, high-density of the stellar interior is at the lower left. The heavy black contour with labeled value 0.3 corresponds to an opacity $\kappa/\kappa_e \approx 2$ that is roughly twice that of the basal value for electron scattering. Indeed, for typical stellar-core temperatures of order several million Kelvin (MK), note that the total opacity is only slightly above this electron scattering value.

---

in the coolest stars (both giants and dwarfs), for which convection dominates the envelope energy transport.

[4] http://opalopacity.llnl.gov/

5 Instabilities in VMS Envelopes and Winds

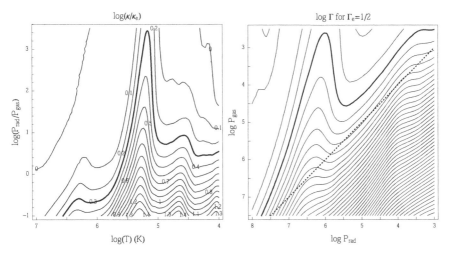

Fig. 5.2 *Left:* OPAL opacity κ, from tables by Grevesse and Noels (1993) for mass fractions $X = 0.7$ and $Z = 0.02$, plotted as contours of $\log(\kappa/\kappa_e)$ in the $\log T$ vs. $\log(P_{\text{rad}}/P_{\text{gas}})$ plane. The peak in contours at $\log T \approx 5.3$ is from the Iron opacity bump. *Right:* For a star with $\Gamma_e = 1/2$, contours of $\Gamma = (\kappa/\kappa_e)\Gamma_e$, now plotted in the $\log P_{\text{rad}}$ vs. $\log P_{\text{gas}}$ plane. In both plots, the contours have log spacings of 0.1, with the *heavy solid contours* representing the case with $\kappa = 2\kappa_e$ and $\Gamma = 1$. In the right panel, the *dotted line* shows the locus where $P_{\text{rad}} = P_{\text{gas}}$

But for temperatures near 1 MK and below, the opacity increases, especially for low values of $P_{\text{rad}}/P_{\text{gas}}$ corresponding to relatively higher densities ρ. Particularly note the strong peak near $T \approx 10^{5.3}$ K $\approx 2 \times 10^5$ K, which is due to a dense "blanketing" of line (bound-bound) opacity from iron group elements, and is thus commonly known as the "*iron opacity bump*". For low $P_{\text{rad}}/P_{\text{gas}}$, and thus *high* density ρ, the opacity can exceed the electron scattering value by an order of magnitude or more. This enhancement decreases at lower density, but only very weakly, requiring several decades decline in ρ (represented here by a several decade increase in $P_{\text{rad}}/P_{\text{gas}}$) to recover the modest factor 2 above electron scattering.

For temperatures along the opacity peak, an approximate fitting relation, normalized about the density $\rho = \rho_2$ that has $\kappa = 2\kappa_e$, is given by the logarithmic form,

$$\frac{2\kappa_e}{\kappa} \approx 1 + \log\left(\frac{\rho_2}{\rho}\right)^{0.2}, \tag{5.26}$$

wherein the 0.2 quantifies the extreme weakness of the dependence on density. For modest deviations about ρ_2, this can alternatively be written in the power-law form,

$$\frac{\kappa}{2\kappa_e} \approx \left(\frac{\rho}{\rho_2}\right)^{0.086}, \tag{5.27}$$

where the small power index $0.086 = 0.2 \log e$ again shows the weak density dependence.

5.3.4 Envelope Inflation and the Iron Bump Eddington Limit

For VMS near the classical Eddington limit, this iron-bump increase in opacity near temperatures $T \approx 150{,}000$–$200{,}000$ K, can have dramatic effects on the near-surface envelope structure, inducing an *inversion* in gas pressure and density, with an associated *inflation* in the surface radius (Petrovic et al. 2006; Gräfener et al. 2012), and possibly triggering strong stellar wind mass loss.

For the specific case of a VMS with $\Gamma_e = 1/2$, the right panel of Fig. 5.2 plots contours of $\log \Gamma$ in the $\log P_{\mathrm{rad}}$ vs. $\log P_{\mathrm{gas}}$ plane, with the heavy solid contour for the Eddington limit value $\Gamma = 1$. Since the combined stellar structure Eq. (5.19) specifies $dP_{\mathrm{gas}}/dP_{\mathrm{rad}}$ in terms of Γ, we can follow the decline of gas pressure from the deep interior, where $\Gamma \approx \Gamma_e = 1/2$, implying from (5.19) that $P_{\mathrm{gas}} \approx P_{\mathrm{rad}}$. But toward the subsurface layer with lower temperature and pressure, the large bump in opacity pushes the integration toward the $\Gamma = 1$ contour, forcing the gas pressure, and thus density, to very low values, in this case to a minimum density $\rho_{min,2} \approx 2.5 \times 10^{-11}$ g cm^{-3} at the peak of the bump, where the opacity is $\kappa = 2\kappa_e$.

A key result here is that, because this iron-peak opacity depends so weakly on density, keeping it limited to the Eddington value $\kappa_{\mathrm{Edd}} \equiv \kappa_e/\Gamma_e$ requires a minimum density that has a *strong inverse* scaling with electron Eddington parameter. Using Eq. (5.26), this can be fit approximately by

$$\log\left(\frac{\rho_{min,2}}{\rho_{min}}\right) \approx 10(\Gamma_e - 1/2). \tag{5.28}$$

Note, for example, that each linear increase of just 0.1 in Γ_e gives an *order magnitude decrease* in ρ_{min}!

For a typical VMS effective temperature $T_{\mathrm{eff}} \sim 60{,}000$ K, the optical depth at the Iron bump temperature $T \sim 180{,}000$ K is, by the diffusion equation (5.17), $\tau \sim (T/T_{\mathrm{eff}})^4 \sim 100$. Because the iron bump region has very low density, and an opacity limited to the Eddington value $\kappa_{\mathrm{Edd}} = \kappa_e/\Gamma_e$, achieving this large $\tau \equiv \int \kappa \rho dr$ requires an extended, or "inflated", range in radius r. Since $\Gamma \approx 1$, we see from the radiative diffusion equation (5.17) that the change in radiation pressures scales as $dP_{\mathrm{rad}}/\rho = GM d(1/r)$. Integration from a "core" radius R_c at the base of the iron bump to an outer "envelope" radius R_e gives

$$\frac{\Delta P}{2\rho_{min}} \approx GM\left(\frac{1}{R_c} - \frac{1}{R_e}\right), \tag{5.29}$$

where $\Delta P \approx 2.2 \times 10^6$ dyn cm^{-2} characterizes the radiation pressure width of the iron bump, and the factor two in the left-side denominator comes from simple trapezoidal rule integration. Equation (5.29) can be readily solved to give a simple analytic scaling for the radius inflation factor in terms of a dimensionless ratio W of the pressure width to gravitational binding energy (Gräfener et al. 2012),

$$\frac{R_e}{R_c} \approx \frac{1}{1-W} \quad ; \quad W \equiv \frac{\Delta P}{2\rho_{min}GM/R_c} = W_{1/2}10^{10(\Gamma_e - 1/2)}. \quad (5.30)$$

Note that when W approaches order unity, the envelope radius can become very large.

Indeed, the radius divergence $R_e \to \infty$ as $W \to 1$ defines a "*Iron Bump Eddington Limit*" (IBEL), for which it is no longer possible to have a radiatively diffusive, hydrostatic envelope. Applying the scaling (5.28) for ρ_{min} within the critical condition $W = 1$, we find the limiting Eddington parameter has the scaling

$$\Gamma_{IBEL} \equiv \Gamma_e(W=1) \approx 0.5 + 0.1 \log \left(\frac{M/M_\odot}{13 R_c/R_\odot} \right). \quad (5.31)$$

To locate this limit on the H-R diagram, let us, for a given luminosity, associate the core radius with an effective temperature $T \sim L^{1/4}/R^{1/2}$. Figure 5.3 shows contours of the stellar mass for this limit, scaled for the given luminosity by the ESM mass of Eq. (5.4), and then plotted in an H-R diagram of $\log T$ vs. $\log(L/L_\odot)$. Here the steep dashed curve represents the locus of the zero-age main sequence (ZAMS), and the shaded regions outline the observational domains for S-Doradus type LBV stars, which typically vary horizontally within the V-shape on times scales of a few years.

Near the ZAMS, stars should roughly follow the ESM M-L scaling, and so the high location of the uppermost, unit-value contour representing the ESM mass indicates that only the most luminous VMS stars will breach the IBEL on the MS. As less-massive MS stars evolve to the right, their luminosities tend to increase while the masses decrease due to mass loss, making M/M_{ESM} decrease and so

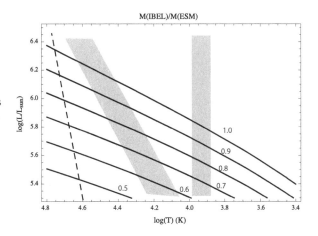

Fig. 5.3 Uppermost part of H-R diagram, showing contours of the limiting mass for the iron bump Eddington limit, scaled by the mass for the Eddington standard model. The *dashed curve* shows the locus of the ZAMS, and the V-shaped shaded regions outline the observational range for S-Doradus LBVs

bringing them up against to the IBEL. In principle, this could be an underlying cause or trigger for the observed variability in surface temperature of S-Doradus LBVs, but further work is needed to flesh out the nature of the time-dependent variations and relaxation processes. As discussed by Petrovic et al. (2006) and Gräfener et al. (2012), the overall inflation effect may also help explain the larger than expected core radii inferred for Wolf-Rayet stars.

A key general unresolved issue is the interplay between this iron-bump inflation and mass loss. Under what conditions might mass loss eliminate inflation, or inflation initiate mass loss? Other complications include the potential roles of mixing, porosity, pulsation, etc. in limiting or disrupting the inflation effect from this idealized 1D, hydrostatic model.

5.4 Basic Formalism for Envelope Instability and Mass Loss

5.4.1 General Time-Dependent Conservation Equations

The dominant role of radiative forces in VMS can lead to both time-dependent instabilities of the stellar interior and strong stellar wind mass loss from their surface. To treat these we need to generalize the above assumption of a hydrostatic equilibrium balance to consider now cases wherein the vector sum of forces acting on the gas is no longer zero, but instead has an imbalance that leads, via Newton's second law, to a net acceleration,

$$\frac{d\mathbf{v}}{dt} = \frac{\partial \mathbf{v}}{\partial t} + \mathbf{v} \cdot \nabla \mathbf{v} = \mathbf{g}_{\rm rad} - g\hat{\mathbf{r}} - \frac{\nabla P_{\rm gas}}{\rho}. \quad (5.32)$$

Here $\mathbf{v}$ is the flow velocity, and $\mathbf{g}_{\rm rad}$ and $-g\hat{\mathbf{r}}$ and are the vector forms for the radiative acceleration and gravity, with $\hat{\mathbf{r}}$ a unit radial vector. The first equality relates the total time derivative d/dt as the sum of intrinsic variation $\partial/\partial t$ and advective changes along a flow gradient, $\mathbf{v} \cdot \nabla$.

The density and velocity are related through the mass conservation relation

$$\frac{\partial \rho}{\partial t} + \nabla \cdot \rho \mathbf{v} = \frac{d\rho}{dt} + \rho \nabla \cdot \mathbf{v} = 0, \quad (5.33)$$

while conservation of energy takes the form

$$\frac{\partial e}{\partial t} + \nabla \cdot e\mathbf{v} = -P_{\rm gas} \nabla \cdot \mathbf{v} - \nabla \cdot \mathbf{F}, \quad (5.34)$$

where the divergence of vector radiative flux $\mathbf{F}$ represents a local source or sink of gas internal energy e. For an ideal gas with ratio of specific heats γ, this is related to the gas pressure through

$$P_{\text{gas}} = \rho a^2 = (\gamma - 1)e, \tag{5.35}$$

where $a \equiv \sqrt{kT/\mu}$ is the isothermal sound speed, with k Boltzmann's constant and μ the mean molecular weight.

Collectively, Eqs. (5.32)–(5.35) represent the general equations for a potentially time-dependent, multi-dimensional flow.

5.4.2 Local Linear Analysis for "Strange-Mode" Instability of Hydrostatic Envelope

Let us first apply these general equations as the basis for a local, linear perturbation analysis of hydrostatic radiative envelopes in VMS with a dynamically significant radiative acceleration g_{rad}. The response of this acceleration to local perturbations, e.g. in the gas density ρ, gives rise to the so-called *strange-mode* instability (Blaes and Socrates 2003; Glatzel and Kiriakidis 1993; Glatzel 1994, 2005). The most general form requires a global analysis of the envelope structure; but in the limit of wavelengths shorter than the local gravitational scale height, we can apply the so-called "WKB approximation" that ignores background gradients and analyzes the effect of localized, small-amplitude perturbations $\delta\rho$, δv, etc. (Since the coupling to radiation keeps the gas nearly isothermal, a simplified analysis can ignore perturbations in temperature.)

For strange modes, the simplest case involves purely radial variations of sinusoidal form $\delta \sim e^{i(kr - \omega t)}$, with k the (real) radial wavenumber and ω the (possibly complex) frequency. To first order in small-amplitude perturbations, the momentum equation (5.32) gives

$$-i\omega \delta v = -ika^2 \frac{\delta\rho}{\rho} + \delta g_{\text{rad}}. \tag{5.36}$$

A similar application to the perturbed continuity equation (5.33) relates the perturbed velocity to density,

$$\delta v = \frac{\omega}{k} \frac{\delta\rho}{\rho}. \tag{5.37}$$

If we further assume that the opacity has density dependence given by the logarithmic derivative

$$\Theta_\rho \equiv \frac{\partial \ln \kappa}{\partial \ln \rho}, \tag{5.38}$$

then the perturbed radiative acceleration can also be expressed in terms of the perturbed density,

$$\delta g_{\text{rad}} = \Theta_\rho g_{\text{rad}} \frac{\delta \rho}{\rho}. \qquad (5.39)$$

The combination of (5.36), (5.37), and (5.39) allows us to solve for a *dispersion relation* for the frequency

$$\omega = \sqrt{a^2 k^2 + ik\Theta_\rho g_{\text{rad}}} \approx \pm ak\left(1 + \frac{i\Theta_\rho g_{\text{rad}}}{2a^2 k}\right) = \pm ak \pm i\Theta_\rho \frac{g_{\text{rad}}}{2a}, \qquad (5.40)$$

where the middle approximation uses a high-wavenumber limit to expand the radical from the first form. For upward-propagating (+) modes, the frequency has a positive imaginary component, implying an instability with growth rate $\eta = \Theta_\rho g_{\text{rad}}/2a = \Theta_\rho \Gamma g/2a$. The associated instability growth time is

$$t_g \equiv \eta^{-1} = \frac{2a}{\Theta_\rho \Gamma g} = 400\,\text{s}\,\frac{a_{20}}{g_4 \Gamma \Theta_\rho}, \qquad (5.41)$$

where $a_{20} \equiv a/(20\,\text{km/s})$ and $g_4 \equiv g/(10^4 \text{cm/s}^2)$.

Figure 5.4a shows that for OPAL opacities Θ_ρ is typically a few tenths in the subsurface regions of a stellar envelope. Taking $\Theta_\rho \approx 0.5$ and typical stellar parameters $g_4 = a_{20} = 1$, we see that instability growth time $t_g \sim 2{,}000\,\text{s}$ is quite short for VMS with Γ a factor few below unity, but is much longer for lower-mass stars with very small Γ.

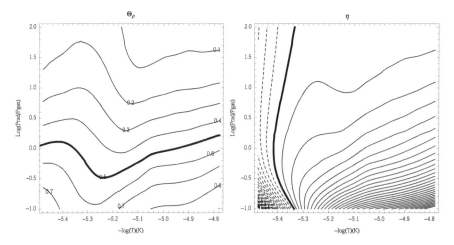

Fig. 5.4 Logarithmic derivative of opacity with density Θ_ρ (*left*) and net instability growth rate η (scaled by $g_{\text{rad}}/2a$; *right*) from Eq. (5.42), plotted as contours in the $\log T$ vs. $\log(P_{\text{rad}}/P_{\text{gas}})$ plane for the case of a hot VMS with $T_{\text{eff}} = 60k K.x$ ($\log T_{\text{eff}} = 4.8$). The contours are in increments of 0.1 and the *heavy thick curves* represent a value of 0.5 for Θ_ρ and 0 for η. The *dashed contours* indicate negative growth rates, and so show deep interior regions have a net damping due to radiation drag effects

The above stability analysis is based solely on mass and momentum conservation, ignoring the gas and radiative energy. A generalization (Blaes and Socrates 2003) that takes proper account of associated perturbations in the radiation field shows that the instability competes against the damping from a "radiation drag" effect, yielding now a net growth rate (cf. Eq. 62 of Blaes and Socrates 2003)

$$\eta = \frac{g_{\rm rad}}{2a}\left(1 + \frac{P_{\rm gas}}{4 P_{\rm rad}}\right)(\Theta_\rho - D), \qquad (5.42)$$

where the radiative drag coefficient term is given by

$$D = \left(1 + \frac{P_{\rm gas}}{4 P_{\rm rad}}\right)\frac{4 P_{\rm rad}}{F} a = \left(1 + \frac{P_{\rm gas}}{4 P_{\rm rad}}\right)\frac{16}{3}\left(\frac{T}{T_{\rm eff}}\right)^4 \frac{a}{c}. \qquad (5.43)$$

As shown in the right panel of Fig. 5.4, in the hot interior regions with $\log T \gtrsim 5.4$, the medium is now stabilized by the net damping from this radiation drag term, but in the subsurface layers where $T \lesssim T_{\rm eff}$, the a/c factor keeps this drag small, thus preserving the instability found above, and so again giving growth rates η that are a few tenths times $g_{\rm rad}/2a$.

The net instability should lead to amplification of upward-propagating sound waves, eventually limited by the steeping into weak or moderate shocks with velocity amplitude on the order of the sound speed, $\delta v \sim a$. For near-surface temperatures $T \sim T_{\rm eff}$ of a few 10^4 K, the sound speed is $a \sim 20$ km/s, much smaller (by factor $\sim 1/30$) than the near-surface escape speed $v_{\rm esc} \equiv \sqrt{2GM/R}$, which is about 600 km/s for a star with a solar value for the ratio of mass to radius. As such, this short-wavelength form of such strange mode instability is not at all suitable to providing the kind of large-scale mass ejection inferred from LBVs. But there have been suggestions (Glatzel and Kiriakidis 1993; Glatzel 1994, 2005) that analogous larger-scale, global modes of strange-mode pulsations might be important in triggering episodic mass loss.

5.4.3 General Equations for Steady, Spherically Symmetric Wind

The general flow conservation Eqs. (5.32)–(5.35) also provide the basis for modeling stellar wind outflows. First-order wind models are commonly based on the simplifying approximations of steady-state ($\partial/\partial t = 0$), spherically symmetric, radial outflow ($\mathbf{v} = v(r)\hat{\mathbf{r}}$). The mass conservation requirement (5.33) then can then be used to define a constant overall mass loss rate,

$$\dot{M} \equiv 4\pi \rho v r^2. \qquad (5.44)$$

Using this and the ideal gas law (5.35) to eliminate the density in the pressure gradient term then gives for the radial equation of motion

$$\left(1 - \frac{a^2}{v^2}\right) v \frac{dv}{dr} = g_{\text{rad}} - \frac{GM}{r^2} + \frac{2a^2}{r} + \frac{da^2}{dr}. \tag{5.45}$$

The gas pressure terms (containing the sound speed a) on the right-hand-side are key to accelerating the hot (MK) coronal-type winds of the sun and other cool, lower-mass stars. But in winds from massive stars – which are kept almost isothermal near the stellar effective temperature by the competition between photoionization heating and radiative cooling – these terms are negligible, since compared to competing terms needed to drive the wind, they are of order $w_s \equiv (a/v_{\text{esc}})^2 \approx 0.001$, where $v_{\text{esc}} \equiv \sqrt{2GM/R}$ is the escape speed from the stellar surface radius R. These gas pressure terms on the right-hand side of (5.45) can thus be quite generally neglected in VMS winds. However, to allow for a smooth mapping of a wind model onto a hydrostatic atmosphere through a subsonic wind base, one can still retain the sound-speed term on the left-hand-side. Transitioning to a supersonic wind then requires $g_{\text{rad}} = GM/r^2$, and so $\Gamma = 1$, at the sonic point $v = a$.

Since overcoming gravity is key, it is convenient to rewrite (5.45) in a dimensionless form that scales all accelerations by gravity,

$$\left(1 - \frac{w_s}{w}\right) w' = \Gamma - 1 \tag{5.46}$$

Here $w \equiv v^2/v_{\text{esc}}^2$ is the flow kinetic energy in terms of the escape energy from the surface radius R, and $w' \equiv dw/dx$ is the change of this scaled energy with scaled the gravitational potential $x \equiv 1 - R/r$ at any radius r.

In characterizing the sonic point as the flow "critical point", it is sometimes suggested (Nugis and Lamers 2002) that conditions for reaching $\Gamma = 1$, for example in the iron opacity bump, set the sonic point density ρ_s and thus the mass loss rate $\dot{M} = 4\pi \rho_s a R^2$ of a steady wind outflow. But it is important to emphasize that, because the flow energy at the sonic point is just a tiny fraction $w_s \approx 0.001$ of what's needed to escape the star's gravitational potential, maintaining a steady wind requires keeping $\Gamma > 1$ over an extended range of the supersonic region. In Sect. 5.6.3 we discuss the flow stagnation that occurs if the opacity or radiative energy flux is insufficient to maintain initial outflow from a limited super-Eddington region. But the next section first reviews the standard "CAK" theory (Castor et al. 1975) for a steady-state wind outflow driven by line-opacity.

5.5 Line-Driven Stellar Winds

As noted in Sects. 5.2.2 and 5.2.3, the resonant nature of line (bound-bound) scattering from metal ions leads to an opacity that is inherently much stronger than from free electrons. In the deep envelope where the radiative transfer is well characterized as a radiative diffusion, the cumulative opacity of lines is given by the Rosseland mean, which can be several times that from pure electron scattering, e.g. near the Iron bump; in VMS approaching the IBEL, this can lead to a strong inflation of the stellar envelope (Sect. 5.3.4).

But in the idealized case that ions are illuminated by an unattenuated continuum – as would occur in the limit of *optically thin* radiative transfer–, the cumulative opacity is given by the *Planck* mean (Sect. 5.2.2), wherein the dominant contribution of lines leads to an opacity enhancement that is a huge factor $\overline{Q} \approx 2{,}000$ larger than from free electrons (Gayley 1995). In terms of the radiative force, this implies the Eddington parameter associated with lines can be enhanced by a similarly large factor over the classical electron scattering value. This suggest that, even in moderately massive stars with electron Eddington parameters $\Gamma_e > 5 \times 10^{-4}$, line opacity could completely overcome gravity and so initiate a sustained stellar wind outflow.

In practice, self-absorption within strong lines limits the line force, giving it a value that is intermediate between that from the Rosseland and Planck means. Modeling such line-driven wind outflows thus requires a treatment of the line radiation transport in regimes between the associated diffusion vs. optically thin limits. Chapter 4 reviews how mass loss rates from VMS are computed from numerical models using Monte-Carlo (MC) treatments of the radiation transport. As complement to this, the next Sect. 5.5.1 reviews the classical Castor et al. (1975, CAK) model, which by using the key Sobolev approximation (Sobolev 1960) for localized line-transport, allows one to obtain fully *analytic* solutions for the steady-state wind, with associated simple scaling forms for the wind mass loss rate and velocity law (Owocki 2013). This provides a basis for a linear perturbation analysis of a strong instability intrinsic to line-driving (Sect. 5.5.2), and for time-dependent numerical hydrodynamical simulations of the resulting instability wind structure (Sect. 5.5.2).

5.5.1 The CAK/Sobolev Model for Steady-State Winds

Sobolev Line-Transfer and Desaturation by Wind Expansion

As illustrated in Fig. 5.5, a key factor in controlling the net strength of the line-force that drives a stellar wind outflow is the desaturation of the lines associated with the variable Doppler shift from the wind acceleration. In the highly supersonic wind, the thermal Doppler broadening of the line, which for heavy ions is set by a thermal speed v_{th} that is a factor several smaller than the sound speed a, is much smaller than the Doppler shift associated with the wind outflow speed $v \gg a$. This allows a localized "Sobolev approximation" (Sobolev 1960) for the line transport, with stellar photons interacting with the wind over a narrow resonance layer, with width set by the Sobolev length, $l_{Sob} = v_{th}/(dv/dr)$, and with associated optical depth proportional to $t \equiv \kappa_e \rho c/(dv/dr) = \Gamma_e \dot{M} c^2/L_* w'$.

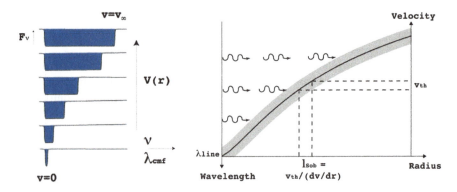

Fig. 5.5 Two perspectives for the Doppler-shifted line-resonance in an accelerating flow. *Right*: Photons with a wavelength just shortward of a line propagate freely from the stellar surface up to a layer where the wind outflow Doppler shifts the line into a resonance over a narrow width (represented here by the shading) equal to the Sobolev length, set by the ratio of thermal speed to velocity gradient, $l_{Sob} \equiv v_{th}/(dv/dr)$. *Left*: Seen from successively larger radii within the accelerating wind, the Doppler-shift sweeps out an increasingly broadened line absorption trough in the stellar spectrum

The CAK Line-Force, Mass Loss Rate, and Wind Velocity Law

Based on this Sobolev treatment of line transport, Castor et al. (1975) developed a powerful and highly useful formalism for treating the cumulative force from a large ensemble of lines by assuming they could be approximated by power-law number distribution in line-strength, with characteristic power index α. A key result is that the cumulative force from the ensemble is reduced by a factor $1/(\overline{Q}t)^\alpha$ from the optically thin value,

$$\Gamma_{CAK} = \frac{\overline{Q}\Gamma_e}{(1-\alpha)(\overline{Q}t)^\alpha} = \Gamma_e k t^{-\alpha} = C(w')^\alpha, \qquad (5.47)$$

where the second equality defines the CAK "force multiplier" $kt^{-\alpha}$, with[5] $k \equiv \overline{Q}^{1-\alpha}/(1-\alpha)$. The last equality relates the line-force to the flow acceleration, with

$$C \equiv \frac{1}{1-\alpha}\left[\frac{L_*}{\dot{M}c^2}\right]^\alpha [\overline{Q}\Gamma_e]^{1-\alpha}. \qquad (5.48)$$

---

[5]Here we use a slight variation of the standard CAK notation in which the artificial dependence on a fiducial ion thermal speed is avoided by simply setting $v_{th} = c$. Back-conversion to CAK notation is achieved by multiplying t by v_{th}/c and k by $(v_{th}/c)^\alpha$. The line normalization $\overline{Q}$ offers the advantages of being a dimensionless measure of line-opacity that is independent of the assumed ion thermal speed, with a nearly constant characteristic value of order $\overline{Q} \sim 10^3$ for a wide range of ionization conditions (Gayley 1995).

Note that, for fixed sets of parameters for the star (L_*, M_*, Γ_e) and line-opacity (α, $\overline{Q}$), this constant scales with the mass loss rate as $C \propto 1/\dot{M}^\alpha$.

Neglecting the small sound-speed term $w_s \approx 0.001 \ll 1$, application of Eq. (5.47) into (5.46) gives the CAK equation of motion,

$$F = w' + 1 - \Gamma_e - C(w')^\alpha = 0. \tag{5.49}$$

For small $\dot{M}$ (large C), there are two solutions, while for large $\dot{M}$ (small C), there are no solutions. The CAK critical solution corresponds to a *maximal* mass loss rate, defined by $\partial F/\partial w' = 0$, for which the $C(w')^\alpha$ is tangent to the line $1 - \Gamma_e + w'$ at a critical acceleration $w'_c = (1 - \Gamma_e)\alpha/(1 - \alpha)$. Since the scaled equation of motion (5.49) has no explicit spatial dependence, this critical acceleration applies throughout the wind, and so can be trivially integrated to yield $w(x) = w'_c x$. In terms of dimensional quantities, this represents a specific case of the general "beta"-velocity-law,

$$v(r) = v_\infty \left(1 - \frac{R_*}{r}\right)^\beta, \tag{5.50}$$

where here $\beta = 1/2$, and the wind terminal speed $v_\infty = v_{esc}\sqrt{\alpha(1 - \Gamma_e)/(1 - \alpha)}$. Similarly, the critical value C_c yields, through Eq. (5.48), the standard CAK scaling for the mass loss rate

$$\dot{M}_{CAK} = \frac{L_*}{c^2} \frac{\alpha}{1 - \alpha} \left[\frac{\overline{Q}\Gamma_e}{1 - \Gamma_e}\right]^{(1-\alpha)/\alpha}. \tag{5.51}$$

Modifications and Limitations of the CAK Mass Loss Scaling

These CAK results strictly apply only under the idealized assumption that the stellar radiation is radially streaming from a point-source. If one takes into account the finite angular extent of the stellar disk, then near the stellar surface the radiative force is reduced by a factor $f_{d*} \approx 1/(1 + \alpha)$, leading to a reduced mass loss rate (Friend and Abbott 1986; Pauldrach et al. 1986)

$$\dot{M}_{fd} = f_{d*}^{1/\alpha} \dot{M}_{CAK} = \frac{\dot{M}_{CAK}}{(1 + \alpha)^{1/\alpha}} \approx \dot{M}_{CAK}/2. \tag{5.52}$$

Away from the star, the correction factor increases back toward unity, which for the reduced base mass flux implies a stronger, more extended acceleration, giving a somewhat higher terminal speed, $v_\infty \approx 3 v_{esc}$, and a flatter velocity law, approximated by replacing the exponent in Eq. (5.50) by $\beta \approx 0.8$.

The effect of a radial change in ionization can be approximately taken into account by correcting the CAK force (5.47) by a factor of the form $(n_e/W)^\delta$, where

n_e is the electron density, $W \equiv 0.5\left(1 - \sqrt{1 - R_*^2/r^2}\right)$ is the radiation "dilution factor", and the exponent has a typical value $\delta \approx 0.1$ (Abbott 1982). This factor introduces an additional density dependence to that already implied by the optical depth factor $1/t^\alpha$ given in Eq. (5.47). Its overall effect can be roughly accounted with the simple substitution $\alpha \to \alpha' \equiv \alpha - \delta$ in the power exponents of the CAK mass loss scaling law (5.51). The general tendency is to moderately increase $\dot{M}$, and accordingly to somewhat decrease the wind speed.

The above scalings also ignore the finite gas pressure associated with a small but non-zero sound-speed parameter w_s. Through a perturbation expansion of the equation of motion (5.46) in this small parameter, it possible to derive simple scalings for the fractional corrections to the mass loss rate and terminal speed (Owocki and ud-Doula 2004)

$$\delta m_s \approx \frac{4\sqrt{1-\alpha}}{\alpha} \frac{a}{v_{esc}} \quad ; \quad \delta v_{\infty,s} \approx \frac{-\alpha \delta m_s}{2(1-\alpha)} \approx \frac{-2}{\sqrt{1-\alpha}} \frac{a}{v_{esc}}. \tag{5.53}$$

For a typical case with $\alpha \approx 2/3$ and $w_s = 0.001$, the net effect is to increase the mass loss rate and decrease the wind terminal speed, both by about 10%.

An important success of these CAK scaling laws is the theoretical rationale they provide for an empirically observed "Wind-Momentum-Luminosity" (WML) relation (Kudritzki et al. 1999). Combining the CAK mass-loss law (5.51) together with the scaling of the terminal speed with the effective escape, we obtain a WML relation of the form,

$$\dot{M} v_\infty \sqrt{R_*} \sim L^{1/\alpha'} \overline{Q}^{1/\alpha'-1} \tag{5.54}$$

wherein we have neglected a residual dependence on $M(1 - \Gamma_e)$ that is generally very weak for the usual case that α' is near $2/3$. Note that the direct dependence $\overline{Q} \sim Z$ provides the scaling of the WML with metalicity Z.

Finally, as a star approaches the classical Eddington limit $\Gamma_e \to 1$, these standard CAK scalings formally predict the mass loss rate to diverge as $\dot{M} \propto 1/(1 - \Gamma_e)^{(1-\alpha)/\alpha}$, but with a vanishing terminal flow speed $v_\infty \propto \sqrt{1 - \Gamma_e}$. The former might appear to provide an explanation for the large mass losses inferred in LBV's, but the latter fails to explain the moderately high inferred ejection speeds, e.g. the 500–800 km/s kinematic expansion inferred for the Homunculus nebula of η Carinae (Smith 2002; Smith et al. 2003).

So one essential point is that line-driving could never explain the extremely large mass loss rates needed to explain the Homunculus nebula. To maintain the moderately high terminal speeds, the $\Gamma_e/(1 - \Gamma_e)$ factor would have to be of order unity. Then for optimal realistic values $\alpha = 1/2$ and $Q \approx 2{,}000$ for the line opacity parameters (Gayley 1995), the maximum mass loss from line driving is given by Smith and Owocki (2006),

$$\dot{M}_{max,lines} \approx 1.4 \times 10^{-4} L_6 \, M_\odot/\text{year}, \tag{5.55}$$

where $L_6 \equiv L/10^6 L_\odot$. Even for peak luminosities of a few times $10^7 L_\odot$ during η Carinae's eruption, this limit is still several orders of magnitude below the mass loss needed to form the Homunculus. Thus, if mass loss during these eruptions occurs via a wind, it must be a super-Eddington wind driven by continuum radiation force (e.g., electron scattering opacity) and not lines (Quinn and Paczynski 1985; Belyanin 1999). Such continuum-driven wind models for LBVs are discussed further in Sect. 5.6.

5.5.2 Non-Sobolev Models of Wind Instability

The above CAK steady-state model depends crucially on the use of the Sobolev approximation to compute the local CAK line force (5.47). Analyses that relax this approximation show that the flow is subject to a strong, "line-deshadowing instability" (LDI) for velocity perturbations on a scale near and below the Sobolev length $l_{Sob} = v_{th}/(dv/dr)$ (Lucy and Solomon 1970; MacGregor et al. 1979; Owocki and Rybicki 1984, 1985; Owocki and Puls 1996). Moreover, the diffuse, scattered component of the line force, which in the Sobolev limit is nullified by the fore-aft symmetry of the Sobolev escape probability (see Fig. 5.6), turns out to have important dynamical effects on the instability through a "diffuse line-drag" (Lucy 1984).

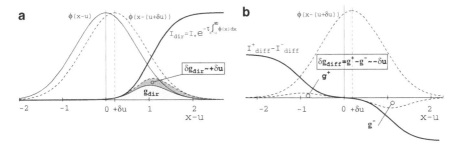

Fig. 5.6 (a) The line profile ϕ and direct intensity plotted vs. comoving frame frequency $x - u = x - v/v_{th}$, with the light shaded overlap area proportional to the net direct line-force g_{dir}. The *dashed profile* shows the effect of the Doppler shift from a perturbed velocity δv, with the resulting extra area in the overlap with the blue-edge intensity giving a perturbed line-force δg that scales in proportion to this perturbed velocity $\delta u = \delta v/v_{th}$. (b) The comoving-frequency variation of the forward (+) and backward (−) streaming parts of the diffuse, scattered radiation. Because of the Doppler shift from the perturbed velocity, the *dashed profile* has a stronger interaction with the backward streaming diffuse radiation, resulting in a diffuse-line-drag force that scales with the negative of the perturbed velocity, and so tends to counter the instability of the direct line-force in part a

Linear Analysis of Line-Deshadowing Instability

For sinusoidal perturbations ($\sim e^{i(kr-wt)}$) with wavenumber k and frequency ω, the linearized momentum equation (5.36) (ignoring the small gas pressure by setting $a = 0$) relating the perturbations in velocity and radiative acceleration implies $\omega = i\frac{\delta g}{\delta v}$, which shows that unstable growth, with $\Im\omega > 0$, requires $\Re(\delta g/\delta v) > 0$. For a purely Sobolev model (Abbott 1980), the CAK scaling of the line-force (5.47) with velocity gradient v' implies $\delta g \sim \delta v' \sim ik\delta v$, giving a purely real ω, and thus a stable wave that propagates inward at phase speed,

$$\frac{\omega}{k} = -\frac{\partial g}{\partial v'} \equiv -U, \qquad (5.56)$$

which is now known as the "Abbott speed". Abbott (1980) showed this is comparable to the outward wind flow speed, and in fact exactly equals it at the CAK critical point.

As illustrated in Fig. 5.6a, instability arises from the deshadowing of the line by the extra Doppler shift from the velocity perturbation, giving $\delta g \sim \delta v$ and thus $\Im\omega > 0$. A general analysis (Owocki and Rybicki 1984) yields a "bridging law" encompassing both effects,

$$\frac{\delta g}{\delta v} \approx \Omega \frac{ik\Lambda}{1 + ik\Lambda}, \qquad (5.57)$$

where $\Omega \approx g_{cak}/v_{th}$ sets the instability growth rate, and the "bridging length" Λ is found to be of order the Sobolev length l_{sob}. As illustrated in Fig. 5.7, in the long-wavelength limit $k\Lambda \ll 1$, we recover the stable, Abbott-wave scalings of the Sobolev approximation, $\delta g/\delta v \approx ik\Omega\Lambda = ikU$; while in the short-wavelength limit $k\Lambda \gg 1$, we obtain the instability scaling $\delta g \approx \Omega\delta v$. The instability growth rate is very large, about the flow rate through the Sobolev length, $\Omega \approx v/l_{Sob}$. Since this

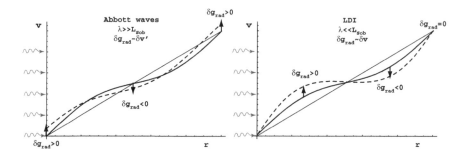

Fig. 5.7 Illustration of the scaling for perturbed line-acceleration δg_{rad} with velocity perturbation δv, showing how this goes from being proportional to the perturbed velocity gradient $\delta v' \sim ik\delta v$ in the long-wavelength Sobolev limit appropriate for Abbott waves, to scaling directly with the perturbed velocity δv in the short-wavelength limit of the Line-Deshadowing Instability (LDI)

is a large factor v/v_{th} bigger than the typical wind expansion rate $dv/dr \approx v/R_*$, a small perturbation at the wind base would, within this lineary theory, be amplified by an enormous factor, of order $e^{v/v_{th}} \approx e^{100}$!

Numerical Simulations of Instability-Generated Wind Structure

Numerical simulations of the nonlinear evolution require a non-Sobolev line-force computation on a spatial grid that spans the full wind expansion over several R_*, yet resolves the unstable structure at small scales near and below the Sobolev length. The first tractable approach (Owocki et al. 1988) focussed on the *absorption* of the *direct* radiation from the stellar core, accounting now for the attenuation from intervening material by carrying out a *nonlocal integral* for the frequency-dependent radial optical depth. Simulations show that because of inward nature of wave propagation implies an anti-correlation between velocity and density variation, the nonlinear growth leads to high-speed rarefactions that steepen into strong *reverse* shocks and compress material into dense clumps (or shells in these 1D models) (Owocki et al. 1988).

The assumption of pure-absorption was criticized by Lucy (1984), who pointed out that the interaction of a velocity perturbation with the background, *diffuse* radiation from line-scattering results in a *line-drag* effect that reduces, and potentially could even eliminate, the instability associated with the direct radiation from the underlying star. The basic effect is illustrated in Fig. 5.6. The fore-aft ($\pm$) symmetry of the diffuse radiation leads to cancellation of the g_+ and g_- force components from the forward and backward streams, as computed from a line-profile with frequency centered on the local comoving mean flow. Panel b shows that the Doppler shift associated with the velocity perturbation δv breaks this symmetry, and leads to stronger forces from the component opposing the perturbation.

Full linear stability analyses accounting for scattering effects (Owocki and Rybicki 1985) show the fraction of the direct instability that is canceled by the line-drag of the perturbed diffuse force depends on the ratio of the scattering source function S to core intensity I_c,

$$s = \frac{r^2}{R_*^2}\frac{2S}{I_c} \approx \frac{1}{1+\mu_*} \quad ; \quad \mu_* \equiv \sqrt{1-R_*^2/r^2}, \tag{5.58}$$

where the latter approximation applies for the optically thin form $2S/I_c = 1-\mu_*$. The net instability growth rate thus becomes

$$\Omega(r) \approx \frac{g_{cak}}{v_{th}}\frac{\mu_*(r)}{1+\mu_*(r)}. \tag{5.59}$$

This vanishes near the stellar surface, where $\mu_* = 0$, but it approaches half the pure-absorption rate far from the star, where $\mu_* \to 1$. This implies that the outer wind is still very unstable, with cumulative growth of ca. $v_\infty/2v_{th} \approx 50$ e-folds.

Most efforts to account for scattering line-drag in simulations of the nonlinear evolution of the instability have centered on a *Smooth Source Function* (SSF) approach (Owocki 1991; Feldmeier 1995; Owocki and Puls 1996, 1999). This assumes that averaging over frequency and angle makes the scattering source function relatively insensitive to flow structure, implying it can be pulled out of the integral in the formal solution for the diffuse intensity. Within a simple *two-stream* treatment of the line-transport, the net diffuse line-force then depends on the *difference* in the *nonlocal* escape probabilities $b_\pm$ associated with forward (+) vs. backward (−) *integrals* of the frequency-dependent line-optical-depth.

In the Sobolev approximation, both the forward and backward integrals give the same form, viz. $b_+ \approx b_-$, leading to the net cancellation of the Sobolev diffuse force. But for perturbations on a spatial scale near and below the Sobolev length, the perturbed velocity breaks the forward/back symmetry (Fig. 5.6b), leading to perturbed diffuse force that now scales in proportion to the *negative* of the perturbed velocity, and thus giving the diffuse line-drag that reduces the net instability by the factors given in (5.58) and (5.59).

The left panel of Fig. 5.8 illustrates the results of a 1D SSF simulation, starting from an initial condition set by smooth, steady-state CAK/Sobolev model (dashed curves). Because of the line-drag stabilization of the driving near the star (Eq. 5.59), the wind base remains smooth and steady. But away from the stellar surface, the net strong instability leads to extensive structure in both velocity and density, roughly straddling the CAK steady-state. Because of the backstreaming component of the diffuse line-force causes any outer wind structure to induce small-amplitude fluctuations near the wind base, the wind structure, once initiated, is "self-excited", arising spontaneously without any explicit perturbation from the stellar boundary.

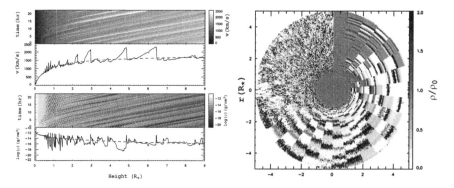

Fig. 5.8 *Left*: Results of 1D Smooth-Source-Function (SSF) simulation of the line-deshadowing instability. The line plots show the spatial variation of velocity (*upper*) and density (*lower*) at a fixed, arbitrary time snapshot. The corresponding grey scales show both the time (*vertical axis*) and height (*horizontal axis*) evolution. *The dashed curve* shows the corresponding smooth, steady CAK model. *Right*: For 2DH+1DR SSF simulation, grayscale representation for the density variations rendered as a time sequence of 2-D wedges of the simulation model azimuthal range $\Delta\phi = 12°$ stacked clockwise from the vertical in intervals of 4,000 s from the CAK initial condition

In the outer wind, the velocity variations become highly nonlinear and non-monotonic, with amplitudes approaching 1,000 km/s, leading to formation of strong shocks. However, these high-velocity regions have very low density, and thus represent only very little material. As noted for the pure-absorption models, this anti-correlation between velocity and density arises because the unstable linear waves that lead to the structure have an *inward* propagation relative to the mean flow. For most of the wind mass, the dominant overall effect of the instability is to concentrate material into dense clumps. As discussed below, this can lead to overestimates in the mass loss rate from diagnostics that scale with the square of the density.

The presence of multiple, embedded strong shocks suggests a potential source for the soft X-ray emission observed from massive star winds; but the rarefied nature of the high-speed gas implies that this self-excited structure actually feeds very little material through the strong shocks needed to heat gas to X-ray emitting temperatures. To increase the level of X-ray emission, Feldmeier et al. (1997), introduced intrinsic perturbations at the wind base, assuming the underlying stellar photosphere has a turbulent spectrum of compressible sound waves characterized by abrupt phase shifts in velocity and density. These abrupt shifts seed wind variations that, when amplified by the line-deshadowing instability, now include substantial velocity variations among the dense clumps. As illustrated in Fig. 5.9, when these dense clumps collide, they induce regions of relatively dense, hot gas which produce localized bursts of X-ray emission. Averaged over time, these localized regions can collectively yield X-ray emission with a brightness and spectrum that is comparable to what is typically observed from such hot stars.

Because of the computational expense of carrying out nonlocal optical depth integrations at each time step, such SSF instability simulations have generally been limited to just 1D. More realistically, various kinds of thin-shell instabilities (Vishniac 1994; Kee et al. 2014) can be expected to break up the structure into a complex, multidimensional form. A first step to modelling both radial and lateral

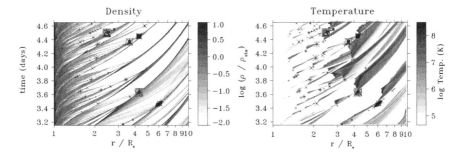

Fig. 5.9 Greyscale rendition of the evolution of wind density and temperature, for time-dependent wind-instability models with structure formation triggered by photospheric perturbations. The *boxed crosses* identify localized region of clump-clump collision that lead to the hot, dense gas needed for a substantial level of soft X-rays emission

structure is to use a restricted "2D-H+1D-R" approach (Dessart and Owocki 2003), extending the hydrodynamical model to 2D in radius and azimuth, but still keeping the 1D-SSF radial integration for the inward/outward optical depth within each azimuthal zone. The right panel of Fig. 5.8 shows the resulting 2D density structure within a narrow (12°) wedge, with the time evolution rendered clockwise at fixed time intervals of 4,000 s starting from the CAK initial condition at the top. The line-deshadowing instability is first manifest as strong radial velocity variations and associated density compressions that initially extend nearly coherently across the full azimuthal range of the computational wedge.

But as these initial "shell" structures are accelerated outward, they become progressively disrupted by Rayleigh-Taylor or thin-shell instabilities that operate in azimuth down to the grid scale $d\phi = 0.2°$. Such a 2DR+1DH approach may well exaggerate the level of variation on small lateral scales. The lack of *lateral* integration needed to compute an azimuthal component of the diffuse line-force means that the model ignores a potentially strong net lateral line-drag that should strongly damp azimuthal velocity perturbations on scales below the lateral Sobolev length $l_0 \equiv r v_{th}/v_r$ (Rybicki et al. 1990). Presuming that this would inhibit development of lateral instability at such scales, then any lateral breakup would be limited to a minimum lateral angular scale of $\Delta\phi_{min} \approx l_0/r = v_{th}/v_r \approx 0.01$ rad $\approx 0.5°$. Further work is needed to address this issue through explicit incorporation of the lateral line-force and the associated line-drag effect.

5.5.3 Clumping, Porosity and Vorosity: Implications for Mass Loss Rates

Both the 1D and 2D SSF simulations thus predict a wind with extensive structure in both velocity and density. A key question then is how such structure might affect the various wind diagnostics that are used to infer the mass loss rate. Historically such wind clumping has been primarily considered for its effect on diagnostics that scale with the square of the density. The strength of such diagnostics is enhanced in a clumped wind, leading to an overestimate of the wind mass loss rate that scales with $\sqrt{f_{cl}}$, where the clumping factor $f_{cl} \equiv \langle \rho^2 \rangle / \langle \rho \rangle^2$, with angle brackets denoting a local averaging over many times the clump scale. For strong density contrast between the clump and interclump medium, this is just inverse of the clump volume filling factor, i.e. $f_{cl} \approx 1/f_{vol}$. 1D SSF simulations by Runacres and Owocki (2002) generally find f_{cl} increasing from unity at the structure onset radius $\sim 1.5 R_*$, peaking at a value $f_{cl} \gtrsim 10$ at $r \approx 10 R_*$, with then a slow outward decline to ~ 5 for $r \sim 100 R_*$.

These thus imply that thermal IR and radio emission formed in the outer wind $r \approx 10$–$100 R_*$ may overestimate mass loss rates by a factor 2–3. The 2D models of Dessart and Owocki (2003, 2005) find a similar variation, but somewhat lower peak value, and thus a lower clumping factor than in 1D models, with a peak value of about $f_{cl} \approx 6$, apparently from the reduced collisional compression from clumps with different radial speeds now being able to pass by each other. But in both 1D and 2D models, the line-drag near the base means that self-excited, intrinsic structure does not appear till $r \gtrsim 1.5$, implying little or no clumping effect on $H\alpha$ line emission formed in this region. It should be stressed, however, that this is not necessarily a very robust result, since turbulent perturbations at the wind base, and/or a modestly reduced diffuse line-drag, might lead to onset of clumping much closer to the wind base.

If clumps remain optically thin, then they have no effect on single-density diagnostics, like the bound-free absorption of X-rays. The recent analysis by Cohen et al. (2010) of the X-ray line-profiles observed by Chandra from ζ-Pup indicates matching the relatively modest skewing of the profile requires mass loss reduction of about a factor 3 from typical density-squared diagnostic value. However, a key issue here is whether the individual clumps might become *optically thick* to X-ray absorption. In this case, the self-shadowing of material within the clump can lead to an overall reduction in the effective opacity of the clumped medium (Owocki et al. 2004; Oskinova et al. 2007),

$$\kappa_{eff} = \kappa \frac{1 - e^{-\tau_{cl}}}{\tau_{cl}}, \qquad (5.60)$$

where κ is the microscopic opacity, and the optical thickness for clumps of size ℓ is $\tau_{cl} = \kappa \rho \ell f_{cl}$. The product $\ell f_{cl} \equiv h$ is known as the *porosity length*, which also represents the *mean-free-path* between clumps. A medium with optically thick clumps is thus porous, with an opacity reduction factor $\kappa_{eff}/\kappa = 1/\tau_{cl} = 1/\kappa \rho h$.

However, it is important to emphasize that getting a significant porosity decrease in the *continuum* absorption of a wind can be quite difficult, since clumps must become optically thick near the radius of the smoothed-wind photosphere, implying a collection of a substantial volume of material into each clump, and so a porosity length on order the local radius. Owocki and Cohen (2006) showed in fact that a substantial porosity reduction the absorption-induced asymmetry of X-ray line profiles required such large porosity lengths $h \sim r$. Since the LDI operates on perturbations at the scale of the Sobolev length $l_{sob} \equiv v_{th}/(dv/dr) \approx (v_{th}/v_\infty) R_* \approx R_*/300$, the resulting structure is likewise very small scale, as illustrated in the 2D SSF simulations in Fig. 5.8. Given the modest clumping factor $f_{cl} \lesssim 10$, it seems clear that the porosity length is quite small, $h < 0.1r$, and thus that porosity from LDI structure is not likely to be an important factor for continuum processes like bound-free absorption of X-rays.

The situation is however quite different for *line* absorption, which can readily be optically thick in even a smooth wind, with *Sobolev optical depth* $\tau_{sob} = \kappa_l \rho v_{th}/(dv/dr) = \kappa_l \rho l_{sob} > 1$. In a simple model with a smooth velocity law but material collected into clumps with volume filling factor $f_{vol} = 1/f_{cl}$, this clump optical depth would be even larger by a factor f_{cl}. As noted by Oskinova et al. (2007), the escape of radiation in the gaps between the thick clumps might then substantially reduce the effective line strength, and so help explain the unexpected weakness of PV lines observed by FUSE (Fullerton et al. 2006), which otherwise might require a substantial, factor-ten or more reduction in wind mass loss rate.

But instead of *spatial* porosity, the effect on lines is better characterized as a kind of velocity porosity, or "*vorosity*", which is now relatively insensitive to the spatial scale of wind structure (Owocki 2008). The left panel of Fig. 5.10 illustrates the typical result of 1D dynamical simulation of the wind instability, plotted here as a time-snapshot of velocity vs. a *mass* coordinate, instead of radius. The intrinsic instability of line-driving leads to a substantial velocity structure, with narrow peaks corresponding to spatially extended, but tenuous regions of high-speed flow; these bracket dense, spatially narrow clumps/shells that appear here as nearly flat, extended velocity plateaus in mass. The right panel of Fig. 5.10 illustrates a simplified, heuristic model of such wind structure for a representative wind section, with the velocity clumping now represented by a simple "staircase" structure, compressing the wind mass into discrete sections of the wind velocity law, while evacuating the regions in between; the structure is characterized by a "velocity clumping factor" f_{vel}, set by the ratio between the internal velocity width δv to the velocity separation Δv of the clumps. The straight line through the steps represents the corresponding smooth wind flow.

The effect of the velocity structure on the line-absorption profile depends on the local Sobolev optical depth, which scales with the inverse of the mass derivative of velocity, $\tau_y \sim 1/(dv/dm)$, evaluated at a resonance location r_s, where the velocity-scaled, observer-frame wavelength $y = -v(r_s)/v_\infty$. In a smooth wind with Sobolev optical depth τ_y, the absorption profile is given simply by $A_y = 1 - e^{-\tau_y}$ (Owocki 2008). In the structured model, the optical thickness of individual clumps is increased by the inverse of the clumping factor $1/f_{vel}$, but they now only cover a fraction f_{vel} of the velocity/wavelength interval. The net effect on the averaged line profile is to *reduce* the net absorption by a factor (Owocki 2008),

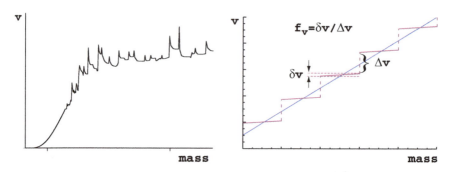

Fig. 5.10 *Left*: Self-excited velocity structure arising in a 1D SSF simulation of the line-driven instability, plotted versus a mass coordinate, $M(r) = \int_R^r 4\pi \rho r'^2 \, dr'$. Note the formation of velocity plateaus in the outer regions of the wind. *Right*: Velocity vs. mass in a wind seqment with structure described by a simplified velocity staircase model with multiple large steps Δv between plateaus of width δv. Here the associated velocity clumping factor $f_{vel} \equiv \delta v / \Delta v = 1/10$. The *straight line* represents the corresponding smooth CAK/Sobolev model

$$R_A(\tau_y, f_{vel}) = f_{vel} \frac{1 - e^{-\tau_y/f_{vel}}}{1 - e^{-\tau_y}}. \tag{5.61}$$

Note that for optically thick lines, $\tau_y \gg 1$, the reduction approaches a fixed value, given in fact by the clumping factor, $R_A \approx f_{vel}$. If the smooth-wind line is optically thin, $\tau_y \ll 1$, then $R_A(\tau_y, f_{vel}) \approx (1 - e^{-\tau_y/f})/(\tau_y/f_{vel})$, which is quite analogous to the opacity reduction for *continuum* porosity (Eq. 5.60), if we just substitute for the clump optical depth, $\tau_c \to \tau_y/f_{vel}$.

But a key point here is that, unlike for the continuum case, the *net reduction in line absorption no longer depends on the spatial scale* of the clumps. Instead one might think of this velocity clumping model as a kind of velocity form of the standard venetian blind, with f_{vel} representing the fractional projected covering factor of the blinds relative to their separation. The $f_{vel} = 1$ case represents closed blinds that effectively block the background light, while small f_{vel} represent cases when the blinds are broadly open, letting through much more light.

Further discussion of the potentially key role of wind clumping and vorosity for determining wind mass loss rates is given by Sundqvist et al. (2011).

5.6 Continuum-Driven Mass Loss from Super-Eddington LBVs

5.6.1 Lack of Self-Regulation for Continuum Driving

Despite the extensive instability-generated structure in the outer regions of line-driven winds, their overall mass loss is quite steady, and can persist throughout the lifetime of even moderately massive stars. But VMS show occasional episodes of much stronger mass loss, known generally as giant eruption LBVs, commonly characterized by a radiative luminosity that exceeds even the classical electron-scattering Eddington limit, and lasting for up to about a decade. The energy source and trigger of such eruptive LBVs is uncertain, and could even have an explosive character seated in the deep stellar interior; but their persistence for much longer than the dynamical time scale of few days suggests they can be at least partly modeled as a quasi-steady stellar wind, though now driven by *continuum* opacity through electron scattering of their super-Eddington luminosity.

A key issue for such continuum-driven wind models is that they lack a natural self-regulation. In line-driven winds, the self-absorption and saturation of line flux defers the onset of line-acceleration to a relatively low-density near-surface layer, thus limiting the associated mass flux to a value that can be sustained to full escape from star by the expansion-desaturated line-driving in the outer wind. For continuum driving by a gray opacity like electron scattering, the bolometric flux does not saturate, keeping the radiative force strong even in dense, optically thick layers well below the photospheric surface. As discussed below, a mass outflow initiated from such deep, dense layers becomes difficult to sustain with the finite energy flux available from the stellar interior, and this can lead to flow stagnation and infall, with extensive variability and spatial structure.

5.6.2 Convective Instability of a Super-Eddington Interior

It should be emphasized, however, that locally exceeding the Eddington limit need *not* necessarily lead to initiation of a mass outflow. As first shown by Joss et al. (1973), in a stellar envelope allowing the Eddington parameter $\Gamma \to 1$ generally implies through the Schwarzschild criterion that material becomes *convectively unstable*. Since convection in such deep layers is highly efficient at transporting the energy, the contribution from the radiative flux is reduced, thereby lowering the associated radiative Eddington parameter away from unity.

This suggests that, even in a star that formally exceeds the Eddington limit, a radiatively driven outflow could only be initiated *outside* the region where convection is *efficient*. An upper bound to the convective energy flux is set by

$$F_{conv} \approx v_{conv} \, l \, dU/dr \lesssim a \, H \, dP/dr \approx a^3 \rho, \tag{5.62}$$

where v_{conv}, l, and U are the convective velocity, mixing length, and internal energy density, and a, H, P, and ρ are the sound speed, pressure scale height, pressure, and mass density. Setting this maximum convective flux equal to the total stellar energy flux $L/4\pi r^2$ yields an estimate for the maximum mass loss rate that can be initiated by radiative driving,

$$\dot{M} \leq \frac{L}{a^2} \equiv \dot{M}_{max,conv}. \tag{5.63}$$

This is a very large rate, generally well in excess of the fundamental limit set by the energy available to lift the material out of the star's gravitational potential. In terms of the escape speed $v_{esc} \equiv \sqrt{GM/R}$ from the stellar surface radius R, this can be written as

$$\dot{M}_{tir} = \frac{L}{v_{esc}^2/2} = \frac{L}{GM/R} = 0.032 \, \frac{M_\odot}{\text{year}} \, \frac{L}{10^6 L_\odot} \, \frac{M_\odot/R_\odot}{M/R}. \tag{5.64}$$

where $L_6 \equiv L/10^6 L_\odot$. As indicated by the subscript, this is commonly referred to as the *photon tiring* limit, since the radiation driving such a mass loss would lose energy, or become "tired", from the work done to lift the material against gravity. Since generally $a \ll v_{esc}$, a mass flux initiated from the radius of inefficient convection would greatly exceed the photon tiring limit, implying again that any such outflow would necessarily stagnate at some finite radius.

5.6.3 Flow Stagnation from Photon Tiring

To account for the reduction in the radiative luminosity $L(r)$ due to the net work done in lifting and accelerating the wind from the stellar radius R to a local radius r with wind speed $v(r)$, we can write

$$L(r) = L_o - \dot{M} \left[\frac{v(r)^2}{2} + \frac{GM}{R} - \frac{GM}{r} \right], \quad (5.65)$$

where $L_o \equiv L(R)$ is the radiative luminosity at the wind base. For the dimensionless equation of motion (5.46),

$$\left(1 - \frac{w_s}{w}\right) w' = \Gamma(x) - 1,$$

the associated Eddington parameter depends on the scaled wind energy w and inverse radius coordinate x,

$$\Gamma(x) = \Gamma_o(x)[1 - m(w + x)], \quad (5.66)$$

where $\Gamma_o(x) \equiv \kappa(x) L_o / 4\pi GMc$. Here the gravitational "tiring number",

$$m \equiv \frac{\dot{M}}{\dot{M}_{tir}} = \frac{\dot{M} GM}{LR} \approx 0.012 \frac{\dot{M}_{-4} V_{1,000}^2}{L_6}, \quad (5.67)$$

characterizes the *fraction* of radiative energy lost in lifting the wind out of the stellar gravitational potential. The last expression allows easy evaluation of the likely importance of photon tiring for characteristic scalings, where $\dot{M}_{-4} \equiv \dot{M}/10^{-4} M_\odot$/year, $L_6 \equiv L/10^6 L_\odot$, and $V_{1,000} \equiv v_{esc}/1,000$ km/s $\approx 0.62 (M/R)/(M_\odot/R_\odot)$. In particular, for the typical CAK wind scalings of line-driven winds, $m < 0.01$, justifying the neglect of photon tiring in the CAK model discussed in Sect. 5.5.

More generally for cases with non-negligible tiring numbers $m \lesssim 1$, the equation of motion (5.66) can be solved using integrating factors, yielding an explicit solution for $w(x)$ in terms of the integral quantity $\bar{\Gamma}_o(x) \equiv \int_0^x dx' \Gamma_o(x')$,

$$w(x) = -x + \frac{1}{m}\left[1 - e^{-m\bar{\Gamma}_o(x)}\right] + w_s, \quad (5.68)$$

where for typical hot-star atmospheres the sonic point boundary value is very small, $w(0) = w_s < 10^{-3}$.

As a simple example, consider the case[6] with $\Gamma_o(x) = 1 + \sqrt{x}$, for which $\bar{\Gamma}_o = x + 2x^{3/2}/3$. Figure 5.11a plots solutions $w(x)$ vs. x from Eq. (5.68) with various

---

[6]The choice of these functions is arbitrary, to illustrate the photon-tiring effect within a simple model. More physically motivated models based on a medium's porosity are presented in Sect. 5.6.5

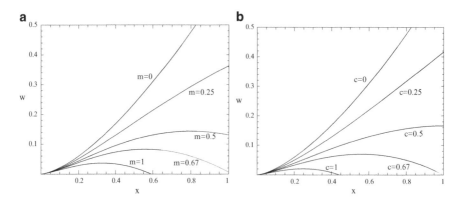

Fig. 5.11 (a) Wind energy w vs. scaled inverse radius $x(\equiv 1 - R/r)$, plotted for Eddington parameter $\Gamma_o(x) = 1 + \sqrt{x}$ with various photon tiring numbers m. (b) Same as (a), except for weak tiring limit $m \ll 1$, and for various constants c in the Eddington parameter scaling $\Gamma_o(x) = 1 + \sqrt{x} - 2cx$

m. For low m, the flow reaches a finite speed at large radii ($x = 1$), but for high m, it curves back, stopping at some finite *stagnation* point x_s, where $w(x_s) \equiv 0$. The latter solutions represent flows for which the mass loss rate is too high for the given stellar luminosity to be able to lift the material to full escape at large radii. In a time-dependent model, such material can be expected to accumulate at this stagnation radius, and eventually fall back to the star (see Sect. 5.6.5 and Fig. 5.13).

Figure 5.11b shows that, even without photon tiring, a similar stagnation can occur from an outward decline in Γ after an initially super-Eddington driving, as might occur, for example, in the region above the iron bump. Again, a general point here is that even if a super-Eddington condition $\Gamma > 1$ initiates an outflow, it does not guarantee that material will escape to large radii in a steady-state wind.

5.6.4 Porosification of VMS Atmospheres by Stagnation and Instabilities

The stagnation implied by the above simple steady 1D models can lead to a complex, time-dependent, 3D pattern of outflows and inflows in actual VMS atmospheres. Moreover, dating back to early work by Spiegel (1976, 1977), there have been speculations that atmospheres supported by radiation pressure would likely exhibit instabilities not unlike that of Rayleigh-Taylor, associated with the support of a heavy fluid by a lighter one, leading to formation of "photon bubbles". Quantitative stability analyses (Spiegel and Tao 1999; Shaviv 2001) indicate that even a simple case of a pure "Thomson atmosphere" – i.e., supported by Thomson scattering of radiation by free electrons – could be subject to intrinsic instabilities for development of lateral inhomogeneities, with many similar properties to the excitation of strange mode pulsations (Glatzel 1994; Papaloizou et al. 1997) discussed in

Sect. 5.4.2. If magnetic fields are introduced, even more instabilities come into play (Arons 1992; Gammie 1998; Begelman 2002; Blaes and Socrates 2003).

The general upshot is that the atmospheres of VMS should have extensive variability and spatial structure, characterized by strong density inhomogeneities over a wide range in length scales. In deeper layers with a high mean density, we can expect that many of the largest and/or densest clumps should individually become optically thick, forcing the radiation flux to preferentially diffuse through relatively low-density channels *between* the clumps.

This is the same spatial "porosity" effect discussed in Sect. 5.5.3, which *reduces* the effective coupling of the gas and radiation in the deeper layers. In the present context the net result is to keep these inner dense layers gravitationally bound even when the radiative flux exceeds the Eddington limit. This defers the onset of a continuum driven wind outflow to a higher, lower-density layer where the clumps are becoming optically thin, resulting in a more moderate mass loss rate that can be more readily sustained.

Figure 5.12 illustrates the expected overall structure of stars with a super-Eddington luminosity, wherein porosity-regulated continuum opacity drives a quasi-steady wind from a stably bound atmospheric base. The distinct physical layers from interior to wind are as follows:

(**A**) As elaborated upon in Sect. 5.6.2, deep inside the star where the density is sufficiently high, any excess flux above the Eddington luminosity is necessarily advected through convection. Thus, we have a bound layer with $L_{rad} < L_{Edd} < L_{tot}$.

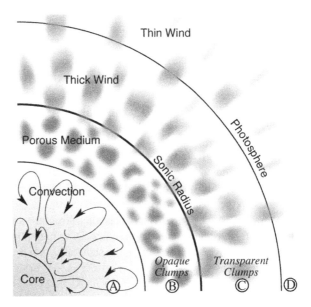

Fig. 5.12 The structure of a super-Eddington star. The labelled regions are described in the text

(B) At lower densities, where convection is inefficient, radiative instabilities necessarily force the atmosphere to become inhomogeneous. This reduces the effective opacity and thus increases the effective Eddington luminosity $L_{\rm eff}$. In other words, this layer is bound, not because the flux is lowered (as it is in the convective regions), but because the opacity is reduced. Thus here we find $L_{\rm Edd} < L_{\rm rad} = L < L_{\rm eff}$.

(C) Opacity reduction can operate only as long as the individual clumps are optically thick. In high layers with lower density, the clumps lose their opaqueness and so the effective opacity recovers the microscopic value (and thus $L_{\rm eff}$ to $L_{\rm Edd}$). A sonic/critical point of a wind will therefore be located where $L = L_{\rm eff} \gtrsim L_{\rm Edd}$.

(D) Since the mass loss rate is large, the inner wind is optically thick and the radiative photosphere resides in the wind itself, at some radius where geometrical dilution eventually makes it become transparent.

5.6.5 Continuum-Driven Winds Regulated by Porous Opacity

To develop quantitative scalings for such porosity-regulated mass loss, let us again first consider a medium in which material has coagulated into discrete clumps of individual optical thickness $\tau_{cl} = \kappa \rho_{cl} \ell$, where ℓ is the clump scale, and the clump density is enhanced compared to the mean density of the medium by a volume filling factor $f = \rho_{cl}/\rho$. The effective overall opacity of this medium can then be approximated by the form given in (5.60),

$$\kappa_{\rm eff} \approx \kappa \, \frac{1 - e^{-\tau_{cl}}}{\tau_{cl}}.$$

Note again that in the limit of optically thin clumps ($\tau_{cl} \ll 1$) this reproduces the usual microscopic opacity ($\kappa_{\rm eff} \approx \kappa$); but in the optically thick limit ($\tau_{cl} \gg 1$), the effective opacity is reduced by a factor of $1/\tau_{cl}$, thus yielding a medium with opacity characterized instead by the clump cross section divided by the clump mass ($\kappa_{\rm eff} = \kappa/\tau_{cl} = \ell^2/m_{cl}$). The critical mean density at which the clumps become optically thin is given by $\rho_o = 1/\kappa h$, where $h \equiv \ell/f$ is a characteristic "porosity length" parameter. A key upshot of this is that the radiative acceleration in such a gray, but spatially porous medium would likewise be reduced by a factor that depends on the mean density.

More realistically, it seems likely that structure should occur with a range of compression strengths and length scales. Drawing upon an analogy with the power-law distribution of line-opacity in the standard CAK model of line-driven winds, let us thereby consider a *power-law-porosity* model in which the associated structure has a broad range of porosity length h. As detailed in Owocki et al. (2004), this leads to an effective Eddington parameter that scales as

$$\Gamma_{\text{eff}} \approx \Gamma \left(\frac{\rho_o}{\rho}\right)^{\alpha_p} \quad ; \quad \rho > \rho_o , \tag{5.69}$$

where α_p is the porosity power index (analogous to the CAK line-distribution power index α), and $\rho_o \equiv 1/\kappa h_o$, with h_o now the porosity-length associated with the *strongest* (i.e. most optically thick) clump. As discussed by Sundqvist et al. (2012), a power index $\alpha_p = 2$ gives the same transport as a simple two-component (clump + void) medium described by Markovian statistics (Levermore et al. 1986; Pomraning 1991).

In rough analogy with the "mixing length" formalism of stellar convection, let us assume the basal porosity length h_o scales with gravitational scale height $H \equiv a^2/g$. Then the requirement that $\Gamma_{\text{eff}} = 1$ at the wind sonic point yields a scaling for the mass loss rate scaling with luminosity. For a canonical case $\alpha_p = 1/2$ (Owocki et al. 2004), we find

$$\dot{M}_{por}(\alpha_p = 1/2) = 4(\Gamma - 1)\frac{L}{ac}\frac{H}{h_o} \tag{5.70}$$

$$= 0.004(\Gamma - 1)\frac{M_\odot}{\text{year}}\frac{L_6}{a_{20}}\frac{H}{h_o} . \tag{5.71}$$

The second equality gives numerical evaluation in terms of characteristic values for the sound speed $a_{20} \equiv a/20\,\text{km/s}$ and luminosity $L_6 \equiv L/10^6 L_\odot$. Comparision with the CAK scalings (5.51) for a line-driven wind shows that the mass loss can be substantially higher from a super-Eddington star with porosity-regulated, continuum driving. Applying the extreme luminosity $L \approx 20 \times 10^6 L_\odot$ estimated for the 1840–1860 outburst of η Carinae, which implies an Eddington parameter $\Gamma \approx 5$, the derived mass loss rate for a canonical porosity length of $h_o = H$ is $\dot{M}_{por} \approx 0.32 M_\odot/\text{year}$, quite comparable to the inferred average $\sim 0.5\,M_\odot/\text{year}$ during this epoch.

For comparison, a Markov model with $\alpha_p = 2$ gives a different, weaker scaling of mass loss with Γ,

$$\dot{M}_{por}(\alpha_p = 2) = \left(1 - \frac{1}{\Gamma}\right)\frac{L}{ac}\frac{H}{h_o} = \frac{\dot{M}_{por}(\alpha_p = 1/2)}{4\Gamma} , \tag{5.72}$$

which saturates to a fixed limit for $\Gamma \gg 1$. To reach the mass loss inferred for η Carianae's giant eruption, such a Markov model would need to have a much smaller porosity length, e.g. $h_o \approx 0.05H$.

But overall, it seems that, together with the ability to drive quite fast outflow speeds (of order the surface escape speed), the extended porosity formalism provides a promising basis for self-consistent dynamical modeling of even the most extreme mass loss outbursts of Luminous Blue Variables, namely those that, like the giant eruption of η Carinae, approach the photon tiring limit.

5.6.6 Simulation of Stagnation and Fallback Above the Tiring Limit

For porosity models in which the base mass flux *exceeds* the photon tiring limit, numerical simulations van Marle et al. (2009) have explored the nature of the resulting complex pattern of infall and outflow. Despite the likely 3-D nature of such flow patterns, to keep the computation tractable, this initial exploration assumes 1-D spherical symmetry, though now allowing a fully time-dependent density and flow speed. The total rate of work done by the radiation on the outflow (or vice versa in regions of inflow) is again accounted for by a radial change of the radiative luminosity with radius,

$$\frac{dL}{dr} = -\dot{m}g_{rad} = -\kappa_{\text{eff}} \, \rho v L/c , \qquad (5.73)$$

where $\dot{m} \equiv 4\pi\rho v r^2$ is the local mass-flux at radius r, which is no longer a constant, or even monotonically positive, in such a time-dependent flow. The latter equality then follows from the definition (5.1) of the radiative acceleration g_{rad} for a gray opacity κ_{eff}, set here by porosity-modified electron scattering. At each time step, Eq. (5.73) is integrated from an assumed lower boundary luminosity $L(R)$ to give the local radiative luminosity $L(r)$ at all radii $r > R$. Using this to compute the local radiative acceleration, the time-dependent equations for mass and momentum conservation are evolved forward to obtain the time and radial variation of density $\rho(r,t)$ and flow speed $v(r,t)$. (For simplicity, the temperature is fixed at the stellar effective temperature.) The base Eddington parameter is $\Gamma = 10$, and the analytic porosity mass flux is 2.3 times the tiring limit.

Figure 5.13 illustrates the flow structure as a function of radius (for $r = 1-15\,R$) and time (over an arbitrary interval long after the initial condition). The left panel grayscale shows the local mass flux, in $M_\odot$/year, with dark shades representing inflow, and light shades outflow. In the right panel, the shading represents the local luminosity in units of the base value, $L(r)/L(R)$, ranging from zero (black) to one (white); in addition, the superposed lines represent the radius and time variation of selected mass shells.

Both panels show the remarkably complex nature of the flow, with positive mass flux from the base overtaken by a hierarchy of infall from stagnated flow above. However, the re-energization of the radiative luminosity from this infall makes the region above have an outward impulse. The shell tracks thus show that, once material reaches a radius $r \approx 5R$, its infall intervals become ever shorter, allowing it eventually to drift outward. The overall result is a net, time-averaged mass loss through the outer boundary that is very close to the photon-tiring limit, with however a terminal flow speed $v_\infty \approx 50$ km/s that is substantially below the surface escape speed $v_{esc} \approx 600$ km/s.

These initial 1-D simulations thus provide an interesting glimpse into this competition below inflow and outflow. Of course, the structure in more realistic

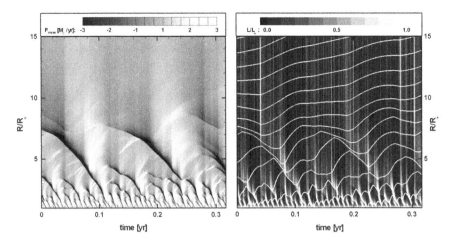

Fig. 5.13 Grayscale plot of radius and time variation of mass flux (*left*) and luminosity (*right*) in a time-dependent simulation of a super-Eddington wind with a porosity-mediated base mass flux above the photon tiring limit. The *white contours* on the right trace the height progression of fixed mass shells

2-D and 3-D models is likely to be even more complex, and may even lead itself to a highly porous medium. But overall, it seems that one robust property of super-Eddington stars may well be mass loss that is of the order of the photon tiring limit.

5.6.7 LBV Eruptions: Enhanced Winds or Explosions?

The previous section has modeled the eruptive, LBV mass loss of VMS in terms of a quasi-steady, continuum-driven wind that results from the stellar luminosity exceeding the Eddington limit. But an alternative paradigm is that such eruptions might in fact be point-time "explosions" that simply did not have sufficient energy to completely disrupt the star.

Both paradigms require an unknown energy source, but one important distinction is that explosions are driven by *gas pressure*, whereas super-Eddington winds are driven by *radiation*. The two have markedly different timescales.

The overpressure from an explosion propagates through the star on a very short dynamical time scale, of order R/a, where a is the sound-speed in the very high temperature gas that is heated by the energy deposition of the explosion. In supernovae, this sound speed is on the order of the mass ejection speed, on the order of 10,000 km/s; even in a "failed" LBV explosion, it would be on the order of the surface escape speed, or a few hundred km/s, implying a dynamical time of order the free fall time, or just a few hours. Of course, the release of radiative energy is tied to the expansion (and later on, radioactive β-decay), and thus peaks on a somewhat longer time of a few days or weeks for supernovae. But it is difficult to see how such a direct gas-pressure-driven explosion could be maintained for the years to decade timescale inferred for LBV eruptions.

This then is perhaps the key argument for a radiation-driven model. If energy is released in the deep interior, its *radiative* signature can take up to a much longer *diffusion* time to reach the surface.[7] This can be long as a few years.

In contrast to the explosive disruption of supernovae, for LBV eruptions the total energy is typically well below the stellar binding energy. Thus even if this energy were released suddenly in the deep interior, the initial dynamical response would quickly stagnate, leaving then radiative diffusion as the fall-back transport. But since massive stars are already close to the Eddington limit, the associated excess luminosity should push it over this limit, leading then to the strong, *radiatively driven* mass loss described above.

Because this time scale is still much longer than any dynamical time in the system, the essential processes can be modeled in terms of a quasi-steady continuum-driven wind during this super-Eddington epoch, as described above.

Perhaps the least understood aspect of LBVs is the mechanism giving rise to the observed eruptions. In supernovae explosions, the energy source is obviously the core-collapse to a neutron star or black hole. But in LBV eruptions, the post-eruption survival of an intact star, and the indication at least some LBVs can undergo multiple giant eruptions, both show that the energy source cannot be a one-time singular event like core collapse. Some other mechanism must provide the energy, but the exact nature of this is still unknown. So it is still unclear why LBVs erupt, or what sets the eruption time, amplitude, and repetition rate. In particular, there is currently no model that predicts these quantities.

Concluding Summary

An overall theme of this chapter is that, because of their very high luminosity, radiative forces play an important, dynamical role in the stability of the envelopes and winds of VMS. A key issue is the nature of the opacity that links the radiation to gas, and in particular the distinction between line vs. continuum processes. Line opacity can in principle be much stronger, but in the stellar envelope the saturation of the radiative flux within the line means that flux-weighted line-force depends on an inverse or harmonic mean (a.k.a. Rosseland mean). This only becomes moderately strong (factor ten above electron scattering) in regions of strong line overlap, most particularly the so-called Iron bump near 150,000 K. This iron bump can cause a strong, even

(continued)

---

[7]Since the luminous stars are likely to be mostly convective (e.g. Sect. 5.6.2), the limiting time scale is that of the convective diffusion's mixing length time in the stellar cores, which due to the high density is much longer than the dynamical time scales.

runaway inflation of the stellar envelope, leading to an Iron-Bump Eddington Limit that might be associated with S-Doradus type LBVs.

Near surface layers, the desaturation of the lines leads to a much stronger line-force that drives a strong stellar wind, with a well defined mass loss rate regulated by the level of line saturation at the sonic point base. Away from the wind base, there develops a strong "line-deshadowing instability" that induces an extensive clumping and associated porosity in the outer wind.

For stars that exceed the classical Eddington limit, much stronger mass loss can be driven by the continuum opacity, even approaching the "photon tiring" limit, in which the full stellar energy flux is expended to lift and accelerate the mass outflow. A key issue here is regulation of the continuum driving by the porosity that develops from instability and flow stagnation of the underlying stellar envelope. For a simple power-law model of the porous structure, the derived mass loss rates seem capable of explaining the giant eruption LBVs, including the 1840s eruption seen in Eta Carinae. Two key remaining issues are the cause of the super-Eddington luminosity, and whether the response might be better modeled as an explosion vs. a quasi-steady mass loss eruption.

Acknowledgements This work was supported in part by NASA ATP grant NNX11AC40G, NASA Chandra grant TM3-14001A, and NSF grant 1312898 to the University of Delaware. I thank M. Giannotti for sharing his Mathematica notebook for the OPAL opacity tables, and N. Shaviv for many helpful discussions and for providing Fig. 5.12. I also acknowledge numerous discussions with G. Graefener, N. Smith, J. Sundqvist, J. Vink and A.J. van Marle.

References

Abbott, D. C. (1980). The theory of radiatively driven stellar winds. I - A physical interpretation. *Astrophysical Journal, 242*, 1183.
Abbott, D. C. (1982). The theory of radiatively driven stellar winds. II - The line acceleration. *Astrophysical Journal, 259*, 282.
Arons, J. (1992). Photon bubbles - Overstability in a magnetized atmosphere. *Astrophysical Journal, 388*, 561.
Begelman, M. C. (2002). Super-eddington fluxes from thin accretion disks? *Astrophysical Journal Letters, 568*, L97.
Belyanin, A. A. (1999). Optically thick super-Eddington winds in galactic superluminal sources. *Astronomy and Astrophysics, 344*, 199.
Blaes, O., & Socrates, A. (2003). Local radiative hydrodynamic and magnetohydrodynamic instabilities in optically thick media. *Astrophysical Journal, 596*, 509.
Castor, J. I., Abbott, D. C., & Klein, R. I. (1975). Radiation-driven winds in of stars. *Astrophysical Journal, 195*, 157.
Cohen, D. H., Leutenegger, M. A., Wollman, E. E., Zsargó, J., Hillier, D. J., Townsend, R. H. D., & Owocki, S. P. (2010). A mass-loss rate determination for ζ Puppis from the quantitative analysis of X-ray emission-line profiles. *Monthly Notices of the Royal Astronomical Society, 405*, 2391.

Crowther, P. A. (2012). In *Death of massive stars: Supernovae and gamma-ray bursts* (Volume 279 of IAU symposium, Environments of massive stars and the upper mass limit, pp. 9–17), Nikkon.

Crowther, P. A., Schnurr, O., Hirschi, R., Yusof, N., Parker, R. J., Goodwin, S. P., & Kassim, H. A. (2010). The R136 star cluster hosts several stars whose individual masses greatly exceed the accepted 150M_{solar} stellar mass limit. *Monthy Notices of the Royal Astronomical Society, 408,* 731.

Dessart, L., & Owocki, S. P. (2003). Two-dimensional simulations of the line-driven instability in hot-star winds. *Astronomy and Astrophysics, 406,* L1.

Dessart, L., & Owocki, S. P. (2005). 2D simulations of the line-driven instability in hot-star winds. II. Approximations for the 2D radiation force. *Astronomy and Astrophysics, 437,* 657.

Eddington, A. S. (1926). *The internal constitution of the stars.* Cambridge: Cambridge University Press.

Feldmeier, A. (1995). Time-dependent structure and energy transfer in hot star winds. *Astronomy and Astrophysics, 299,* 523.

Feldmeier, A., Puls, J., & Pauldrach, A. W. A. (1997). The X-ray emission from shock cooling zones in O star winds. *Astronomy and Astrophysics, 322,* 878.

Figer, D. F. (2005). An upper limit to the masses of stars. *Nature, 434,* 192.

Friend, D. B., & Abbott, D. C. (1986). The theory of radiatively driven stellar winds. III - Wind models with finite disk correction and rotation. *Astrophysical Journal, 311,* 701.

Fullerton, A. W., Massa, D. L., & Prinja, R. K. (2006). The discordance of mass-loss estimates for galactic O-type stars. *Astrophysical Journal, 637,* 1025.

Gammie, C. F. (1998). Photon bubbles in accretion discs. *Monthy Notices of the Royal Astronomical Society, 297,* 929.

Gayley, K. G. (1995). An improved line-strength parameterization in hot-star winds. *Astrophysical Journal, 454,* 410.

Glatzel, W. (1994). On the origin of strange modes and the mechanism of related instabilities. *Monthy Notices of the Royal Astronomical Society, 271,* 66.

Glatzel, W. (2005). In R. Humphreys & K. Stanek (Eds.) *The fate of the most massive stars* (Volume 332 of Astronomical Society of the Pacific conference series, Instabilities in the most massive evolved stars, p. 22), Jackson Hole, WY.

Glatzel, W., & Kiriakidis, M. (1993). Stability of massive stars and the humphreys / davidson limit. *Monthy Notices of the Royal Astronomical Society, 263,* 375.

Gräfener, G., Owocki, S. P., & Vink, J. S. (2012). Stellar envelope inflation near the Eddington limit. Implications for the radii of Wolf-Rayet stars and luminous blue variables. *Astronomy and Astrophysics, 538,* A40.

Grevesse, N., & Noels, A. (1993). Atomic data and the spectrum of the solar photosphere. *Physica Scripta T47,* 133.

Humphreys, R. M., Davidson, K. (1979). Studies of luminous stars in nearby galaxies. III - Comments on the evolution of the most massive stars in the milky way and the large magellanic cloud. *Astrophysical Journal, 232,* 409.

Iglesias, C. A., & Rogers, F. J. (1996). Updated opal opacities. *Astrophysical Journal, 464,* 943.

Joss, P. C., Salpeter, E. E., & Ostriker, J. P. (1973). On the "critical luminosity" in stellar interiors and stellar surface boundary conditions. *Astrophysical Journal, 181,* 429.

Kee, N. D., Owocki, S., & ud-Doula, A. (2014). Suppression of X-rays from radiative shocks by their thin-shell instability. *Monthy Notices of the Royal Astronomical Society, 438,* 3557.

Kippenhahn, R., Weigert, A., & Weiss, A. (2013). *Stellar structure and evolution: Astronomy and astrophysics library.* Berlin/Heidelberg: Springer.

Kudritzki, R. P., Puls, J., Lennon, D. J., Venn, K. A., Reetz, J., Najarro, F., McCarthy, J. K., & Herrero, A. (1999). The wind momentum-luminosity relationship of galactic A- and B-supergiants. *Astronomy and Astrophysics, 350,* 970.

Levermore, C. D., Pomraning, G. C., Sanzo, D. L., & Wong, J. (1986). Linear transport theory in a random medium. *Journal of Mathematical Physics, 27,* 2526.

Lucy, L. B. (1984). Wave amplification in line-driven winds. *Astrophysical Journal, 284,* 351.

Lucy, L. B., & Solomon, P. M. (1970). Mass loss by hot stars. *Astrophysical Journal, 159,* 879.

MacGregor, K. B., Hartmann, L., & Raymond, J. C. (1979). Radiative amplification of sound waves in the winds of O and B stars. *Astrophysical Journal, 231*, 514.

Nugis, T., & Lamers, H. J. G. L. M. (2002). The mass-loss rates of Wolf-Rayet stars explained by optically thick radiation driven wind models. *Astronomy and Astrophysics, 389*, 162.

Oey, M. S., & Clarke, C. J. (2005). Statistical confirmation of a stellar upper mass limit. *Astrophysical Journal Letters, 620*, L43.

Oskinova, L. M., Hamann, W.-R., & Feldmeier, A. (2007). Neglecting the porosity of hot-star winds can lead to underestimating mass-loss rates. *Astronomy and Astrophysics, 476*, 1331.

Owocki, S. P. (1991). In: L. Crivellari, I. Hubeny, & D. G. Hummer (Eds.) NATO ASIC proceedings 341: Stellar atmospheres – beyond classical models (A smooth source function method for including scattering in radiatively driven wind simulations, p. 235), Trieste.

Owocki, S. P. (2008). In W.-R. Hamann, A. Feldmeier, L. M. Oskinova (Eds.), *Clumping in hot-star winds* (Dynamical simulation of the "velocity-porosity" reduction in observed strength of stellar wind lines, p. 121). Germany: Universitätsverlag Potsdam.

Owocki, S. P. (2013). In T. D. Oswalt & M. A. Barstow (Eds.), *Planets, stars and stellar systems*. (Volume 4 of Stellar structure and evolution stellar winds, p. 735). Dordrecht/New York: Springer.

Owocki, S. P., Castor, J. I., & Rybicki, G. B. (1988). Time-dependent models of radiatively driven stellar winds. I - Nonlinear evolution of instabilities for a pure absorption model. *Astrophysical Journal, 335*, 914.

Owocki, S. P., & Cohen, D. H. (2006). The effect of porosity on X-ray emission-line profiles from hot-star winds. *Astrophysical Journal, 648*, 565.

Owocki, S. P., Gayley, K. G., & Shaviv, N. J. (2004). A porosity-length formalism for photon-tiring-limited mass loss from stars above the eddington limit. *Astrophysical Journal, 616*, 525.

Owocki, S. P., & Puls, J. (1996). Nonlocal escape-integral approximations for the line force in structured line-driven stellar winds. *Astrophysical Journal, 462*, 894.

Owocki, S. P., & Puls, J. (1999). Line-driven stellar winds: The dynamical role of diffuse radiation gradients and limitations to the sobolev approach. *Astrophysical Journal, 510*, 355.

Owocki, S. P., & Rybicki, G. B. (1984). Instabilities in line-driven stellar winds. I - Dependence on perturbation wavelength. *Astrophysical Journal, 284*, 337.

Owocki, S. P., & Rybicki, G. B. (1985). Instabilities in line-driven stellar winds. II - Effect of scattering. *Astrophysical Journal, 299*, 265.

Owocki, S. P., & ud-Doula, A. (2004). The effect of magnetic field tilt and divergence on the mass flux and flow speed in a line-driven stellar wind. *Astrophysical Journal, 600*, 1004.

Papaloizou, J. C. B., Alberts, F., Pringle, J. E., & Savonije, G. J. (1997). On the nature of strange modes in massive stars. *Monthy Notices of the Royal Astronomical Society, 284*, 821.

Pauldrach, A., Puls, J., & Kudritzki, R. P. (1986). Radiation-driven winds of hot luminous stars - Improvements of the theory and first results. *Astronomy and Astrophysics, 164*, 86.

Petrovic, J., Pols, O., & Langer, N. (2006). Are luminous and metal-rich Wolf-Rayet stars inflated? *Astronomy and Astrophysics, 450*, 219.

Pomraning, G. C. (1991). *Linear kinetic theory and particle transport in stochastic mixtures*. Singapore/New Jersey: World Scientific.

Quinn, T., & Paczynski, B. (1985). Stellar winds driven by super-Eddington luminosities. *Astrophysical Journal, 289*, 634.

Runacres, M. C., & Owocki, S. P. (2002). The outer evolution of instability-generated structure in radiatively driven stellar winds. *Astronomy and Astrophysics, 381*, 1015.

Rybicki, G. B., Owocki, S. P., & Castor, J. I. (1990). Instabilities in line-driven stellar winds. IV - Linear perturbations in three dimensions. *Astrophysical Journal, 349*, 274.

Shaviv, N. J. (1998). The eddington luminosity limit for multiphased media. *Astrophysical Journal Letters, 494*, L193.

Shaviv, N. J. (2000). The porous atmosphere of η carinae. *Astrophysical Journal Letters, 532*, L137.

Shaviv, N. J. (2001). The nature of the radiative hydrodynamic instabilities in radiatively supported thomson atmospheres. *Astrophysical Journal, 549*, 1093.

Smith, N. (2002). Dissecting the Homunculus nebula around Eta Carinae with spatially resolved near-infrared spectroscopy. *Monthy Notices of the Royal Astronomical Society, 337*, 1252.

Smith, N., Davidson, K., Gull, T. R., Ishibashi, K., & Hillier, D. J. (2003). *Astrophysical Journal, 586*, 432.

Smith, N., & Owocki, S. P. (2006). Latitude-dependent effects in the stellar wind of η Carinae. *Astrophysical Journal Letters, 645*, L45.

Sobolev, V. V. (1960). *Moving envelopes of stars*. Cambridge: Harvard University Press.

Spiegel, E. A. (1976). In: R. Cayrel & M. Steinberg (Eds.) *Physique des Mouvements dans les Atmospheres* (Photohydrodynamic instabilities of hot stellar atmospheres, p. 19). Paris: Editions du Centre National de la Recherche Scientifique.

Spiegel, E. A. (1977). In: E. A. Spiegel & J.-P. Zahn (Eds.) *Problems of Stellar Convection* (Volume 71 of Lecture Notes in Physics; Photoconvection, pp. 267–283). Berlin: Springer.

Spiegel, E. A., & Tao, L. (1999). Photofluid instabilities of hot stellar envelopes. *Physics Reports, 311*, 163.

Sundqvist, J. O., Owocki, S. P., Cohen, D. H., Leutenegger, M. A., & Townsend, R. H. D. (2012). A generalized porosity formalism for isotropic and anisotropic effective opacity and its effects on X-ray line attenuation in clumped O star winds. *Monthy Notices of the Royal Astronomical Society, 420*, 1553.

Sundqvist, J. O., Puls, J., Feldmeier, A., & Owocki, S. P. (2011). The nature and consequences of clumping in hot, massive star winds. *Astronomy and Astrophysics, 528*, A64.

van Marle, A. J., Owocki, S. P., & Shaviv, N. J. (2009). On the behaviour of stellar winds that exceed the photon-tiring limit. *Monthy Notices of the Royal Astronomical Society, 394*, 595.

Vishniac, E. T. (1994). Nonlinear instabilities in shock-bounded slabs. *Astrophysical Journal, 428*, 186.

Chapter 6
Evolution and Nucleosynthesis of Very Massive Stars

Raphael Hirschi

Abstract In this chapter, after a brief introduction and overview of stellar evolution, we discuss the evolution and nucleosynthesis of very massive stars (VMS: $M > 100\,M_\odot$) in the context of recent stellar evolution model calculations. This chapter covers the following aspects: general properties, evolution of surface properties, late central evolution, and nucleosynthesis including their dependence on metallicity, mass loss and rotation. Since very massive stars have very large convective cores during the main-sequence phase, their evolution is not so much affected by rotational mixing, but more by mass loss through stellar winds. Their evolution is never far from a homogeneous evolution even without rotational mixing. All VMS at metallicities close to solar end their life as WC(-WO) type Wolf-Rayet stars. Due to very important mass loss through stellar winds, these stars may have luminosities during the advanced phases of their evolution similar to stars with initial masses between 60 and 120 $M_\odot$. A distinctive feature which may be used to disentangle Wolf-Rayet stars originating from VMS from those originating from lower initial masses is the enhanced abundances of neon and magnesium at the surface of WC stars. At solar metallicity, mass loss is so strong that even if a star is born with several hundred solar masses, it will end its life with less than 50 $M_\odot$ (using current mass loss prescriptions). At the metallicity of the LMC and lower, on the other hand, mass loss is weaker and might enable stars to undergo pair-instability supernovae.

6.1 Introduction

For a long time, the evolution of VMS was considered only in the framework of Pop III stars. Indeed, it was expected that, only in metal free environments, could such massive stars be formed, since the absence of dust, an efficient cooling agent, would prevent a strong fragmentation of the proto-stellar cloud (Bromm et al. 1999;

R. Hirschi (✉)
Astrophysics, Lennard-Jones Labs 2.09, EPSAM, Keele University, ST5 5BG, Staffordshire, UK

Kavli Institute for the Physics and Mathematics of the Universe (WPI), University of Tokyo, 5-1-5 Kashiwanoha, Kashiwa, 277-8583, Japan
e-mail: r.hirschi@keele.ac.uk

© Springer International Publishing Switzerland 2015
J.S. Vink (ed.), *Very Massive Stars in the Local Universe*, Astrophysics and Space Science Library 412, DOI 10.1007/978-3-319-09596-7_6

Abel et al. 2002).[1] It came therefore as a surprise when it was discovered that the most metal-poor low-mass stars, likely formed from a mixture between the ejecta of these Pop III stars and pristine interstellar medium, did not show any signature of the peculiar nucleosynthesis of the VMS (Heger and Woosley 2002; Umeda and Nomoto 2002; Christlieb et al. 2002; Frebel et al. 2005). While such observations cannot rule out the existence of these VMS in Pop III generations (their nucleosynthetic signature may have been erased by the more important impact of stars in other mass ranges), it seriously questions the importance of such object for understanding the early chemical evolution of galaxies. Ironically, when the importance of VMS in the context of the first stellar generations fades, they appear as potentially interesting objects in the framework of present day stellar populations.

For a long time, observations favored a present-day upper mass limit for stars around 150 $M_\odot$ (Figer 2005; Oey and Clarke 2005). Recently, however, Crowther et al. (2010) have re-assessed the properties of the brightest members of the R136a star cluster, revealing exceptionally high luminosities (see Chaps. 1 and 2 for more details). The comparison between main sequence evolutionary models for rotating and non-rotating stars and observed spectra resulted in high current ($\leq 265\ M_\odot$) and initial ($\leq 320\ M_\odot$) masses for these stars. The formation scenarios for these VMS are presented in Chap. 3.

The above observations triggered a new interest in the evolution of very massive stars. However, since VMS are so rare, only a few of them are known and we have to rely on stellar evolution models in order to study their properties and evolution. In this chapter, the evolution of VMS will be discussed based on stellar evolution models calculated using the Geneva stellar evolution code (Eggenberger et al. 2007) including the modifications implemented to follow the advanced stages as described in Hirschi et al. (2004a). Models at solar ($Z = 0.014$), Large Magellanic Cloud (LMC, $Z = 0.006$) and Small Magellanic Cloud (SMC, $Z = 0.002$) metallicities will be presented (see Yusof et al. 2013, for full details about these models). These models will also be compared to models of normal massive stars calculated with the same input physics at solar metallicity ($Z = 0.014$) presented in Ekström et al. (2012) and Georgy et al. (2012) as well as their extension to lower metallicities (Georgy et al. 2013).

In Sect. 6.2, we review the basics of stellar evolution models and their key physical ingredients. The general properties and early evolution of VMS are presented in Sect. 6.3. The Wolf-Rayet stars originating from VMS are discussed in Sect. 6.4. The late evolution and possible fates of the VMS is the subject of Sect. 6.5. The nucleosynthesis and contribution to chemical evolution of galaxies is discussed in Sect. 6.6. A summary and conclusions are given in section "Summary and Conclusion".

---

[1]Note, however, that recent star formation simulations find lower-mass stars forming in groups, similarly to present-day star formation (Stacy et al. 2010; Greif et al. 2010).

6.2 Stellar Evolution Models

The evolution of VMS is similar enough to more common massive stars that the same stellar evolution codes can be used to study their evolution and corresponding nucleosynthesis. Stellar evolution models require a wide range of input physics ranging from nuclear reaction rates to mass loss prescriptions. In this section, we review the basic equations that govern the structure and evolution of stars as well as some of the key input physics with a special emphasis on mass loss, rotation and magnetic fields.

6.2.1 Stellar Structure Equations

There are four equations describing the evolution of the structure of stars: the mass, momentum and energy conservation equations and the energy transport equations, which we recall below. On top of that, the equations of the evolution of chemical elements abundances are to be followed. These equations are discussed in the next section. In the Geneva stellar evolution code (GENEC, see Eggenberger et al. 2007), which we base our presentation on in this section, the problem is treated in one dimension (1D) and the equations of the evolution of chemical elements abundances are calculated separately from the structure equations, as in the original version of Kippenhahn and Weigert (Kippenhahn et al. 1967; Kippenhahn and Weigert 1990). In GENEC, rotation is included and spherical symmetry is no longer assumed. The effective gravity (sum of the centrifugal force and gravity) can in fact no longer be derived from a potential and the case is said to be non–conservative. The problem can still be treated in 1D by assuming that the angular velocity is constant on isobars. This assumes that there is a strong horizontal (along isobars) turbulence which enforces constant angular velocity on isobars (Zahn 1992). The case is referred to as "shellular" rotation and using reasonable simplifications described in Meynet and Maeder (1997), the usual set of four structure equations (as used for non-rotating stellar models) can be recovered:

- Energy conservation:

$$\frac{\partial L_P}{\partial M_P} = \epsilon_{nucl} - \epsilon_\nu + \epsilon_{grav} = \epsilon_{nucl} - \epsilon_\nu - c_P \frac{\partial \overline{T}}{\partial t} + \frac{\delta}{\rho}\frac{\partial P}{\partial t} \qquad (6.1)$$

Where L_P is the luminosity, M_P the Lagrangian mass coordinate, and ϵ_{nucl}, ϵ_ν, and ϵ_{grav} are the energy generation rates per unit mass for nuclear reactions, neutrinos and gravitational energy changes due to contraction or expansion, respectively. T is the temperature, c_P the specific heat at constant pressure, t the time, P the pressure, ρ the density and $\delta = -\partial \ln\rho/\partial \ln T$.

- Momentum equation:

$$\frac{\partial P}{\partial M_P} = -\frac{GM_P}{4\pi r_P^4} f_P \qquad (6.2)$$

Where r_P is the radius of the shell enclosing mass M_P and G the gravitational constant.

- Mass conservation (continuity equation):

$$\frac{\partial r_P}{\partial M_P} = \frac{1}{4\pi r_P^2 \overline{\rho}} \qquad (6.3)$$

- Energy transport equation:

$$\frac{\partial \ln \overline{T}}{\partial M_P} = -\frac{GM_P}{4\pi r_P^4 P} f_P \min[\nabla_{ad}, \nabla_{rad}\frac{f_T}{f_P}] \qquad (6.4)$$

where

$$\nabla_{ad} = \left(\frac{\partial \ln \overline{T}}{\partial \ln P}\right)_{ad} = \frac{P\delta}{\overline{T}\overline{\rho}c_P} \quad \text{(convective zones)},$$

$$\nabla_{rad} = \frac{3}{64\pi\sigma G} \frac{\kappa L_P P}{M_P \overline{T}^4} \quad \text{(radiative zones)},$$

where κ is the total opacity and σ is the Stefan-Boltzmann constant.

$$f_P = \frac{4\pi r_P^4}{GM_P S_P} \frac{1}{<g^{-1}>},$$

$$f_T = \left(\frac{4\pi r_P^2}{S_P}\right)^2 \frac{1}{<g><g^{-1}>},$$

$<x>$ is x averaged on an isobaric surface, $\overline{x}$ is x averaged in the volume separating two successive isobars and the index P refers to the isobar with a pressure equal to P. g is the effective gravity and S_P is the surface of the isobar (see Meynet and Maeder 1997, for more details). The implementation of the structure equations into other stellar evolution codes are presented for example in Paxton et al. (2011) and Chieffi et al. (1998).

6.2.2 Mass Loss

Mass loss strongly affects the evolution of very massive stars as we shall describe below. Mass loss is already discussed in Chap. 4 but here we will recall the different mass loss prescriptions used in stellar evolution calculations and how they relate to each other. In the models presented in this chapter, the following prescriptions were used. For main-sequence stars, the prescription for radiative line driven winds from Vink et al. (2001a) was used, which compare rather well with observations (Crowther et al. 2010; Muijres et al. 2011). For stars in a domain not covered by the Vink et al. prescription, the de Jager et al. (1988a) prescription was applied to models with $\log(T_{\text{eff}}) > 3.7$. For $\log(T_{\text{eff}}) \leq 3.7$, a linear fit to the data from Sylvester et al. (1998) and van Loon et al. (1999) (see Crowther 2001) was performed. The formula used is given in Eq. 2.1 in Bennett et al. (2012).

In the stellar evolution simulations, the stellar wind is not simulated self-consistently and a criterion is used to determine when a star becomes a WR star. Usually, a star becomes a WR when the surface hydrogen mass fraction, X_s, becomes inferior to 0.3 (sometimes when it is inferior to 0.4) and the effective temperature, $\log(T_{\text{eff}})$, is greater than 4.0. The mass loss rate used during the WR phase depends on the WR sub-type. For the eWNL phase (when $0.3 > X_s > 0.05$), the Gräfener and Hamann (2008) recipe was used (in the validity domain of this prescription, which usually covers most of the eWNL phase). In many cases, the WR mass-loss rate of Gräfener and Hamann (2008) is lower than the rate of Vink et al. (2001a), in which case, the latter was used. For the eWNE phase – when $0.05 > X_s$ and the ratio of the mass fractions of $(^{12}\text{C}+{}^{16}\text{O})/{}^4\text{He} < 0.03$ – and WC/WO phases – when $(^{12}\text{C}+{}^{16}\text{O})/{}^4\text{He} > 0.03$ – the corresponding prescriptions of Nugis and Lamers (2000a) were used. Note also that both the Nugis and Lamers (2000a) and Gräfener and Hamann (2008) mass-loss rates account for clumping effects (Muijres et al. 2011).

As is discussed below, the mass loss rates from Nugis and Lamers (2000a) for the eWNE phase are much larger than in other phases and thus the largest mass loss occurs during this phase. In Crowther et al. (2010), the mass loss prescription from Nugis and Lamers (2000a) was used for both the eWNL and eWNE phases (with a clumping factor, $f = 0.1$). The models presented in this chapter thus lose less mass than those presented in Crowther et al. (2010) during the eWNL phase.

The metallicity dependence of the mass loss rates is commonly included in the following way. The mass loss rate used at a given metallicity, $\dot{M}(Z)$, is the mass loss rate at solar metallicity, $\dot{M}(Z_\odot)$, multiplied by the ratio of the metallicities to the power of α: $\dot{M}(Z) = \dot{M}(Z_\odot)(Z/Z_\odot)^\alpha$. α was set to 0.85 for the O-type phase and WN phase and 0.66 for the WC and WO phases; and for WR stars the initial metallicity rather than the actual surface metallicity was used in the equation above following Eldridge and Vink (2006). α was set to 0.5 for the de Jager et al. (1988a) prescription.

For rotating models, the correction factor described below in Eq. 6.5 is applied to the radiative mass-loss rate.

6.2.3 Rotation and Magnetic Mields

The physics of rotation included in stellar evolution codes has been developed extensively over the last 20 years. A recent review of this development can be found in Maeder and Meynet (2012). The effects induced by rotation can be divided into three categories.

1. **Hydrostatic effects**: The centrifugal force changes the hydrostatic equilibrium of the star. The star becomes oblate and the equations describing the stellar structure have to be modified as described above.
2. **Mass loss enhancement and anisotropy**: Mass loss depends on the opacity and the effective gravity (sum of gravity and centrifugal force) at the surface. The larger the opacity, the larger the mass loss. The higher the effective gravity, the higher the radiative flux (von Zeipel 1924) and effective temperature. Rotation, via the centrifugal force, reduces the surface effective gravity at the equator compared to the pole. As a result, the radiative flux of the star is larger at the pole than at the equator. In massive hot stars, since the opacity is dominated by the temperature–independent electron scattering, rotation enhances mass loss at the pole. If the opacity increases when the temperature decreases (in cooler stars), mass loss can be enhanced at the equator when the bi-stability is reached (see mass loss chapter for more details).

For rotating models, the mass loss rates can be obtained by applying a correction factor to the radiative mass loss rate as described in Maeder and Meynet (2000):

$$\dot{M}(\Omega) = F_\Omega \cdot \dot{M}(\Omega = 0) = F_\Omega \cdot \dot{M}_{\rm rad}$$

$$\text{with} \quad F_\Omega = \frac{(1-\Gamma)^{\frac{1}{\alpha}-1}}{\left[1 - \frac{\Omega^2}{2\pi G \rho_{\rm m}} - \Gamma\right]^{\frac{1}{\alpha}-1}} \quad (6.5)$$

where $\Gamma = L/L_{\rm Edd} = \kappa L/(4\pi c GM)$ is the Eddington factor (with κ the total opacity), and α the $T_{\rm eff}$–dependent force multiplier parameter. Enhancement factors (F_Ω) are generally close to one but they may become very large when $\Gamma \gtrsim 0.7$ or $\Omega/\Omega_{\rm crit} > 0.9$ (see Maeder and Meynet 2000; Georgy et al. 2011, for more details). If critical rotation, where the centrifugal force balances gravity at the equator, is reached, mechanical mass loss may occur and produce a decretion disk (see Krtička et al. 2011, for more details). In most stellar evolution codes, the mass loss is artificially enhanced when $\Omega/\Omega_{\rm crit} \gtrsim 0.95$ to ensure that the ratio does not become larger than unity but multi-dimensional simulations are required to provide new prescriptions to use in stellar evolution codes.

For mass loss rates, $\dot{M}(\Omega = 0)$, the following prescriptions are commonly used: Vink et al. (2001b) for radiatively driven wind of O-type stars, Nugis and Lamers (2000b) for Wolf-Rayet stars and de Jager et al. (1988b) for cooler stars not covered by the other two prescriptions and for which dust and pulsation could play a role in the driving of the wind.

3. **Rotation driven instabilities**: The main rotation driven instabilities are horizontal turbulence, meridional circulation and dynamical and secular shear (see Maeder 2009, for a comprehensive description of rotation-induced instabilities).

Horizontal turbulence corresponds to turbulence along the isobars. If this turbulence is strong, rotation is constant on isobars and the situation is usually referred to as "shellular rotation" (Zahn 1992). The horizontal turbulence is expected to be stronger than the vertical turbulence because there is no restoring buoyancy force along isobars (see Maeder 2003, for recent development on this topic).

Meridional circulation, also referred to as Eddington–Sweet circulation, arises from the local breakdown of radiative equilibrium in rotating stars. This is due to the fact that surfaces of constant temperature do not coincide with surfaces of constant pressure. Indeed, since rotation elongates isobars at the equator, the temperature on the same isobar is lower at the equator than at the pole. This induces large scale circulation of matter, in which matter usually rises at the pole and descends at the equator (see Fig. 6.1).

In this situation, angular momentum is transported inwards. It is however also possible for the circulation to go in the reverse direction and, in this second case, angular momentum is transported outwards. Circulation corresponds to an advective process, which is different from diffusion because the latter can only erode gradients. Advection can either build or erode angular velocity gradients (see Maeder and Zahn 1998, for more details).

Dynamical shear occurs when the excess energy contained in differentially rotating layers is larger then the work that needs to be done to overcome the buoyancy force. The criterion for stability against dynamical shear instability is the

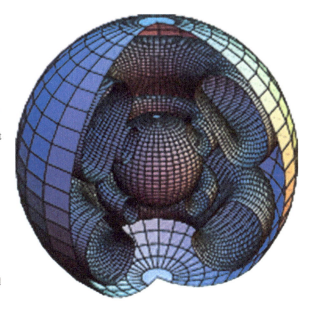

Fig. 6.1 Streamlines of meridional circulation in a rotating $20\,M_\odot$ model with solar metallicity and $v_{\text{ini}} = 300\,\text{km}\,\text{s}^{-1}$ at the beginning of the H–burning phase. The streamlines are in the meridian plane. In the upper hemisphere on the right section, matter is turning counterclockwise along the outer streamline and clockwise along the inner one. The outer sphere is the star surface and has a radius equal to $5.2\,R_\odot$. The inner sphere is the outer boundary of the convective core. It has a radius of $1.7\,R_\odot$ (Illustration from Meynet and Maeder 2002)

Richardson criterion:

$$Ri = \frac{N^2}{(\partial U/\partial z)^2} > \frac{1}{4} = Ri_c, \qquad (6.6)$$

where U is the horizontal velocity, z the vertical coordinate and N^2 the Brunt–Väisälä frequency.

The critical value of the Richardson criterion, $Ri_c = 1/4$, corresponds to the situation where the excess kinetic energy contained in the differentially rotating layers is equal to the work done against the restoring force of the density gradient (also called buoyancy force). It is therefore used by most authors as the limit for the occurrence of the dynamical shear. However, studies by Canuto (2002) show that turbulence may occur as long as $Ri \lesssim Ri_c \sim 1$. This critical value is consistent with numerical simulations done by Brüggen and Hillebrandt (2001) where they find shear mixing for values of Ri greater than $1/4$ (up to about 1.5).

Different dynamical shear diffusion coefficients, D, can be found in the literature. The one used in GENEC is:

$$D = \frac{1}{3}vl = \frac{1}{3}\frac{v}{l}l^2 = \frac{1}{3}r\frac{d\Omega}{dr}\Delta r^2 = \frac{1}{3}r\Delta\Omega\,\Delta r \qquad (6.7)$$

where r is the mean radius of the zone where the instability occurs, $\Delta\Omega$ is the variation of Ω over this zone and Δr is the extent of the zone. The zone is the reunion of consecutive shells where $Ri < Ri_c$ (see Hirschi et al. 2004b, for more details and references).

If the differential rotation is not strong enough to induce dynamical shear, it can still induce the secular shear instability when thermal turbulence reduces the effect of the buoyancy force. The secular shear instability occurs therefore on the thermal time scale, which is much longer than the dynamical one. Note that the way the inhibiting effect of the molecular weight (μ) gradients on secular shear is taken into account impacts strongly the efficiency of the shear. In some work, the inhibiting effect of μ–gradients is so strong that secular shear is suppressed below a certain threshold value of differential rotation (Heger et al. 2000). In other work (Maeder 1997), thermal instabilities and horizontal turbulence reduce the inhibiting effect of the μ–gradients. As a result, shear is not suppressed below a threshold value of differential rotation but only decreased when μ–gradients are present.

There are other minor instabilities induced by rotation: the GSF instability (Goldreich and Schubert 1967; Fricke 1968; Hirschi and Maeder 2010), the ABCD instability (Knobloch and Spruit 1983; Heger et al. 2000) and the Solberg–Høiland instability (Kippenhahn and Weigert 1990). The GSF instability is induced by axisymmetric perturbations. The ABCD instability is a kind of horizontal convection. Finally, Solberg–Høiland stability criterion is the criterion that should be used instead of the Ledoux or Schwarzschild criterion in rotating stars. However, including the dynamical shear instability also takes into account the Solberg–Høiland instability (Hirschi et al. 2004b).

Transport of Angular Momentum

For shellular rotation, the equation of transport of angular momentum (Zahn 1992) in the vertical direction is (in lagrangian coordinates):

$$\rho \frac{d}{dt}\left(r^2 \Omega\right)_{M_r} = \frac{1}{5r^2}\frac{\partial}{\partial r}\left(\rho r^4 \Omega U(r)\right) + \frac{1}{r^2}\frac{\partial}{\partial r}\left(\rho D r^4 \frac{\partial \Omega}{\partial r}\right), \tag{6.8}$$

where $\Omega(r)$ is the mean angular velocity at level r, $U(r)$ the vertical component of the meridional circulation velocity and D the diffusion coefficient due to the sum of the various turbulent diffusion processes (convection, shears and other rotation induced instabilities apart from meridional circulation). Note that angular momentum is conserved in the case of contraction or expansion. The first term on the right hand side, corresponding to meridional circulation, is an *advective* term. The second term on the right hand side, which corresponds to the diffusion processes, is a *diffusive* term. The correct treatment of advection is very costly numerically because Eq. 6.8 is a fourth order equation (the expression of $U(r)$ contains third order derivatives of Ω, see Zahn 1992). This is why some research groups treat meridional circulation in a diffusive way (see for example Heger et al. 2000) with the risk of transporting angular momentum in the wrong direction (in the case meridional circulation builds gradients).

Transport of Chemical Species

The transport of chemical elements is also governed by a diffusion–advection equation like Eq. 6.8. However, if the horizontal component of the turbulent diffusion is large, the vertical advection of the elements (but not that of the angular momentum) can be treated as a simple diffusion (Chaboyer and Zahn 1992) with a diffusion coefficient D_{eff},

$$D_{\text{eff}} = \frac{|rU(r)|^2}{30 D_h}, \tag{6.9}$$

where D_h is the coefficient of horizontal turbulence (Zahn 1992). Equation 6.9 expresses that the vertical advection of chemical elements is severely inhibited by the strong horizontal turbulence characterized by D_h. The change of the mass fraction X_i of the chemical species i is simply

$$\left(\frac{dX_i}{dt}\right)_{M_r} = \left(\frac{\partial}{\partial M_r}\right)_t \left[(4\pi r^2 \rho)^2 D_{\text{mix}}\left(\frac{\partial X_i}{\partial M_r}\right)_t\right] + \left(\frac{dX_i}{dt}\right)_{\text{nuclear}}, \tag{6.10}$$

where the second term on the right accounts for composition changes due to nuclear reactions. The coefficient D_{mix} is the sum $D_{\text{mix}} = D + D_{\text{eff}}$, where D is the term appearing in Eq. 6.8 and D_{eff} accounts for the combined effect of advection and horizontal turbulence.

Interaction Between Rotation and Magnetic Fields

Circular spectro-polarimetric surveys have obtained evidence for the presence of magnetic field at the surface of OB stars (see e.g. the review by Walder et al. 2011, and references therein). The origin of these magnetic fields is still unknown. It might be fossil fields or fields produced through a dynamo mechanism.

The central question for the evolution of massive stars is whether a dynamo is at work in internal radiative zones. This could have far reaching consequences concerning the mixing of the elements and the loss of angular momentum. In particular, the interaction between rotation and magnetic fields in the stellar interior strongly affects the angular momentum retained in the core and thus the initial rotation rate of pulsars and which massive stars could die as long & soft gamma-ray bursts (GRBs), see Vink et al. (2011a) and the discussion in Sect. 6 in Georgy et al. (2012, and references therein).

The interplay between rotation and magnetic field has been studied in stellar evolution calculations using the Tayler–Spruit dynamo (Spruit 2002; Maeder and Meynet 2005). Some numerical simulations confirm the existence of a magnetic instability, however the existence of the dynamo is still controversial (Braithwaite 2006; Zahn et al. 2007).

The Tayler-Spruit dynamo is based on the fact that a purely toroidal field $B_\varphi(r, \vartheta)$, even very weak, in a stable stratified star is unstable on an Alfvén timescale $1/\omega_A$. This is the first magnetic instability to appear. It is non–axisymmetric of type $m = 1$ (Spruit 2002), occurs under a wide range of conditions and is characterized by a low threshold and a short growth time. In a rotating star, the instability is also present, however the growth rate σ_B of the instability is, if $\omega_A \ll \Omega$,

$$\sigma_B = \frac{\omega_A^2}{\Omega}, \qquad (6.11)$$

instead of the Alfvén frequency ω_A, because the growth rate of the instability is reduced by the Coriolis force (Spruit 2002). One usually has the following ordering of the different frequencies, $N \gg \Omega \gg \omega_A$. In the Sun, one has $N \approx 10^{-3}\,\text{s}^{-1}$, $\Omega = 3 \times 10^{-6}\,\text{s}^{-1}$ and a field of 1 kG would give an Alfvén frequency as low as $\omega_A = 4 \times 10^{-9}\,\text{s}^{-1}$ (where N^2 is the Brunt–Väisälä frequency).

This theory enables us to establish the two quantities that we are mainly interested in for stellar evolution: the magnetic viscosity ν, which expresses the mechanical coupling due to the magnetic field **B**, and the magnetic diffusivity η, which expresses the transport by a magnetic instability and thus also the damping of the instability. The parameter η also expresses the vertical transport of the chemical elements and enters Eq. 6.10, while the viscosity ν determines the vertical transport

6 Evolution and Nucleosynthesis of Very Massive Stars

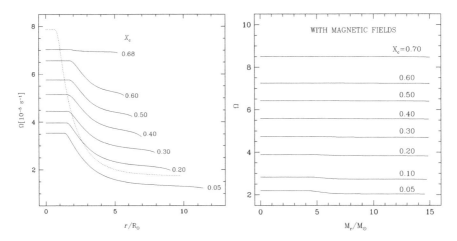

Fig. 6.2 *Left*: evolution of the angular velocity Ω as a function of the distance to the center in a 20 $M_\odot$ star with $v_{\text{ini}} = 300 \text{ km s}^{-1}$. X_c is the hydrogen mass fraction at the center. The *dotted line* shows the profile when the He–core contracts at the end of the H–burning phase. *Right*: rotation profiles at various stages of evolution (labelled by the central H content X_c) of a 15 $M_\odot$ model with $X = 0.705$, $Z = 0.02$, an initial velocity of 300 km s^{-1} and magnetic field from the Tayler–Spruit dynamo (Taken from Maeder and Meynet (2005))

of the angular momentum by the magnetic field and enters the second term on the right-hand side of Eq. 6.8.

Figure 6.2 shows the differences in the internal Ω–profiles during the evolution of a 20 $M_\odot$ star with and without magnetic field created by the Tayler–Spruit dynamo. Without magnetic field, the star has a significant differential rotation, while Ω is almost constant when a magnetic field created by the dynamo is present. It is not perfectly constant, otherwise there would be no dynamo. In fact, the rotation rapidly adjusts itself to the minimum differential rotation necessary to sustain the dynamo. One could then assume that the mixing of chemical elements is suppressed by magnetic fields. This is, however, not the case since the interplay between magnetic fields and the meridional circulation tend to lead to more mixing in models including magnetic fields compared to models not including magnetic fields (Maeder and Meynet 2005). Fast rotating models of GRB progenitors calculated by Yoon et al. (2006) also experience a strong chemical internal mixing leading to the stars undergoing quasi-chemical homogeneous evolution. The study of the interaction between rotation and magnetic fields is still under development (see e.g. Potter et al. 2012, for a different rotation-magnetic field interaction theory, the $\alpha - \Omega$ dynamo, and its impact on massive star evolution) and the next 10 years will certainly provide new insights on this important topic.

Other Input Physics

The other key input physics that are essentials for the computation of stellar evolution models are: nuclear reactions, mass loss prescriptions (discussed above), the equation of state, opacities and neutrino losses. Stellar evolution codes are now able to include larger and more flexible nuclear reaction network (see e.g. Frischknecht et al. 2010, for a description of the implementation of a flexible network in GENEC). Nuclear physics and other inputs are described for other codes for example in Paxton et al. (2011) and Chieffi et al. (1998).

In this chapter, the evolution of single stars is discussed. We refer the reader to Langer (2012) for a review of the impact of binarity on massive star evolution and to Schneider et al. (2014) for the possible impact of a binary companion on the properties of VMS. Note that the mass transfer efficiency prescriptions used in binary model represent an important uncertainty, especially for VMS with high luminosities.

6.3 General Properties and Early Evolution of VMS

6.3.1 VMS Evolve Nearly Homogeneously

Probably the main characteristic that makes VMS quite different from their lower mass siblings is the fact that they possess very large convective cores during the MS phase. They therefore evolve quasi-chemically homogeneously even if there is no mixing (due e.g. by rotation) in radiative zones as discussed in Maeder (1980). To illustrate this last point, Fig. 6.3 shows the convective core mass fraction for

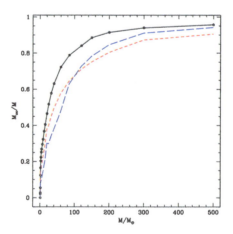

Fig. 6.3 Mass fraction of the convective core in non-rotating solar metallicity models. This figure and all the following figures are taken from Yusof et al. (2013). Models with initial masses superior or equal to 150 M are from Yusof et al. (2013). Models for lower initial masses are from Ekström et al. (2012). The *continuous line* corresponds to the ZAMS, the *short-dashed line* to models when the mass fraction of hydrogen at the centre, X_c, is 0.35, and the *long-dashed line* to models when X_c is equal to 0.05

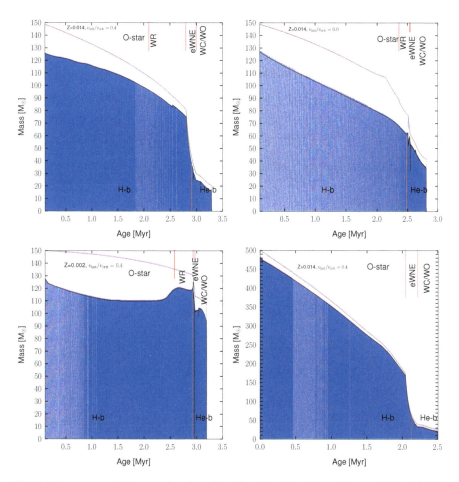

Fig. 6.4 Structure evolution as a function of age for selected models: solar metallicity 150 $M_\odot$ rotating (*top-left*) and non-rotating (*top-right*) models, rotating SMC metallicity 150 $M_\odot$ model (*bottom-left*) and rotating solar metallicity 500 $M_\odot$ model (*bottom-right*). The *blue zones* represent the convective regions. The *top solid black line* indicates the total mass of the star and *vertical red markers* are given for the different phases (O-type, WR = eWNL, eWNE and WC/WO) at the top of the plots. The transition between H- and He-burning phases is indicated by the *red vertical line* at the *bottom* of the plots

non-rotating massive stars at solar metallicity. It is apparent that the convective cores for masses above 150 $M_\odot$ extend over more than 75 % of the total mass of the star.

Figure 6.4 shows how age, metallicity and rotation influence this mass fraction. Comparing the *top-left* and *bottom-left* panels showing the rotating 150 $M_\odot$ models at solar and SMC metallicities (Z), respectively, we can see that the convective core occupies a very slightly larger fraction of the total mass at SMC metallicity on the ZAMS. As for lower-mass massive stars, this is due to a lower CNO content leading to higher central temperature. This effect is counterbalanced by the lower opacity (especially at very low metallicities) and the net change in convective core size is small. As the evolution proceeds mass loss is weaker at lower Z and thus the total mass decreases slower than the convective core mass. This generally leads to a smaller fraction of the total mass occupied by the convective core in the SMC models.

We can see the impact of rotation by comparing the rotating (*top-left*) and non-rotating (*top-right*) 150 $M_\odot$ models. The convective core size remains higher in the rotating model due to the additional mixing in radiative zones. We can see that rotation induced mixing can even lead to an increase of the convective core size as is the case for the SMC model (*bottom-left*). This increase is typical of quasi-chemically homogeneous evolution also found in previous studies (see Yoon et al. 2012, and citations therein). The rotating 500 $M_\odot$ model (*bottom-right* panel) evolves quasi-homogeneously throughout its entire evolution, even with an initial ratio of the velocity to the critical velocity of 0.4.

These features, very large convective cores and quasi-chemi homogeneous evolution, are key factors governing their evolution as is discussed below.

6.3.2 Evolutionary Tracks

In Figs. 6.5 and 6.6, we present the evolutionary tracks of models with initial masses between 150 and 500 $M_\odot$ at various metallicities. Other properties of VMS models at the end of H- and He- burning stages are given in Tables 6.1 and 6.2, respectively.

A very first striking feature is that these massive stars evolve vertically in the HR diagram (HRD) covering only very restricted ranges in effective temperatures but a very large range in luminosities. This is typical of an evolution mainly governed by mass loss and also by a strong internal mixing (here due to convection).

Let us now describe in more details the evolution of the non-rotating 500 $M_\odot$ model at solar metallicity (see Fig. 6.5). In general, the luminosity of stars increases during the MS phase. Here we have that during that phase, the luminosity decreases slightly by about 0.1 dex. This is the consequence of very high mass loss rates (of the order of 7×10^{-5} $M_\odot$ per year) already at very early evolutionary stages.

At an age of 1.43 million years, the mass fraction of hydrogen at the surface becomes inferior to 0.3, the star enters into the WR phase and has an actual mass decreased by about 40 % with respect to its initial value. At that time the mass fraction of hydrogen in the core is 0.24. Thus this star enters the WR phase while still burning hydrogen in its core and having nearly the same amount of hydrogen at the centre and at the surface, illustrating the nearly homogeneous nature of its evolution (see also Maeder 1980). Typically for this model, the convective core encompasses nearly 96 % of the total mass on the ZAMS (see also Fig. 6.3).

At an age equal to 2.00 Myr, the mass fraction of hydrogen is zero in the core ($X_c = 0$). The star has lost a huge amount of mass through stellar winds and has at this stage an actual mass of 55.7 $M_\odot$. So, since the entrance into the WR phase, the

6 Evolution and Nucleosynthesis of Very Massive Stars 171

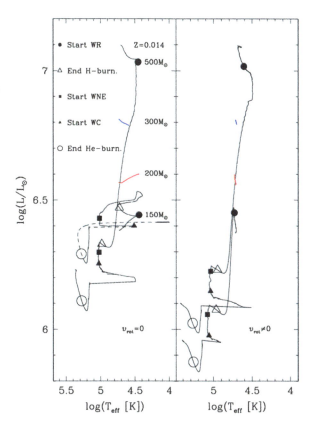

Fig. 6.5 HR diagram from 150 up to 500 $M_\odot$ at solar metallicity for non-rotating (*left*) and rotating (*right*) models, respectively. Key stages are indicated along the tracks. Only the first portion (up to start of WR phase) of the tracks for the 200 and 300 $M_\odot$ are shown

star has lost about 245 $M_\odot$, i.e. about half of its total mass. This strong mass loss episode translates into the HR diagram by a very important decrease in luminosity. Note that when X_c is zero, the convective core still encompass 80 % of the total stellar mass!

The core helium burning phase last for about 0.3 Myr, that means slightly more than 15 % of the MS lifetime. At the end of the core He burning phase, the actual mass of the star is 29.82 $M_\odot$, its age is 2.32 My, the mass fraction of helium at the surface is 0.26. The total WR phase lasts for 0.88 My, that means about 38 % of the total stellar lifetime.

It is interesting to compare the evolution of the 500 $M_\odot$ stellar model with that of the 150 $M_\odot$ model. In contrast to the 500 $M_\odot$ model, the 150 $M_\odot$ increases in luminosity during the MS phase. Looking at the HRD we see that the O-type star phases of the 150 and 500 $M_\odot$ models cover more or less the same effective temperature range. This illustrates the well known fact that the colors of stars for this mass range does not change much with the initial mass.

When the stars enters into the WR phase, in contrast to the case of the 500 $M_\odot$ where the luminosity decreases steeply, the luminosity of the 150 $M_\odot$ model continues to increase a little. The luminosities of the two models when the hydrogen

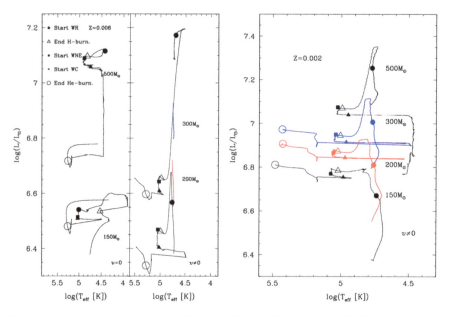

Fig. 6.6 Same as Fig. 6.5 for *right*: LMC models ($Z = 0.006$) and *right*: SMC rotating models ($Z = 0.002$)

mass fraction at the surface becomes inferior to 10^{-5} differ by just a little more than 0.1 dex. The effective temperatures are similar. Thus one expects stars from very different initial masses to occupy similar positions in the HRD (the 500 $M_\odot$ star being slightly less luminous than the 150 $M_\odot$ during the WR phase). We note that after the end of the core He-burning phase, the star evolves to the red and terminate its lifetime around an effective temperature of Log T_{eff} equal to 4. This comes from the core contraction at the end of core He-burning which releases energy and leads to an envelope expansion akin to the expansion of the envelope at the end of the MS (see also Yoshida and Umeda 2011).

The duration of the core H-burning phase of the 150 $M_\odot$ model is not much different from the one of the 500 $M_\odot$ model being 2.5 My instead of 2 My. The core He-burning lifetime lasts for 0.3 My as for the 500 $M_\odot$. The total duration of the WR phase is 0.45 My, thus about half of the WR duration for the 500 $M_\odot$.

The 200 $M_\odot$ model has an evolution similar to the 150 $M_\odot$ model, while the 300 $M_\odot$ has an evolution similar to the 500 $M_\odot$.

Let us now consider how rotation changes the picture. The right panel of Fig. 6.5 shows the evolutionary tracks of the $Z = 0.014$ rotating models in a similar way to the tracks of the non-rotating models in the left panel. The changes brought by rotation are modest. This is expected because of two facts: first, in this high mass range, the evolution is more impacted by mass loss than by rotation, second, stars are already well mixed by the large convective cores. One notes however a few differences between the non-rotating and rotating models. One of the most

6 Evolution and Nucleosynthesis of Very Massive Stars

Table 6.1 Properties of the hydrogen burning phase: initial properties of stellar models (columns 1–3), lifetime of H-burning and O-type star phase (4–5), average MS surface velocity (6) and properties at the end of the core H-burning phase (7–15). Masses are in solar masses, velocities are in km s^{-1}, lifetimes are in 10^6 years and abundances are surface abundances in mass fractions. The luminosity, L, is in log$_{10}(L/L_\odot)$ unit and the effective temperature, T_{eff}, is in log$_{10}$ [K]. Note that the effective temperature given here includes a correction for WR stars to take into account the fact that their winds are optically thick as in Meynet and Maeder (2005)

M_{ini}	Z_{ini}	$\frac{v_{\mathrm{ini}}}{v_{\mathrm{crit}}}$	τ_H	τ_O	$\langle v_{\mathrm{MS}} \rangle$	$M_{\mathrm{H.b.}}^{\mathrm{end}}$	^{1}H	^{4}He	^{12}C	^{14}N	^{16}O	T_{eff}	L	Γ_{Edd}
120	0.014	0.0	2.671	2.592	0.0	63.7	2.04e-1	7.82e-1	8.58e-5	8.15e-3	1.06e-4	4.405	6.334	0.627
150	0.014	0.0	2.497	2.348	0.0	76.3	1.35e-1	8.51e-1	9.26e-5	8.15e-3	9.91e-5	4.413	6.455	0.657
200	0.014	0.0	2.323	2.095	0.0	95.2	7.51e-2	9.11e-1	9.93e-5	8.14e-3	9.23e-5	4.405	6.597	0.687
300	0.014	0.0	2.154	1.657	0.0	65.2	1.24e-3	9.85e-1	1.31e-4	8.11e-3	7.93e-5	4.267	6.401	0.595
500	0.014	0.0	1.990	1.421	0.0	56.3	2.20e-3	9.84e-1	1.26e-4	8.12e-3	8.03e-5	4.301	6.318	0.568
120	0.014	0.4	3.137	2.270	116.71	34.6	1.56e-3	9.85e-1	1.33e-4	8.10e-3	8.48e-5	4.400	6.018	0.463
150	0.014	0.4	2.909	2.074	101.24	37.1	1.80e-3	9.85e-1	1.30e-4	8.11e-3	8.41e-5	4.387	6.062	0.479
200	0.014	0.4	2.649	1.830	89.33	40.0	1.41e-3	9.85e-1	1.33e-4	8.10e-3	8.30e-5	4.372	6.110	0.495
300	0.014	0.4	2.376	1.561	61.16	43.2	1.85e-3	9.85e-1	1.33e-4	8.10e-3	8.23e-5	4.356	6.157	0.511
500	0.014	0.4	2.132	1.377	24.55	48.1	1.24e-3	9.85e-1	1.38e-4	8.10e-3	8.08e-5	4.332	6.221	0.531
120	0.006	0.0	2.675	2.682	0.0	79.0	4.03e-1	5.91e-1	3.29e-5	3.50e-3	4.47e-5	4.441	6.391	0.672
150	0.006	0.0	2.492	2.499	0.0	96.1	3.28e-1	6.67e-1	3.58e-5	3.50e-3	4.25e-5	4.483	6.524	0.709
500	0.006	0.0	1.904	1.636	0.0	238.8	2.56e-2	9.69e-1	5.12e-5	3.48e-3	3.18e-5	4.032	7.094	0.819
120	0.006	0.4	3.140	2.479	208.55	64.0	1.70e-3	9.92e-1	6.06e-5	3.47e-3	3.04e-5	4.387	6.395	0.597
150	0.006	0.4	2.857	2.172	198.19	71.3	9.76e-4	9.93e-1	6.33e-5	3.47e-3	2.97e-5	4.365	6.455	0.615
200	0.006	0.4	2.590	1.894	193.05	80.7	1.22e-3	9.93e-1	6.29e-5	3.47e-3	2.95e-5	4.339	6.525	0.638
300	0.006	0.4	2.318	1.619	173.47	85.8	1.32e-3	9.93e-1	6.30e-5	3.47e-3	2.93e-5	4.327	6.559	0.649
500	0.006	0.4	2.077	1.419	116.76	101.7	1.37e-3	9.93e-1	6.37e-5	3.47e-3	2.89e-5	4.291	6.650	0.676
150	0.002	0.4	2.921	2.567	318.92	128.8	1.67e-3	9.96e-1	2.13e-5	1.16e-3	8.09e-6	4.394	6.780	0.720
200	0.002	0.4	2.612	2.168	333.43	152.2	1.31e-3	9.97e-1	2.26e-5	1.16e-3	7.82e-6	4.363	6.867	0.743
300	0.002	0.4	2.315	1.801	347.32	176.2	1.10e-3	9.97e-1	2.32e-5	1.16e-3	7.68e-6	4.279	7.067	0.763

Table 6.2 Properties of the helium burning phase: initial properties of stellar models (columns 1–3), age of star at the end of He-burning (4), average He-b. surface velocity (5) and properties at the end of the core He-burning phase (6–15). Abundances are given for the surface, except for $^{12}C_c$, which represents the central C abundance. Same units as in Table 6.1. $\mathcal{L}_{CO}$ [$10^{50}\,\frac{\text{g cm}^2}{\text{s}}$] is the angular momentum contained in the CO core (Note that at this stage the angular velocity is constant in the CO core due to convective mixing)

M_{ini}	Z_{ini}	$\frac{v_{\text{ini}}}{v_{\text{crit}}}$	$\text{age}_{\text{He-b.}}^{\text{end}}$	$\langle v_{\text{He-b.}} \rangle$	$M_{\text{He-b.}}^{\text{end}}$	^{4}He	^{12}C	^{12}C$_c$	^{16}O	^{22}Ne	T_{eff}	L	Γ_{Edd}	$\mathcal{L}_{\text{CO}}$
120	0.014	0.0	3.003	0.00	30.9	0.242	0.458	0.150	0.281	1.081e-02	4.819	6.117	0.650	0
150	0.014	0.0	2.809	0.00	41.3	0.234	0.436	0.126	0.312	1.003e-02	4.822	6.278	0.706	0
200	0.014	0.0	2.622	0.00	49.4	0.207	0.408	0.112	0.366	8.811e-03	4.807	6.377	0.737	0
300	0.014	0.0	2.469	0.00	38.2	0.234	0.443	0.133	0.305	1.029e-02	4.825	6.236	0.691	0
500	0.014	0.0	2.314	0.00	29.8	0.261	0.464	0.152	0.257	1.110e-02	4.811	6.095	0.640	0
120	0.014	0.4	3.513	1.58	18.8	0.292	0.492	0.195	0.198	1.196e-02	4.806	5.814	0.533	1.91
150	0.014	0.4	3.291	1.18	20.3	0.286	0.488	0.187	0.208	1.184e-02	4.808	5.863	0.551	1.91
200	0.014	0.4	3.020	0.50	22.0	0.277	0.484	0.180	0.221	1.172e-02	4.812	5.912	0.570	1.37
300	0.014	0.4	2.733	0.13	24.0	0.270	0.479	0.172	0.233	1.151e-02	4.814	5.965	0.589	0.75
500	0.014	0.4	2.502	0.03	25.9	0.269	0.473	0.164	0.239	1.140e-02	4.811	6.010	0.606	0.28
120	0.006	0.0	2.993	0.00	54.2	0.229	0.391	0.098	0.372	3.701e-03	4.860	6.424	0.753	0
150	0.006	0.0	2.845	0.00	59.7	0.241	0.370	0.086	0.380	3.597e-03	4.844	6.474	0.767	0
500	0.006	0.0	2.182	0.00	94.7	0.251	0.392	0.078	0.349	3.318e-03	4.833	6.711	0.834	0
120	0.006	0.4	3.472	6.84	39.3	0.294	0.457	0.132	0.241	4.709e-03	4.387	6.395	0.692	16.2
150	0.006	0.4	3.164	3.67	45.7	0.310	0.451	0.122	0.231	4.701e-03	4.824	6.329	0.767	14.7
200	0.006	0.4	2.904	1.33	51.1	0.303	0.444	0.114	0.245	4.547e-03	4.825	6.390	0.738	9.98
300	0.006	0.4	2.625	0.35	54.1	0.291	0.439	0.110	0.262	4.433e-03	4.830	6.421	0.748	5.18
500	0.006	0.4	2.387	0.13	74.9	0.330	0.425	0.090	0.237	4.356e-03	4.790	6.590	0.798	4.83
150	0.002	0.4	3.193	64.94	106.7	0.809	0.153	0.074	0.035	1.730e-03	4.743	6.766	0.841	412.5
200	0.002	0.4	2.889	29.88	129.3	0.880	0.109	0.066	0.009	1.777e-03	4.789	6.861	0.863	355.6
300	0.002	0.4	2.585	5.10	149.8	0.938	0.058	0.060	0.001	1.798e-03	4.833	6.933	0.880	156.8

6 Evolution and Nucleosynthesis of Very Massive Stars

striking differences is the fact that the models during their O-type phase evolve nearly vertically when rotation is accounted for. This is the effect of rotational mixing which keeps the star more chemically homogeneous than in the non-rotating cases (although, as underlined above, already in models with no rotation, due to the importance of the convective core, stars are never very far from chemical homogeneity). As was the case in the non-rotating tracks, the O-type star phase corresponds to an upward displacement when time goes on in the HR diagram for the 150 $M_\odot$ model, while, it corresponds to a downwards displacement for the three more massive models. One notes finally that lower luminosities are reached by the rotating models at the end of their evolution (decrease by about 0.3 dex in luminosity, thus by a factor 2). This comes mainly because the rotating models enter earlier into their WR phase and thus lose more mass.

How does a change in metallicity alter the picture? When the metallicity decreases to Z=0.006 (see Fig. 6.6, *left*), as expected, tracks are shifted to higher luminosities and effective temperatures. In this metallicity range, all models evolve upwards during their O-type star phase in the HR diagram. This is an effect of the lower mass loss rates.

As was already the case at Z=0.014, rotation makes the star evolve nearly vertically in the HR diagram. One notes in this metallicity range, much more important effects of rotation than at Z=0.014, which is also expected, since at these lower metallicity, mass loss rates are smaller and rotational mixing more efficient. We note that most of the decrease in luminosity in the 500 $M_\odot$ solar mass model occurs during the WC phase in the Z=0.006 non-rotating model, while it occurs during the WNL phase in the rotating one. This illustrates the fact that rotational mixing, by creating a much larger H-rich region in the star, tends to considerably increase the duration of the WNL phase. One notes also that while the 150 $M_\odot$ model enters the WR phase only after the MS phase, the rotating model becomes a WR star before the end of the MS phase.

At the metallicity of the SMC (see Fig. 6.6, *right*), except for the 500 $M_\odot$, the tracks evolve horizontally after the end of the core H-burning phase (triangle in Fig. 6.6, *right*). The much lower mass loss rates are responsible for this effect.

6.3.3 Lifetimes and Mass-Luminosity Relation

In Tables 6.1 and 6.2, we provide ages at the end of core hydrogen burning and core helium burning, respectively. We see that the MS lifetime of non-rotating models at solar metallicity ranges from 2.67 to 1.99 Myr for initial masses ranging from 120 to 500 $M_\odot$ showing the well known fact that VMS have a very weak lifetime dependence on their initial mass.

The mass-luminosity relation on the ZAMS for rotating massive stars at solar composition is shown in Fig. 6.7. The relation ($L \propto M^\alpha$) is steep for low and intermediate-mass stars ($\alpha \sim 3$ for $10 < M/M_\odot < 20$) and flattens for VMS ($\alpha \sim 1.3$ for $200 < M/M_\odot < 500$). This flattening is due to the increased radiation

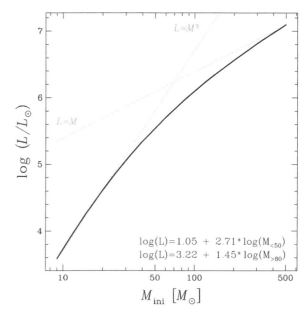

Fig. 6.7 Mass-luminosity relation on the ZAMS for rotating models at solar metallicity. The formulae in the bottom right corner are linear fits for the mass ranges: 9–50 $M_\odot$ and 80–500 $M_\odot$. The non-rotating models have very similar properties on the ZAMS

pressure relative to gas pressure in massive stars. Since the lifetime of a star is roughly M/L, we get that for VMS $\tau \propto M/L \propto M^{-0.3}$.

The H-burning (and total) lifetimes of VMS are lengthened by rotation as in lower mass stars. Differences in the H-burning lifetimes of rotating and non-rotating 150 $M_\odot$ models at solar metallicity are ∼14 %. The effects of metallicity on the lifetimes are generally very small. The small differences in total lifetimes are due to different mass loss at different metallicities.

6.3.4 Mass Loss by Stellar Winds

Mass loss by stellar winds is a key factor governing the evolution of VMS. This comes from the very high luminosities reached by these objects. For example, the luminosity derived for R136a1 is about 10 million times that of our sun.

For such luminous objects, winds will be very powerful at all evolutionary stages, so while early main-sequence VMS are formally O-type stars from an evolutionary perspective, their spectral appearance may be closer to Of or Of/WN at early phases (Crowther et al. 2012).

Table 6.3 gives the total mass at the start and end of the evolution[2] as well as at the transitions between the different WR phases in columns 1 to 5. The average

---

[2]The models have been evolved beyond the end of core He-burning and usually until oxygen burning, thus very close to the end of their life (see Yusof et al. 2013, for full details)

6 Evolution and Nucleosynthesis of Very Massive Stars

Table 6.3 Mass loss properties: Total mass of the models at various stages (columns 1–5), and average mass loss rates $\langle \dot{M} \rangle$ during the O-type and eWNE phases (6,7). Masses are in solar mass units and mass loss rates are given in $M_\odot$ year^{-1}

ZAMS	start eWNL	start eWNE	start WC	final	$\langle \dot{M}_{\text{Vink}} \rangle$	$\langle \dot{M}_{\text{eWNE}} \rangle$
	$Z=0.014$, $v/v_{crit}=0.0$					
120	69.43	52.59	47.62	30.81	2.477e-05	3.638e-04
150	88.86	66.87	61.20	41.16	3.274e-05	6.107e-04
200	121.06	91.20	83.85	49.32	4.618e-05	1.150e-03
300	184.27	130.47	52.05	38.15	8.047e-05	8.912e-04
500	298.79	169.50	45.14	29.75	1.736e-04	9.590e-04
	$Z=0.014$, $v/v_{crit}=0.4$					
120	88.28	69.54	27.43	18.68	1.675e-05	2.057e-04
150	106.64	80.88	29.49	20.22	2.467e-05	2.640e-04
200	137.52	98.75	31.84	21.93	3.985e-05	3.564e-04
300	196.64	129.10	34.45	23.93	7.559e-05	5.160e-04
500	298.42	174.05	38.30	25.83	1.594e-04	7.901e-04
	$Z=0.006$, $v/v_{crit}=0.0$					
120	74.30	57.91	56.91	54.11	2.140e-05	3.272e-04
150	94.18	74.20	71.75	59.59	2.839e-05	5.038e-04
500	332.68	250.64	197.41	94.56	1.304e-04	3.334e-03
	$Z=0.006$, $v/v_{crit}=0.4$					
120	100.57	90.78	54.43	39.25	9.429e-06	3.219e-04
150	125.79	111.84	60.75	45.58	1.367e-05	4.418e-04
200	166.81	144.86	66.25	51.02	2.180e-05	6.257e-04
300	247.07	207.10	73.11	54.04	4.166e-05	9.524e-04
500	397.34	315.51	86.10	74.75	9.194e-05	1.685e-03
	$Z=0.002$, $v/v_{crit}=0.4$					
150	135.06	130.46	113.51	106.50	6.661e-06	4.485e-04
200	181.42	174.18	137.90	129.21	9.902e-06	6.631e-04
300	273.18	260.81	156.14	149.70	1.730e-05	1.040e-03

mass loss rates during the O-type and eWNE phases (the phase during which the mass loss rates are highest) are given in columns 6 and 7, respectively.

The evolution of the mass loss rates for various models are shown in Fig. 6.8. Following the evolution from left to right for the 150 $M_\odot$ model at solar metallicity (solid-black), mass loss rates slowly increase at the start of the O-type phase with mass loss rates between 10^{-5} $M_\odot$ year^{-1} (absolute values for the mass loss rates, $-\dot{M}$, are quoted in this paragraph) and $10^{-4.5}$ $M_\odot$ year^{-1}. If a bi-stability limit is encountered during the MS phase, as is the case in the non-rotating 150 $M_\odot$ model, mass loss rates can vary significantly over a short period of time and mass loss peaks reach values higher than 10^{-4} $M_\odot$ year^{-1}. The highest mass loss rate is encountered at the start of the eWNE phase (star symbols) with values in excess of 10^{-3} $M_\odot$ year^{-1} (note that the mass loss rate in the non-rotating model has a peak at the end of the H-burning phase. phase due to the star reaching temporarily

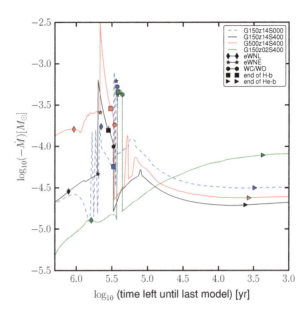

Fig. 6.8 Evolution of the mass loss rate as a function of time left until last model (log scale) for the rotating 500 $M_\odot$ model (*solid-red*), the rotating 150 $M_\odot$ model (*solid-black*), the non-rotating (*dashed*) 150 $M_\odot$ model at solar metallicity, and the rotating 150 $M_\odot$ model at SMC metallicity (*solid-green*). The diamonds indicate the start of the eWNL phase, the stars the start of the eWNE phase and hexagons the start of the WC/WO phase. The *squares* and *triangles* indicate the end of H-b. and He-b. phases, respectively

cooler effective temperatures). Such high mass loss rates quickly reduce the mass and luminosity of the star and thus the mass loss rate also decreases quickly during the eWNE phase. During the WC/WO phase, mass loss rates are of the same order of magnitude as during the O-type phase.

Comparing the rotating 500 and 150 $M_\odot$ model at solar metallicity (solid black and red), we see that more massive stars start with higher mass loss rates but converge later on to similar mass loss rates since the total mass of the models converges to similar values (see Table 6.3).

Comparing the SMC and solar metallicity 150 $M_\odot$ rotating models, we can clearly see the metallicity effect during the O-type star phase. During the eWNE phase, mass loss rates are similar and in the WC/WO, mass loss rates in the SMC model are actually higher since the total mass in that model remained high in contrast with solar metallicity models.

Table 6.3 also shows the relative importance of the mass lost during the various phases and how their importance changes as a function of metallicity. Even though mass loss is the strongest during the eWNE phase, significant amount of mass is lost in all phases.

6.3.5 Mass Loss Rates and Proximity of the Eddington Limit

Vink et al. (2011b) suggest enhanced mass-loss rates (with respect to Vink et al. 2001a, used in the models presented here) for stars with high Eddington parameters (($\Gamma_e \geq 0.7$), see Eq. 1 in Vink et al. 2011b, for the exact definition of Γ_e) that they

6 Evolution and Nucleosynthesis of Very Massive Stars

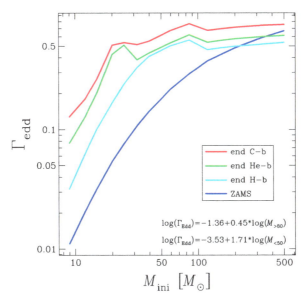

Fig. 6.9 Eddington parameter, $\Gamma_{\rm edd}$ for rotating models at solar metallicity. $\Gamma_{\rm edd}$ is plotted on the ZAMS (*blue line*) and the end of H-(*light blue*), He-(*green*) and C-burning (*red*) phases. Except for the 300 and 500 $M_\odot$ models, $\Gamma_{\rm edd}$ increases throughout the evolution. At solar metallicity, the highest value (close to 0.8) is actually reached by the 85 $M_\odot$ model at the end of its evolution. This could lead to significant mass loss shortly before the final explosion in a model that ends as a WR star and potentially explain supernova surrounded by a thick circumstellar material without the need for the star to be in the luminous variable phase. The formulae in the bottom right corner are linear fits for the mass ranges: 9–50 $M_\odot$ and 80–500 $M_\odot$. The non-rotating models have very similar properties on the ZAMS

attribute to the Wolf-Rayet stage. In order to know whether higher mass loss rates near the Eddington limit could have an impact on the present result, we discuss here the proximity of our models to the Eddington limit.

Figure 6.9 shows the Eddington parameter, $\Gamma_{\rm Edd} = L/L_{\rm Edd} = \kappa L/(4\pi cGM)$, as a function of the initial mass of our models at key stages. Since the Eddington parameter, $\Gamma_{\rm Edd}$ scales with L/M, the curve for $\Gamma_{\rm Edd}$ also flattens for VMS. The ZAMS values for $\Gamma_{\rm Edd}$ range between 0.4 and 0.6, so well below the Eddington limit, $\Gamma_{\rm Edd} = 1$, and below the limiting value of $\Gamma_{\rm e} = 0.7$ where enhanced mass-loss rates are expected according to Vink et al. (2011b).

How does $\Gamma_{\rm Edd}$ change during the lifetime of VMS? Figure 6.10 presents the evolution of $\Gamma_{\rm Edd}$ for a subset of representative models. The numerical values for each model are given at key stages in Tables 6.1 and 6.2. Since $\Gamma_{\rm Edd} \propto \kappa L/M$, an increase in luminosity and a decrease in mass both lead to higher $\Gamma_{\rm Edd}$. Changes in effective temperature and chemical composition affect the opacity and also lead to changes in $\Gamma_{\rm Edd}$.

In rotating models at solar metallicity, $\Gamma_{\rm Edd}$ slowly increases until the start of the eWNE phase. This is mainly due to the increase in luminosity and decrease in mass

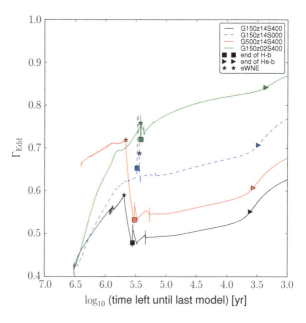

Fig. 6.10 Evolution of the Eddington parameter, Γ_{Edd}, as a function of time left until last model (log scale) for the rotating 500 $M_\odot$ model (*solid-red*), the rotating 150 $M_\odot$ model (*solid-black*), the non-rotating (*dashed*) 150 $M_\odot$ model at solar metallicity, and the rotating 150 $M_\odot$ model at SMC metallicity (*solid-green*). The stars indicate the start of the eWNE phase. The squares and triangles indicate the end of H-b. and He-b. phases, respectively

of the model. At the start of the eWNE phase, mass loss increases significantly. This leads to a strong decrease in the luminosity of the model and as a result Γ_{Edd} decreases sharply.

During the WC/WO phase, mass loss rates being of similar values as during the O-type star phase, Γ_{Edd} increases again gradually.

We can see that, at solar metallicity, Γ_{Edd} rarely increases beyond 0.7 even in the 500 $M_\odot$ model. There are nevertheless two interesting cases in which values above 0.7 are reached. The first case is during the advanced stages. At this stage, mass loss does not have much time to change the total mass of the star (it is mostly changes in effective temperature and to a minor extent in luminosity that influence the increase in Γ_{Edd}). This may nevertheless trigger instabilities resulting in strong mass loss episodes. This may have consequences for the type of SN event that such a star will produce and may be a reason why the explosion of VMS may look like as if they had happened in an environment similar to those observed around Luminous Blue Variable. The second case is at low metallicity, as highlighted by the 150 $M_\odot$ model at SMC metallicity. Indeed, values above 0.7 are reached before the end of the MS (square symbol). Mass loss prescriptions such as the ones of Vink et al. (2011b) and Maeder et al. (2012) may thus play an important role on the fate of VMS. The non-rotating model has a different mass loss history (see Fig. 6.8), which explains the slightly different evolution of Γ_{Edd} near the end of the main sequence.

6.3.6 Evolution of the Surface Velocity

The surface velocity of stars is affected by several processes. Contraction or expansion of the surface respectively increases and decreases the surface velocity due to the conservation of angular momentum. Mass loss removes angular momentum and thus decreases the surface velocity. Finally internal transport of angular momentum generally increases the surface velocity. As shown in Fig. 6.11 (left panel), at solar metallicity, the surface velocity rapidly decreases during the main sequence due to the strong mass loss over the entire mass range of VMS. At SMC metallicity, mass loss is weaker than at solar metallicity and internal transport of angular momentum initially dominates over mass loss and the surface velocity increases during the first half of the MS phase. During this time, the ratio of surface velocity to critical velocity also increases up to values close to 0.7 (note that the models presented include the effect of the luminosity of the star when determining the critical rotation as described in Maeder and Meynet 2000). However, as the evolution proceeds, the luminosity increases and mass loss eventually starts to dominate and the surface velocity and its ratio to critical rotation both decrease for the rest of the evolution. SMC stars thus never reach critical rotation. The situation at very low and zero metallicities has been studied by several groups (see Hirschi 2007; Ekström et al. 2008; Yoon et al. 2012; Chatzopoulos and Wheeler 2012, and references therein). If mass loss becomes negligible, then the surface velocity reaches critical rotation for a large fraction of its lifetime, which probably leads to mechanical mass loss along the equator. The angular momentum content in the core of VMS stars is discussed further in Sect. 6.5.4.

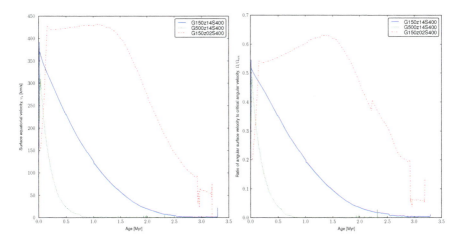

Fig. 6.11 Evolution of surface equatorial velocity (*left*) and ratio of the surface angular velocity to the critical angular velocity (*right*) for the rotating solar metallicity 150 and 500 $M_\odot$ and SMC 150 $M_\odot$ models as a function of age of the star

6.4 WR Stars from VMS

Figure 6.12 presents the evolution of the surface abundances as a function of the total mass for the solar metallicity rotating models of 150 and 60 $M_\odot$. This figure shows how the combined effects of mass loss and internal mixing change their surface composition. Qualitatively there are no big differences between the 60 and 150 $M_\odot$ models. Since the 150 $M_\odot$ has larger cores, the transition to the various WR stages occurs at larger total masses compared to the 60 $M_\odot$ model. It thus confirms the general idea that a more massive (thus more luminous) WR star originates from a more massive O-type star. Figure 6.12 shows that all abundances and abundance ratios are very similar for a given WR phase. It is therefore not easy to distinguish a WR originating from a VMS from its surface chemical composition (however see below).

We present in Table 6.4 the lifetimes of the different WR phases through which all our VMS models evolve. At solar metallicity, the WR phase of non-rotating stellar models for masses between 150 and 500 $M_\odot$ covers between 16 and 38 % of the total stellar lifetime. This is a significantly larger proportion than for masses between 20 and 120 $M_\odot$, where the WR phase covers only 0–13 % of the total stellar lifetimes. At the LMC metallicity, the proportion of the total stellar lifetime spent

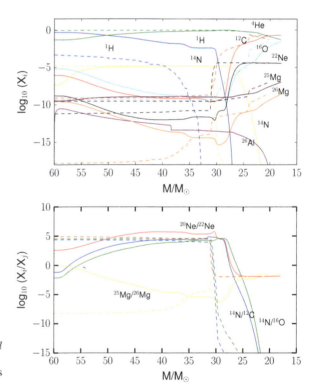

Fig. 6.12 Evolution of surface abundances of the solar metallicity rotating 150 $M_\odot$ (*solid*) and 60 $M_\odot$ (*dashed*) rotating solar Z models as a function of total mass (evolution goes from left to right since mass loss peels off the star and reduces the total mass). The *top panel* shows individual abundances while the *bottom panel* shows abundance ratios

Table 6.4 Lifetimes of the various phases in units of years

M_{ini}	Z_{ini}	$\frac{v_{\text{ini}}}{v_{\text{crit}}}$	O-star	WR	WNL	WNE	WN/WC	WC (WO)
120	0.014	0	2.151e06	3.959e05	1.150e05	9.390e03	2.675e02	2.715e05
150	0.014	0	2.041e06	4.473e05	1.777e05	5.654e03	7.120e02	2.639e05
200	0.014	0	1.968e06	5.148e05	2.503e05	1.773e03	4.576e02	2.626e05
300	0.014	0	1.671e06	8.014e05	5.051e05	9.217e03	2.735e03	2.870e05
500	0.014	0	1.286e06	8.848e05	5.804e05	1.079e04	3.279e03	2.935e05
120	0.014	0.4	2.289e06	1.227e06	8.790e05	4.118e04	4.008e03	3.076e05
150	0.014	0.4	2.105e06	1.189e06	8.567e05	2.579e04	3.649e03	3.068e05
200	0.014	0.4	1.860e06	1.164e06	8.375e05	2.242e04	3.153e03	3.042e05
300	0.014	0.4	1.585e06	1.152e06	8.315e05	1.897e04	2.897e03	3.015e05
500	0.014	0.4	1.422e06	1.083e06	7.663e05	1.830e04	2.899e03	2.990e05
120	0.006	0	2.222e06	2.964e05	2.043e05	1.302e02	6.025e02	9.202e04
150	0.006	0	2.028e06	3.320e05	1.579e05	1.211e03	2.921e02	1.728e05
500	0.006	0	1.388e06	5.362e05	2.690e05	5.211e03	1.350e03	2.620e05
120	0.006	0.4	2.513e06	9.624e05	6.776e05	1.601e04	3.386e03	2.687e05
150	0.006	0.4	2.188e06	9.789e05	6.912e05	2.172e04	2.336e03	2.660e05
200	0.006	0.4	1.922e06	9.848e05	7.073e05	1.347e04	2.757e03	2.640e05
300	0.006	0.4	1.644e06	9.838e05	7.033e05	1.600e04	9.744e02	2.644e05
500	0.006	0.4	1.461e06	9.283e05	6.647e05	9.312e03	6.853e02	2.542e05
150	0.002	0.4	2.583e06	6.119e05	3.691e05	8.459e03	4.874e03	2.343e05
200	0.002	0.4	2.196e06	6.926e05	4.524e05	1.019e04	2.709e03	2.300e05
300	0.002	0.4	1.827e06	7.602e05	5.186e05	1.317e04	1.289e03	2.283e05

as a WR phase for VMS decreases to values between 12% (150 $M_\odot$) and 25% (500 $M_\odot$).

Figure 6.13 shows how these lifetimes vary as a function of mass for non-rotating and rotating solar metallicity models. Looking first at the non-rotating models (Fig. 6.13, *left*), we see that the very massive stars (above 150 $M_\odot$) have WR lifetimes between 0.4 and nearly 1 My. The longest WR phase is the WNL phase since these stars spend a large fraction of H-burning in this phase. The duration of the WC phases of VMS is not so much different from those of stars in the mass range between 50 and 120 $M_\odot$.

Rotation significantly increases the WR lifetimes. Typically, the WR phase of rotating stellar models for masses between 150 and 500 $M_\odot$ covers between 36 and 43 % of the total stellar lifetime. The increase is more important for the lower mass range plotted in the figures. This reflects the fact that for lower initial mass stars, mass loss rates are weaker and thus the mixing induced by rotation has a greater impact. We see that this increase is mostly due to longer durations for the WNL phase, the WC phase duration remaining more or less constant for the whole mass range between 50 and 500 $M_\odot$ as was the case for the non rotating models. Rotation has qualitatively similar effects at the LMC metallicities.

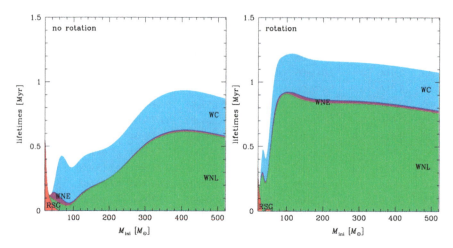

Fig. 6.13 Lifetimes of the RSG phase and of the different WR phases for the solar metallicity non-rotating (*left*) and rotating (*right*) models. Lifetimes are piled up. For example, the lifetime of the WNE phase extent corresponds to the height of the purple area

Would the account of the VMS stars in the computation of the number ratios of WR to O-type stars and on the WN/WC ratios have a significant effect? The inclusion of VMS is marginal at solar metallicity, since the durations are only affected by a factor 2. Convoluted with the weighting of the initial mass function (IMF), WR stars originating from VMS only represent $\sim 10\%$ of the whole population of WR stars (using a Salpeter 1955, IMF) originating from single stars. However, the situation is different at SMC metallicity. Due to the weakness of the stellar winds, single stellar models below $120\,M_\odot$ at this Z do not produce any WC or WO stars (Georgy et al. in prep.). In that case, we expect that the few WC/WO stars observed at low metallicity come from VMS, or from the binary channel (Eldridge et al. 2008). In starburst regions, the detection of WR stars at very young ages would also be an indication that they come from VMS, as these stars enter the WR phase before their less massive counterparts, and well before WRs coming from the binary channel.

We see in Fig. 6.14 that VMS models well fit the most luminous WNL stars. On the other hand, they predict very luminous WC stars. Of course the fact that no such luminous WC stars has ever been observed can simply come from the fact that such stars are very rare and the lifetime in the WC phase is moreover relatively short.

Fig. 6.14 The positions of WR stars observed by Hamann et al. (2006) and Sander et al. (2012) are indicated with the rotating evolutionary tracks taken from Ekström et al. (2012) for masses up to 120 $M_\odot$ and from Yusof et al. (2013) for VMS

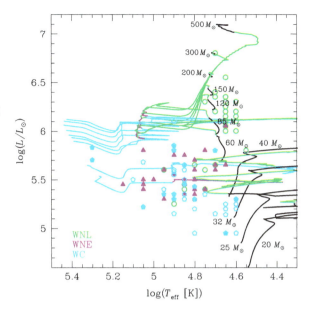

6.5 Late Evolution and Pre-SN Properties of Very Massive Stars

The next chapter discusses the explosion that will take place at the end of VMS life but whether or not a star produces a pair-instability supernova (PISN, aka pair-creation SN, PCSN) can be reasonably estimated from the mass of its carbon-oxygen (CO) core as demonstrated by the similar fate for stars with the same CO core found in various studies of VMS in the early Universe (Bond et al. 1984; Heger and Woosley 2002; Chatzopoulos and Wheeler 2012; Dessart et al. 2013), even if their prior evolution is different. In this section, we will thus use the CO core mass to estimate the fate of the models discussed in the previous sections.[3] We will only briefly discuss the supernova types that these VMS may produce in this chapter as this is discussed in Chaps. 7 and 8.

---

[3]Note that for lower-mass massive stars ($\lesssim 50\,M_\odot$), the CO core mass alone is not sufficient to predict the fate of the star and other factors like compactness, rotation and the central carbon abundance at the end of helium burning also play a role (see e.g. Chieffi and Limongi 2013).

6.5.1 Advanced Phases, Final Masses and Masses of Carbon-Oxygen Cores

In Fig. 6.15, the structure evolution diagrams are drawn as a function of the log of the time left until the last model calculated (as opposed to age in Fig. 6.4). This choice of x-axis allows one to see the evolution of the structure during the advanced stages. In the *left* panel, we can see that, at solar metallicity, VMS have an advanced evolution identical to lower mass stars (see e.g. Fig. 12 in Hirschi et al. 2004a) with a radiative core C-burning followed by a large convective C-burning shell, radiative neon burning and convective oxygen and silicon burning stages. All the solar metallicity models will eventually undergo core collapse after going through the usual advanced burning stages. As presented in Table 6.2 (column 9), the central mass fraction of ^{12}C is very low in all VMS models and is anti-correlated with the total mass at the end of helium burning (column 6): the higher the total mass, the lower the central ^{12}C mass fraction. This is due to the higher temperature in more massive cores leading to a more efficient ^{12}C$(\alpha, \gamma)^{16}$O relative to 3α.

The similarities between VMS and lower mass stars at solar metallicity during the advanced stages can also be seen in the central temperature versus central density diagram (see Fig. 6.16). Even the evolution of the 500 $M_\odot$ rotating model is close to that of the 60 $M_\odot$ model. The non-rotating models lose less mass as described above and thus their evolutionary track is higher (see e.g. the track for the non-rotating 150 $M_\odot$ model in Fig. 6.16). Non-rotating models nevertheless stay clear of the pair-instability region ($\Gamma < 4/3$, where Γ is the adiabatic index) in the centre.

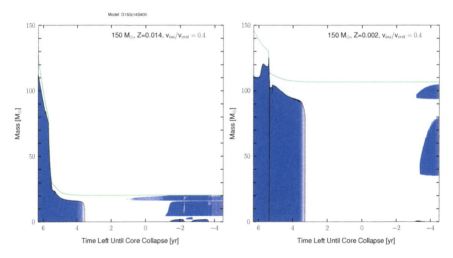

Fig. 6.15 Structure evolution diagram for rotating 150 $M_\odot$ at solar and SMC metallicities as a function of the log of the time left until the last model. The *blue zones* represent the convective regions and the *top solid line* the total mass

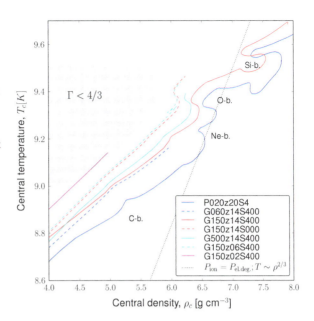

Fig. 6.16 Evolution of the central temperature T_c versus central density ρ_c for the rotating 20 (from Hirschi et al. 2004a), 60 (from Ekström et al. 2012), 150 and 500 $M_\odot$ models and non-rotating 150 $M_\odot$ model at solar metallicity as well as the rotating 150 $M_\odot$ model at SMC metallicity. The *gray shaded area* is the pair-creation instability region ($\Gamma < 4/3$, where Γ is the adiabatic index). The *additional dotted line* corresponds to the limit between non-degenerate and degenerate electron gas

The situation is quite different at SMC metallicity (see Fig. 6.15, *right* panel). Mass loss is weaker and thus the CO core is very large (93.5 $M_\odot$ for this 150 $M_\odot$ model). Such a large core starts the advanced stages in a similar way: radiative core C-burning followed by a large convective C-burning shell and radiative neon burning. The evolution starts to diverge from this point onwards. As can be seen in T_c vs ρ_c plot, the SMC 150 $M_\odot$ model enters the pair-instability region. These models will thus have a different final fate than those at solar metallicity (see below).

Figure 6.17 (see also Table 6.3) shows the final masses of VMS as a function of the initial masses. All models at solar Z, rotating or not, end with a small fraction of their initial mass due to the strong mass loss they experience. Rotation enhances mass loss by allowing the star to enter the WR phase earlier during the MS (see *top* panels of Fig. 6.4) and the final mass of non-rotating models is generally higher than that of rotating models. At low metallicities, due to the metallicity dependence of radiatively-driven stellar winds in both O-type stars (Vink et al. 2001a) and WR stars (Eldridge and Vink 2006), final masses are larger.

Figure 6.18 shows how the CO core masses vary as a function of the initial mass, rotation and metallicity. The CO core (M_{CO}) is here defined as the core mass for which the mass fraction of C+O is greater than 75 %. Since the CO core mass is so close to the total mass, the behavior is the same as for the total mass and for the same reasons. For the rotating solar metallicity models, mass loss is so strong that all models end with roughly the same CO core mass around 20 $M_\odot$. As the metallicity decreases, so does mass loss and thus the LMC and SMC models have higher final CO core masses and the CO core mass does depend on the initial mass in a monotonous way. Finally, non-rotating models lose less mass than their rotating

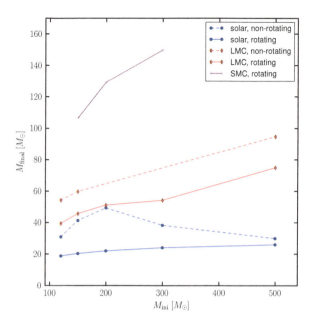

Fig. 6.17 Final mass versus initial mass for all rotating (*solid lines*) and non-rotating (*dashed line*) models

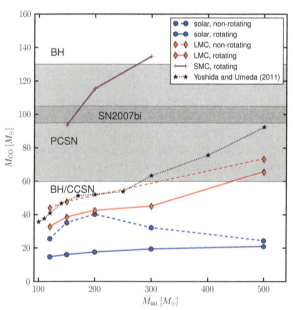

Fig. 6.18 Mass of carbon-oxygen core of all the models as a function of the initial mass. The *light grey shaded area* represents the range of M_{CO}, for which the estimated fate is a PISN. The *thin dark grey shaded area* corresponds to the estimated M_{CO} of the progenitor of SN2007bi assuming it is a PISN (see text for more details). The points linked by the *dotted black line* are from the models of Yoshida and Umeda (2011) at $Z = 0.004$, case A

counterpart since they enter the WR phase later and also have less hot surface. Simulations at $Z = 0.004$ from Yoshida and Umeda (2011) (case A) are also plotted in Fig. 6.18. The CO core masses they obtain are consistently slightly larger than for the LMC ($Z = 0.006$) models.

6.5.2 Do VMS Produce PISNe?

As mentioned above, the core masses, especially the CO core masses, can be used to estimate whether or not models produce a Pair Instability SuperNova (PISN) by using the results of previous studies, which follow the explosion of such massive cores and knowing that VMS with the same CO core masses have similar core evolution from carbon burning onwards. Heger and Woosley (2002) calculated a grid of models and found that stars with helium cores (M_α) between 64 and 133 $M_\odot$ produce PISNe and that stars with more massive M_α will collapse to a BH without explosion, confirming the results of previous studies, such as Bond et al. (1984). The independent results of Chatzopoulos and Wheeler (2012) also confirm the CO core mass range that produce PISNe.

PISNe occur when very massive stars (VMS) experience an instability in their core during the neon/oxygen burning stage due to the creation of electron-positron pairs out of two photons. The creation of pairs in their oxygen-rich core softens the equation of state, leading to further contraction. This runaway collapse is predicted to produce a very powerful explosion, in excess of 10^{53} erg, disrupting the entire star and leaving no remnant (Bond et al. 1984; Fryer et al. 2001).

Heger and Woosley (2002) also find that stars with M_α between roughly 40 and 63 $M_\odot$ will undergo violent pulsations induced by the pair-instability leading to strong mass loss but which will not be sufficient to disrupt the core. Thus these stars will eventually undergo core collapse as lower mass stars. Since in our models, the CO core masses are very close to M_α (equal to the final total mass in our models, see Table 6.5), in this chapter we assume that models will produce a PISN if 60 $M_\odot \leq M_{CO} \leq 130\,M_\odot$. In Fig. 6.18, the light grey shaded region corresponds to the zone where one would expect a PISN, the dark shaded region shows the estimated range of the carbon oxygen core of the progenitor of SN2007bi (see Yusof et al. 2013, for more details).

We see in Fig. 6.18 that at solar metallicity none of the models is expected to explode as a PISN. At the metallicity of the LMC, only stars with initial masses above 450 for the rotating models and above about 300 $M_\odot$ for the non-rotating case are expected to explode as a PISN. At the SMC metallicity, the mass range for the PISN progenitors is much more favorable. Extrapolating the points obtained from our models we obtain that all stars in the mass range between about 100 $M_\odot$ and 290 $M_\odot$ could produce PISNe. Thus these models provide support for the occurrence of PISNe in the nearby (not so metal poor) universe.

Table 6.5 presents for each of the models, the initial mass ($M_{\rm ini}$), the amount of helium left in the star at the end of the calculation ($M_{\rm He}^{\rm env}$), and final total mass as well as the estimated fate in terms of the explosion type: PISN or core-collapse supernova and black hole formation with or without mass ejection (CCSN/BH). The helium core mass (M_α) is not given since it is always equal to the final total mass, all the models having lost the entire hydrogen-rich layers.

Table 6.5 Initial masses, mass content of helium in the envelope, mass of carbon-oxygen core, final mass in solar masses and fate of the models estimated from the CO core mass

M_{ini}	Non-rotating				Rotating			
	M_{He}^{env}	M_{co}	M_{final}	Fate	M_{He}^{env}	M_{co}	M_{final}	Fate
Z=0.014								
120	0.4874	25.478	30.8	CCSN/BH	0.5147	18.414	18.7	CCSN/BH
150	0.6142	35.047	41.2	CCSN/BH	0.5053	19.942	20.2	CCSN/BH
200	0.7765	42.781	49.3	CCSN/BH	0.5101	21.601	21.9	CCSN/BH
300	0.3467	32.204	38.2	CCSN/BH	0.4974	19.468	23.9	CCSN/BH
500	0.3119	24.380	29.8	CCSN/BH	0.5675	20.993	25.8	CCSN/BH
Z=0.006								
120	1.2289	43.851	54.2	CCSN/BH	0.5665	32.669	39.2	CCSN/BH
150	1.1041	47.562	59.7	CCSN/BH	0.7845	38.436	45.6	CCSN/BH
200	–	–	–	CCSN/BH	0.5055	42.357	51.0	CCSN/BH
300	–	–	–	CCSN/BH	0.5802	44.959	54.0	CCSN/BH
500	1.6428	92.547	94.7	PISN	0.7865	73.145	74.8	PISN
Z=0.002								
150	–	–	–	–	2.3353	93.468	106.5	PISN
200	–	–	–	–	3.3022	124.329	129.2	PISN
300	–	–	–	–	5.5018	134.869	149.7	BH

6.5.3 Supernova Types Produced by VMS

Let us recall that, in VMS, convective cores are very large. It is larger than 90 % above $200\,M_\odot$ at the start of the evolution and even though it decreases slightly during the evolution, at the end of core H-burning, the convective core occupies more than half of the initial mass in non-rotating models and most of the star in rotating models. This has an important implication concerning the type of supernovae that these VMS will produce. Indeed, even if mass loss is not very strong in SMC models, all the models calculated have lost the entire hydrogen rich layers long before the end of helium burning. Thus the models predict that all VMS stars in the metallicity range studied will produce either a type Ib or type Ic SN but no type II.

6.5.4 GRBs from VMS?

The evolution of the surface velocity was described in Sect. 6.3.6. Only models at SMC retain a significant amount of rotation during their evolution (see angular momentum contained in the CO core at the end of helium burning in the last column of Table 6.2) but do they retain enough angular momentum for rotation to affect the fate of the star? The angular momentum profile of the SMC models is presented in Fig. 6.19. Note that the models presented in this section do not include the

6 Evolution and Nucleosynthesis of Very Massive Stars 191

Fig. 6.19 Specific angular momentum profile, j_m, as a function of the Lagrangian mass coordinate in the core of the SMC rotating 150, 200, 300 $M_\odot$ models, plotted at the end of the calculations (*solid line*). The *dash-dotted line* is $j_{Kerr,lso} = r_{LSO}\, c$ (Shapiro and Teukolsky 1983, p. 428), where the radius of the last stable orbit, r_{LSO}, is given by r_{ms} in formula (12.7.24) from Shapiro and Teukolsky (1983, p. 362) for circular orbit in the Kerr metric. $j_{Kerr,lso}$ is the minimum specific angular momentum necessary to form an accretion disc around a rotating black hole. $j_{Schwarzschild} = \sqrt{12}\, Gm/c$ (*dotted line*) is the minimum specific angular momentum necessary for a non-rotating *black hole*, for reference

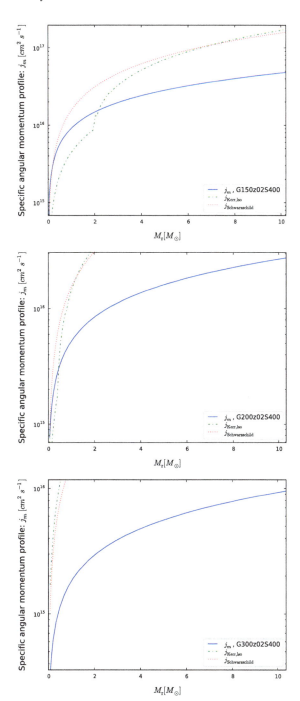

Tayler-Spruit dynamo so represent the most optimistic (highest possible) prediction concerning the angular momentum in the core of these models. Mass loss in the 300 $M_\odot$ model is too strong for the core to retain enough angular momentum for rotation to impact the death of this model. In the 200 $M_\odot$ model, and even more so in the 150 $M_\odot$ model, however, the central part of the core retain a significant amount of angular momentum that could potentially affect the death of the star. Since the role of rotation is very modest from carbon until just after the end of core silicon burning, even for extremely fast rotators (see e.g. Hirschi et al. 2005; Chieffi and Limongi 2013), we do not expect rotation to affect significantly the fate of stars that are predicted to explode as PISN during neon-oxygen burning. However, as discussed in Yoon et al. (2012, and references therein), the large angular momentum content is most interesting for the stars that just fall short of the minimum CO core mass for PISN (since fast rotation plays an important role during the early collapse Ott et al. 2004; O'Connor and Ott 2011; Chieffi and Limongi 2013). Indeed, without rotation, these stars would produce a BH following a possible pulsation pair-creation phase, whereas with rotation, these stars could produce energetic asymmetric explosions (GRBs or magnetars). Since the 150 $M_\odot$ model is predicted to explode as a PISN, we thus do not expect the models presented in this grid to produce GRBs or magnetars but such energetic asymmetric explosions are likely to take place in lower mass and lower metallicity stars (see Hirschi et al. 2005; Yoon and Langer 2005; Woosley and Heger 2006).

Yoon et al. (2012) calculated a grid of zero-metallicity rotating stars, including the Tayler-Spruit dynamo for the interaction between rotation and magnetic fields. They find that fast rotating stars with an initial mass below about 200 $M_\odot$ retain enough angular momentum in their cores in order to produce a collapsar ($j > j_{Kerr,Iso}$ Woosley 1993) or a magnetar (see e.g. Wheeler et al. 2000; Burrows et al. 2007; Dessart et al. 2012). Thus some VMS that do not produce PISNe might produce GRBs instead.

6.6 The Final Chemical Structure and Contribution to Galactic Chemical Evolution

Figure 6.20 shows the chemical structure at the last time steps calculated, which is the end of the carbon burning phase in the case of the 40 $M_\odot$, and the end of the core oxygen-burning phase in the case of the 150 and 500 $M_\odot$ models. A few interesting points come out from considering this figure. First, in all cases, some helium is still present in the outer layers. Depending on how the final stellar explosion occurs, this helium may or may not be apparent in the spectrum, as discussed in Yusof et al. (2013). Second, just below the He-burning shell, products of the core He-burning, not affected by further carbon burning are apparent. This zone extends between about 4 and 10 $M_\odot$ in the 40 $M_\odot$ model, between about 32 and 35 $M_\odot$ in the 150 $M_\odot$ model and in a tiny region centered around 24 $M_\odot$

6 Evolution and Nucleosynthesis of Very Massive Stars 193

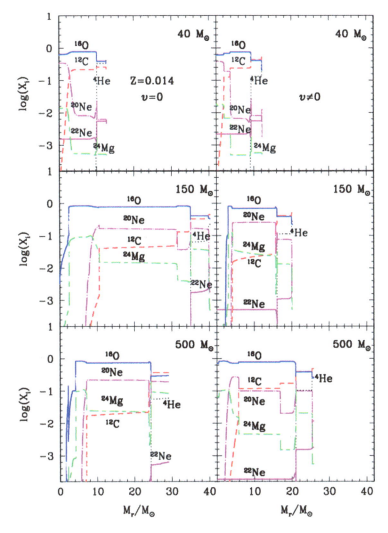

Fig. 6.20 Chemical structure of 40, 150 and 500 $M_\odot$ non-rotating (*left*) and rotating (*right*) models at $Z = 0.014$ at the end of the calculations. Note that the rotating 500 M model is shown at an earlier evolutionary stage than the corresponding non-rotating model

in the 500 $M_\odot$ model. We therefore see that this zone decreases in importance when the initial mass increases. Interestingly, the chemical composition in this zone present striking differences if we compare for instance the 40 $M_\odot$ and the 500 $M_\odot$ model. We can see that the abundance of ^{20}Ne is much higher in the more massive model. This comes from the fact that in more massive stars, due to higher central temperatures during the core He-burning phase the reaction ^{16}O(α, γ)^{20}Ne is more active, building thus more ^{20}Ne. Note that ^{24}Mg is also more abundant, which is natural since the reaction ^{20}Ne(α, γ)^{24}Mg reaction will also be somewhat active in

VMS for the same reasons. While in the case of the $150\,M_\odot$, due to the mass loss history, the ^{20}Ne and ^{24}Mg-rich layers are not uncovered, they are uncovered in the $500\,M_\odot$ model. This implies that strong overabundances of these two isotopes at the surface of WC stars can be taken as a signature for an initially very massive stars as the progenitor of that WC star. It also means that, contrary to what occurs at the surface of WC stars originating from lower initial mass stars, neon is no longer present mainly in the form of ^{22}Ne (and thus be a measure of the initial CNO content since resulting from the transformation of nitrogen produced by CNO burning during the H-burning phase) but will mainly be present in the form of ^{20}Ne.

Rotation does not change much this picture (see right panel of Fig. 6.20), except that, due to different mass loss histories, the rotating models lose much more mass and end their evolution with smaller cores. This is particularly striking for the $150\,M_\odot$ model. Qualitatively the situation is not much different at lower metallicities.

As stars in the mass range 50–100 $M_\odot$ (see e.g. Meynet and Maeder 2005; Chieffi and Limongi 2013), VMS eject copious amount of H-burning products through their stellar winds and to a lesser extent He-burning products. The potential difference between VMS and stars in the mass range 50–100 $M_\odot$ is how they explode (or not) at the end of their life, which is discussed in detail in the next chapter. If they collapse to a black hole the contribution from the supernova will be negligible whereas if they explode as PISNe they will produce large amounts of iron and other heavy elements.

Do VMS contribute to the chemical enrichment of galaxies or are VMS so rare that whatever their evolution, their impact on energy and mass outputs will anyway be very low? Considering a Salpeter IMF, the number of stars with masses between 120 and $500\,M_\odot$ corresponds to only about 2 % of the total number of stars with masses between 8 and $500\,M_\odot$. So they are indeed only very few! On the other hand, one explosion can release a great amount of energy and mass into the interstellar medium. Typically a $200\,M_\odot$ star releases about ten times more mass than a $20\,M_\odot$ star. If we roughly suppose that for hundred $20\,M_\odot$ stars there are only two $200\,M_\odot$ star, this means that the $200\,M_\odot$ stars contribute to the release of mass at a level corresponding to about 20 % of the release of mass by $20\,M_\odot$, which is by far not negligible. Of course this is a rough estimate but, as a rule of thumb we can say that any quantity released by a VMS $\sim$ tenfold intensity compared to that of a typical, $20\,M_\odot$ star will make a non-negligible difference in the overall budget of this quantity at the level of a galaxy. For instance, the high bolometric luminosities, stellar temperatures and mass loss rates of VMS imply that they will contribute significantly to the radiative and mechanical feedback from stars in high mass clusters at ages prior to the first supernovae (Crowther et al. 2010). Core-collapse SNe produce of the order of $0.05\,M_\odot$ (ejected masses) of iron, $1\,M_\odot$ of each of the α−elements. According to the production factors in Table 6.4 in Heger and Woosley (2002), PISN produce up to $40\,M_\odot$ of iron, of the order of $30\,M_\odot$ of oxygen and silicon and of the order of 5–$10\,M_\odot$ of the other α−elements. Considering that PISN may occur up to SMC metallicity and represent 2 % of SNe at a given metallicity, their contribution to the chemical enrichment of galaxies may be significant, especially in the case of iron, oxygen and silicon. If the IMF is top heavy at low metallicities, the impact of VMSs would be even larger.

Summary and Conclusion

In this chapter, we have discussed the evolution of very massive stars based on stellar evolution models at various metallicities. The main properties of VMS are the following:

- VMS possess very large convective cores during the MS phase. Typically, in a 200 $M_\odot$ model on the ZAMS the convective core extends over more than 90 % of the total mass.
- Since the mass-luminosity relation flattens above 20 $M_\odot$, VMS have lifetimes that are not very sensitive to their initial mass and range between 2 and 3.5 million years.
- Even in models with no rotation, due to the importance of the convective core, VMS evolve nearly chemically homogeneously.
- Most of the very massive stars (all at solar Z) remain in blue regions of the HR diagram and do not go through a luminous blue variable phase.
- They all enter into the WR phase and their typical evolution is Of – WNL - WNE - WC/WO.
- Due to increasing mass loss rates with the mass, very different initial mass stars end with similar final masses. As a consequence very different initial masses may during some of their evolutionary phases occupy very similar positions in the HRD.
- A significant proportion of the total stellar lifetimes of VMS is spent in the WR phase (about a third).
- A WC star with high Ne (^{20}Ne) and Mg (^{24}Mg) abundances at the surface has necessarily a VMS as progenitor.
- At solar metallicity VMS are not expected to explode as PISNe because mass loss rates are too high.
- Whether or not some VMS models retain enough mass to produce a PISN at low metallicity is strongly dependent on mass loss. As discussed above, models that retain enough mass at SMC metallicity (and below) also approach very closely the Eddington limit after helium burning and this might trigger a strong enough mass loss in order to prevent any VMS from producing a PISN.
- Most VMS lose the entire hydrogen rich layers long before the end of helium burning. Thus most VMS stars near solar metallicity are expected to produce either a type Ib or type Ic SN but no type II.
- Models near solar metallicity are not expected to produce GRBs or magnetars. The reason for that is that either they lose too much angular momentum by mass loss or they avoid the formation of a neutron star or

(continued)

BH because they explode as PISN. Lower mass stars at low metallicities ($Z \lesssim 0.002$), however, may retain enough angular momentum as in metal free stars (see Yoon et al. 2012; Chatzopoulos and Wheeler 2012) for rotation (and magnetic fields) to play a significant role in their explosion.

Even though VMS are rare, their extreme luminosities and mass loss will still contribute significantly to the light and chemistry budget of their host galaxies. And although many VMS will die quietly and form a black hole, some VMS may die as PISNe or GRBs. Thus the extreme properties of VMS compensate for their rarity and they are worth studying and considering in stellar population and galactic chemical evolution studies. As discussed above, the main uncertainty that strongly affects their evolution and their fate is the uncertainty in mass loss, especially for stars near the Eddington limit. Thus, although the discussions and conclusions presented in this chapter will remain qualitatively valid, quantitative results will change as our knowledge of mass loss in these extreme stars improves.

Acknowledgements The author thanks his collaborators at the University of Keele (C. Georgy), Geneva (G. Meynet, A. Maeder and Sylvia Ekström) and Malaysia (N. Yusof and H. Kassim) for their significant contributions to the results presented in this chapter. R. Hirschi acknowledges support from the World Premier International Research Center Initiative (WPI Initiative), MEXT, Japan and from the Eurogenesis EUROCORE programme. The research leading to these results has received funding from the European Research Council under the European Union's Seventh Framework Programme (FP/2007–2013)/ERC Grant Agreement n. 306901.

References

Abel, T., Bryan, G. L., & Norman, M. L. (2002). *Science, 295*, 93
Bennett, M. E., Hirschi, R., Pignatari, M., et al. (2012). *Monthy Notices of the Royal Astronomical Society, 420*, 3047
Bond, J. R., Arnett, W. D., & Carr, B. J. (1984). *Astrophysics Journal, 280*, 825
Brüggen, M., & Hillebrandt, W. (2001). *Monthy Notices of the Royal Astronomical Society, 323*, 56
Braithwaite, J. (2006). *Astronomy and Astrophysics, 449*, 451
Bromm, V., Coppi, P. S., & Larson, R. B. (1999). *Astrophysics Journal Letters, 527*, L5
Burrows, A., Dessart, L., Livne, E., Ott, C. D., & Murphy, J. (2007). *Astrophysics Journal, 664*, 416
Canuto, V. M. (2002). *Astronomy and Astrophysics, 384*, 1119
Chaboyer, B., & Zahn, J.-P. (1992). *Astronomy and Astrophysics, 253*, 173
Chatzopoulos, E., & Wheeler, J. C. (2012). *Astrophysics Journal, 748*, 42
Chieffi, A., & Limongi, M. (2013). *Astrophysics Journal, 764*, 21
Chieffi, A., Limongi, M., & Straniero, O. (1998). *Astrophysics Journal, 502*, 737
Christlieb, N., Bessell, M. S., Beers, T. C., et al. (2002). *Nature, 419*, 904

Crowther, P. A. (2001). In D. Vanbeveren (Ed.), *Astrophysics and Space Science Library* (The Influence of Binaries on stellar population studies, Vol. 264, p. 215). ISBN: 0792371046, Springer.
Crowther, P. A., Hirschi, R., Walborn, N., & N., Y. (2012). In L. Drissen, C. Robert, N. St-Louis, & A. Moffat (Eds.) *Four decades of research on massive stars* (Astronomical Society of the Pacific, San Fransisco, conference series)
Crowther, P. A., Schnurr, O., Hirschi, R., et al. (2010). *Monthy Notices of the Royal Astronomical Society, 408*, 731
de Jager, C., Nieuwenhuijzen, H., & van der Hucht, K. A. (1988a). *Astronomy and Astrophysics Supplement, 72*, 259
de Jager, C., Nieuwenhuijzen, H., & van der Hucht, K. A. (1988b). *Astronomy and Astrophysics Supplement, 72*, 259
Dessart, L., O'Connor, E., & Ott, C. D. (2012). *Astrophysics Journal, 754*, 76
Dessart, L., Waldman, R., Livne, E., Hillier, D. J., & Blondin, S. (2013). *Monthy Notices of the Royal Astronomical Society, 428*, 3227
Eggenberger, P., Meynet, G., Maeder, A., et al. (2007). *Astrophysics and Space Science, 263*
Ekström, S., Georgy, C., Eggenberger, P., et al. (2012). *Astronomy and Astrophysics, 537*, A146
Ekström, S., Meynet, G., Chiappini, C., Hirschi, R., & Maeder, A. (2008). *Astronomy and Astrophysics, 489*, 685
Eldridge, J. J., Izzard, R. G., & Tout, C. A. (2008). *Monthy Notices of the Royal Astronomical Society, 384*, 1109
Eldridge, J. J., & Vink, J. S. (2006). *Astronomy and Astrophysics, 452*, 295
Figer, D. F. (2005). *Nature, 434*, 192
Frebel, A., Aoki, W., & Christlieb, N., e. a. (2005). *Nature, 434*, 871
Fricke, K. (1968). *Zeitschrift fur Astrophysics, 68*, 317
Frischknecht, U., Hirschi, R., Meynet, G., et al. (2010). *Astronomy and Astrophysics, 522*, A39
Fryer, C. L., Woosley, S. E., & Heger, A. (2001). *Astrophysics Journal, 550*, 372
Georgy, C., Ekström, S., Meynet, G., et al. (2012). *Astronomy and Astrophysics, 542*, A29
Georgy, C., Ekström, S., Eggenberger, P., et al. (2013). *Astronomy and Astrophysics, 558*, A103
Georgy, C., Meynet, G., & Maeder, A. (2011) *Astronomy and Astrophysics, 527*, A52
Goldreich, P., & Schubert, G. (1967). *Astrophysics Journal, 150*, 571
Gräfener, G., & Hamann, W.-R. (2008). *Astronomy and Astrophysics, 482*, 945
Greif, T. H., Glover, S. C. O., Bromm, V., & Klessen, R. S. (2010). *Astrophysics Journal, 716*, 510
Hamann, W.-R., Gräfener, G., & Liermann, A. (2006). *Astronomy and Astrophysics, 457*, 1015
Heger, A., Langer, N., & Woosley, S. E. (2000). *Astrophysics Journal, 528*, 368
Heger, A. & Woosley, S. E. (2002). *Astrophysics Journal, 567*, 532
Hirschi, R. (2007). *Astronomy and Astrophysics, 461*, 571
Hirschi, R. & Maeder, A. (2010). *Astronomy and Astrophysics, 519*, A16
Hirschi, R., Meynet, G., & Maeder, A. (2004a). *Astronomy and Astrophysics, 425*, 649
Hirschi, R., Meynet, G., & Maeder, A. (2004b). *Astronomy and Astrophysics, 425*, 649
Hirschi, R., Meynet, G., & Maeder, A. (2005). *Astronomy and Astrophysics, 443*, 581
Kippenhahn, R., & Weigert, A. (1990). *Stellar structure and evolution*. Berlin/New York, Springer.
Kippenhahn, R., Weigert, A., & Hofmeister, E. (1967). In B. Alder, S. Fernbach, & M. Rotenberg (Eds.), *Methods in Computational Physics* (Vol. 7). New York: Academic Press.
Knobloch, E., & Spruit, H. C. (1983). *Astronomy and Astrophysics, 125*, 59
Krtička, J., Owocki, S. P., & Meynet, G. (2011). *Astronomy and Astrophysics, 527*, A84
Langer, N. (2012). *ARA&A, 50*, 107
Maeder, A. (1980). *Astronomy and Astrophysics, 92*, 101
Maeder, A. (1997). *Astronomy and Astrophysics, 321*, 134
Maeder, A. (2003). *Astronomy and Astrophysics, 399*, 263
Maeder, A. (2009). *Physics, formation and evolution of rotating stars*. Berlin/Heidelberg, Springer.
Maeder, A., Georgy, C., Meynet, G., & Ekström, S. (2012). *Astronomy and Astrophysics, 539*, A110
Maeder, A., & Meynet, G. (2000). *Astronomy and Astrophysics, 361*, 159

Meynet, G., & Maeder, A. (2002). *Astronomy and Astrophysics, 390*, 561
Maeder, A., & Meynet, G. (2005). *Astronomy and Astrophysics, 440*, 1041
Maeder, A., & Meynet, G. (2012). *Reviews of Modern Physics, 84*, 25
Maeder, A., & Zahn, J. (1998). *Astronomy and Astrophysics, 334*, 1000
Meynet, G., & Maeder, A. (1997). *Astronomy and Astrophysics, 321*, 465
Meynet, G., & Maeder, A. (2005). *Astronomy and Astrophysics, 429*, 581
Muijres, L. E., de Koter, A., Vink, J. S., et al. (2011). *Astronomy and Astrophysics, 526*, A32
Nugis, T., & Lamers, H. J. G. L. M. (2000a). *Astronomy and Astrophysics, 360*, 227
Nugis, T., & Lamers, H. J. G. L. M. (2000b). *Astronomy and Astrophysics, 360*, 227
O'Connor, E., & Ott, C. D. (2011). *Astrophysics Journal, 730*, 70
Oey, M. S., & Clarke, C. J. (2005). *Astrophysics Journal Letters, 620*, L43
Ott, C. D., Burrows, A., Livne, E., & Walder, R. (2004). *Astrophysics Journal, 600*, 834
Paxton, B., Bildsten, L., Dotter, A., et al. (2011). *Astrophysics Journal Supplement Series, 192*, 3
Potter, A. T., Chitre, S. M., & Tout, C. A. (2012). *Monthy Notices of the Royal Astronomical Society, 424*, 2358
Salpeter, E. E. (1955). *Astrophysics Journal, 121*, 161
Sander, A., Hamann, W.-R., & Todt, H. (2012). *Astronomy and Astrophysics, 540*, A144
Schneider, F. R. N., Izzard, R. G., de Mink, S. E., et al. (2014). *Astrophysics Journal, 780*, 117
Shapiro, S. L. & Teukolsky, S. A. (1983). *Black Holes, White Dwarfs and Neutron Stars: The Physics of Compact Objects* (Research supported by the National Science Foundation, p. 663) New York: Wiley-Interscience. http://eu.wiley.com/WileyCDA/WileyTitle/productCd-0471873160.html
Spruit, H. C. (2002). *Astronomy and Astrophysics, 381*, 923
Stacy, A., Greif, T. H., & Bromm, V. (2010). *Monthy Notices of the Royal Astronomical Society, 403*, 45
Sylvester, R. J., Skinner, C. J., & Barlow, M. J. (1998). *Monthy Notices of the Royal Astronomical Society, 301*, 1083
Umeda, H., & Nomoto, K. (2002). *Astrophysics Journal, 565*, 385
van Loon, J. T., Groenewegen, M. A. T., de Koter, A., et al. (1999). *Astronomy and Astrophysics, 351*, 559
Vink, J. S., de Koter, A., & Lamers, H. J. G. L. M. (2001a). *Astronomy and Astrophysics, 369*, 574
Vink, J. S., de Koter, A., & Lamers, H. J. G. L. M. (2001b). *Astronomy and Astrophysics, 369*, 574
Vink, J. S., Gräfener, G., & Harries, T. J. (2011a). *Astronomy and Astrophysics, 536*, L10
Vink, J. S., Muijres, L. E., Anthonisse, B., de Koter, A., Gräfener, G., Langer, N., (2011b). Wind modelling of very massive stars up to 300 solar masses. *Astronomy and Astrophysics, 531*, A132. doi: 10.1051/0004-6361/201116614. Provided by the SAO/NASA Astrophysics Data System http://adsabs.harvard.edu/abs/2011A26A...531A.132V
von Zeipel, H. (1924). *Monthy Notices of the Royal Astronomical Society, 84*, 665
Walder, R., Folini, D., & Meynet, G. (2011). *Space Science Reviews, 125*
Wheeler, J. C., Yi, I., Höflich, P., & Wang, L. (2000). *Astrophysics Journal, 537*, 810
Woosley, S. E. (1993). *Astrophysics Journal, 405*, 273
Woosley, S. E. & Heger, A. (2006). *Astrophysics Journal, 637*, 914
Yoon, S.-C., Dierks, A., & Langer, N. (2012). *Astronomy and Astrophysics, 542*, A113
Yoon, S.-C. & Langer, N. (2005). *Astronomy and Astrophysics, 443*, 643
Yoon, S.-C., Langer, N., & Norman, C. (2006). *Astronomy and Astrophysics, 460*, 199
Yoshida, T. & Umeda, H. (2011). *Monthy Notices of the Royal Astronomical Society, 412*, L78
Yusof, N., Hirschi, R., Meynet, G., et al. (2013). *Monthy Notices of the Royal Astronomical Society, 433*, 1114
Zahn, J., Brun, A. S., & Mathis, S. (2007). *Astronomy and Astrophysics, 474*, 145
Zahn, J.-P. (1992). *Astronomy and Astrophysics, 265*, 115

Chapter 7
The Deaths of Very Massive Stars

Stan. E. Woosley and Alexander Heger

Abstract The theory underlying the evolution and death of stars heavier than 10 M$_\odot$ on the main sequence is reviewed with an emphasis upon stars much heavier than 30 M$_\odot$. These are stars that, in the absence of substantial mass loss, are expected to either produce black holes when they die, or, for helium cores heavier than about 35 M$_\odot$, encounter the pair instability. A wide variety of outcomes is possible depending upon the initial composition of the star, its rotation rate, and the physics used to model its evolution. These stars can produce some of the brightest supernovae in the universe, but also some of the faintest. They can make gamma-ray bursts or collapse without a whimper. Their nucleosynthesis can range from just CNO to a broad range of elements up to the iron group. Though rare nowadays, they probably played a disproportionate role in shaping the evolution of the universe following the formation of its first stars.

7.1 Introduction

Despite their scarcity, massive stars illuminate the universe disproportionately. They light up regions of star formation and stir the media from which they are born. They are the fountains of element creation that make life possible. The neutron stars and black holes that they make are characterized by extreme physical conditions that can never be attained on the earth. They are thus unique laboratories for nuclear physics, magnetohydrodynamics, particle physics, and general relativity. And they are never quite so fascinating as when they die.

Here we briefly review some aspects of massive star death. The outcomes can be crudely associated with three parameters – the star's mass, metallicity, and rotation rate. In the simplest case of no rotation and no mass loss, one can delineate five outcomes and assign approximate mass ranges (in some cases *very* approximate

---

S.E. Woosley (✉)
Department of Astronomy and Astrophysics, UCSC, Santa Cruz, CA 95064, USA
e-mail: woosley@ucolick.org

A. Heger
School of Mathematical Sciences, Monash University, Victoria 3800, Australia
e-mail: alexander.heger@monash.edu

mass ranges) for each. These masses then become the section heads for the first part of this chapter. (1) From 8 to 30 $M_\odot$ on the main sequence (presupernova helium core masses up to 12 $M_\odot$), stars mostly produce iron cores that collapse to neutron stars leading to explosions that make most of today's observable supernovae and heavy elements. Within this range there are probably islands of stars that either do not explode or explode incompletely and make black holes, especially for helium cores from 7 to 10 $M_\odot$. (2) From 30 to 80 $M_\odot$ (helium core mass 10–35 $M_\odot$), black hole formation is quite likely. Except for their winds, stars in this mass range may be nucleosynthetically barren. Again though there will be exceptions, especially when the effects of rotation during core collapse are included. (3) 80 to (very approximately) 150 $M_\odot$ (helium cores 35–63 $M_\odot$), pulsational-pair instability supernovae. Violent nuclear-powered pulsations eject the star's envelope and, in some cases, part of the helium core, but no heavy elements are ejected and a massive black hole of about 40 $M_\odot$ is left behind. (4) 150–260 $M_\odot$ (again very approximate for the main sequence mass range, but helium core 63–133 $M_\odot$), pair instability supernovae of increasing violence and heavy element synthesis. No gravitationally bound remnant is left behind. (5) Over 260 $M_\odot$ (133 $M_\odot$ of helium), with few exceptions, a black hole consumes the whole star. Rotation generally shifts the main sequence mass ranges (but not the helium core masses) downwards for each outcome. Mass loss complicates the relation between initial main sequence mass and final helium core mass.

The latter part of the paper deals with some possible effects of rapid rotation on the outcome. In the most extreme cases, gamma-ray bursts are produced, but even milder rotation can have a major affect on the light curve and hydrodynamics if a magnetar is formed.

7.2 The Deaths of Stars 8–80 $M_\odot$

7.2.1 Compactness as a Guide to Outcome

The physical basis for distinguishing stars that become supernovae rather than planetary nebulae, and that are therefore, in some sense, "massive", is the degeneracy of the carbon-oxygen (CO) core following helium core burning. Stars with dense, degenerate CO cores develop thin helium shells and eject their envelopes leaving behind stable white dwarfs, while heavier stars go on to burn carbon and heavier fuels. A mass around 8 $M_\odot$ is usually adopted for the transition point. The effects of degeneracy linger, however, on up to at least 30 $M_\odot$ at oxygen ignition, and to still heavier masses for silicon burning. Even at 80 $M_\odot$, the center of a massive star has become degenerate by silicon depletion.

Were the core fully degenerate and composed of nuclei with equal numbers of neutrons and protons, its maximum mass would be the cold Chandrasekhar mass, 1.38 $M_\odot$. This cold Chandrasekhar mass is altered however, both by electron capture reactions, which tend to reduce it, and the high temperatures necessary to burn oxygen and silicon, which increase it (Chandrasekhar 1939; Hoyle and Fowler 1960; Timmes et al. 1996). For main sequence stars from 8 to 80 $M_\odot$, the iron core mass at the time it collapses varies from about 1.3–2.3 $M_\odot$ (baryonic mass), with the larger values appropriate for more massive stars. Surrounding this degenerate core is a nested structure of shells that cause adjustments to the density structure. For very degenerate cores with energetic shells at their edges, the presupernova structure resembles that of an asymptotic giant branch star – a compact core surrounded by thin burning shells and a low density envelope with little gravitational binding energy. The matter outside of the iron core is easily ejected in such stars, and it is easy to make a supernova out of them, even with an inefficient energy source like neutrinos. Heavier stars with less degenerate cores and shells farther out, on the other hand, have a density that declines more slowly. These mantles of heavy elements, where ultimately most of the nucleosynthesis occurs, are more tightly bound and the star is more difficult to blow up.

O'Connor and Ott (2011) have defined a "compactness parameter", $\xi_{2.5} = 2.5/R_{2.5}$, that is a quantitative measure of this density fall off. Here $R_{2.5}$ is the radius, in units of 1,000 km, of the mass shell in the presupernova star that encloses 2.5 $M_\odot$. The fiducial mass is taken to be well outside the iron core but deep enough in to sample the density structure around that core. It makes little difference whether this compactness is evaluated at the onset of hydrodynamical instability or at core bounce (Sukhbold and Woosley 2014). Figure 7.1 shows $\xi_{2.5}$ as a function of main sequence mass for stars of solar metallicity. O'Connor and Ott and Ugliano et al. (2012) have both shown that it becomes difficult to explode the star by neutrino transport alone if $\xi_{2.5}$ becomes very large. The critical value is not certain and may vary with other properties of the star, but in Ugliano's study is usually 0.20–0.30. By this criterion, it may be difficult to explode stars in the 22–24 $M_\odot$ range (at least) as well as all stars above about 30 $M_\odot$ that do not lose substantial mass along the way to their deaths. The latter especially includes stars with very subsolar metallicity.

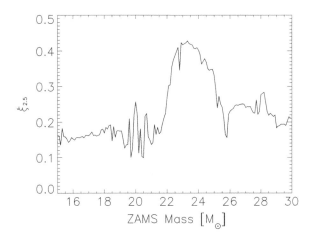

Fig. 7.1 Compactness parameter for presupernova stars of solar metallicity as a function of main sequence mass (Sukhbold and Woosley 2014). Stars with smaller $\zeta_{2.3}$ explode more easily

There are a number of caveats that go along with this speculation. The structure of a presupernova star is not fully represented by a single number and its compactness is sensitive to a lot of stellar physics, including the treatment of semiconvection and convective overshoot mixing and mass loss and the nuclear reaction rates employed (Sukhbold and Woosley 2014). Rotation and magnetic fields will change both the presupernova structure and its prospects for explosion by non-neutrino processes. Finally, the surveys of how neutrino-powered explosions depend on compactness have, so far, been overly simple and mostly in 1D, though see recent work by Janka et al. (2012), Janka (2012), Müller et al. (2012). Still the simplification introduced by this parametrization is impressive and reasonably consistent with what we know about the systematics of supernova progenitors.

Figure 7.1 suggests that stars below $22\,M_\odot$ should be, for the most part, easy to explode using neutrinos alone and no rotation. This is consistent with the observational limits that Smartt (2009) and Smartt et al. (2009) placed upon about a dozen presupernova progenitor masses as well as the estimated mass of SN 1987A. It also is a minimal set of masses if the solar abundances are to be produced (Brown and Woosley 2013). The compactness of stars between 22 and about $35\,M_\odot$ is highly variable though due to the migration outwards of the carbon and oxygen burning shells (Sukhbold and Woosley 2014). For a standard choice of stellar physics, there exists an island of compact cores between 26 and $30\,M_\odot$ that might allow for islands of "explodability". This would help with nucleosynthesis and also possibly have implications for the properties of the Cas A supernova remnant. Cas A, like SN 1993J and 2001gd (Chevalier and Soderberg 2010), is thought to be the remnant of a relatively massive single star that lost most of its hydrogenic envelope either to a wind or a binary companion, yet its remnant contains a neutron star. If the mass loss was to a companion, as is currently thought, then the progenitor mass was probably less than $20\,M_\odot$, but if a star of $30\,M_\odot$ could explode after losing most of its envelope, this might provide an alternate, solitary star explanation.

On the other hand, binary x-ray sources exist and the black holes in them are thought to be quite massive (Özel et al. 2010; Wiktorowicz et al. 2014). Stars above $35\,M_\odot$ either make black holes if their mass loss during the Wolf-Rayet stage is small, or some variant of Type Ibc supernovae if it is large and shrinks the carbon oxygen core below about $6\,M_\odot$.

Probably the greatest omission here is the effect of rotation and the need to produce gamma-ray bursts in a subset of stars. We also have said nothing about the fate of stars over $80\,M_\odot$. Both topics will be covered in later sections.

7.2.2 $8\text{--}30\,M_\odot$; Today's Supernovae and Element Factories

For reasonable choices of initial mass function, stars in this mass range are responsible for most of the supernovae we see today and for the synthesis of most of the heavy elements. This does not preclude many of these stars from making black holes, but the supernovae we see are in this range. Baring binary interaction, including mergers, or low metallicity, such stars are, at death, red supergiants, and so the most common supernovae are Type IIp. Explosion energies range from 0.5 to 4×10^{51} erg with a typical value of 9×10^{50} erg (Kasen and Woosley 2009). These values are the kinetic energy of all ejecta at infinity and the actual energy requirement for the central engine may be larger, especially for more massive stars with large binding energies in their mantles. The light curves and spectra of the models are consistent with observations, to the extent that models for SN IIp can even be used as "standard candles" based upon the expanding photosphere method.

Including binary interactions, one can account for the remainder of common (non-thermonuclear) supernovae, including Type Ib, Ic, IIb, etc. (Dessart et al. 2011, 2012). These events typically come from massive stars in the 12–18 $M_\odot$ range that lose their binary envelopes and die as stripped down helium cores of 3–4 $M_\odot$. On the low end, the explosion ejects too little ^{56}Ni to be a bright optical event. Heavier stars are rarer and may not explode. If they do their light curves are broader and fainter than typical Ib and Ic supernovae.

The nucleosynthesis produced by solar metallicity stars in this mass range has been explored many times (Woosley and Weaver 1995; Woosley et al. 2002; Woosley and Heger 2007; Thielemann et al. 1996; Nomoto et al. 2006; Limongi et al. 2000; Chieffi and Limongi 2004, 2013; Hirschi et al. 2005; Nomoto et al. 2013). While the results from the different groups studying the problem vary depending upon the treatment of critical reaction rates, mass loss, semiconvection, convective overshoot, and rotationally induced mixing, some general conclusions may be noted.

- The majority of the elements and their isotopes from carbon (Z = 6) through strontium (Z = 38) are made in solar proportions in supernovae with an average production factor of around 15 (IMF averaged yield expressed as a mass fraction and divided by the corresponding solar mass fraction). The iron group, Ti through Ni, is underproduced in massive stars by a factor of several, which is consistent with the premise that most of the solar abundances of these species were made recently in thermonuclear (Type Ia) supernovae. In the distant past, the oxygen to iron ratio was larger, and massive stars probably produced the iron group in very low metallicity stars.
- For a reasonable choice for the critical ^{22}Ne(α, n)^{25}Mg reaction rate, the light s-process up to A = 90 is made well in massive stars, but only if the upper bound for the masses of stars that explode is not too low (Brown and Woosley 2013). The heavy component of the p-process above A = 130 is also produced in massive stars, but the production of the lighter p-process isotopes (A = 90 – 130) remains a mystery, especially the origin of the abundant closed shell nucleus ^{92}Mo (Z = 42, N = 50).
- While oxygen is definitely a massive star product, the elemental yield of carbon (^{12}C) is sensitive to how mass loss is treated and requires for its production the inclusion of the winds of stars heavier than 30 $M_\odot$. Red giant winds, AGB mass loss, and planetary nebulae also produce ^{12}C, perhaps most of it, as well as all of

^{13}C and ^{14}N. ^{15}N and ^{17}O are not sufficiently produced in massive stars and may be made in classical novae.
- ^{11}B and about one-third of ^{19}F are made by neutrino spallation in massive star supernovae. ^{6}Li, ^{9}Be, and ^{10}B do not appear to be substantially made, and probably owe their origin to cosmic ray spallation in the interstellar medium. Some but not all of 7Li is made by neutrino spallation.
- Certain select nuclei like ^{44}Ca, ^{48}Ca, and ^{64}Zn are underproduced and may require alternate synthesis.

In addition to the previously mentioned uncertainties affecting presupernova evolution, assumptions about the explosion mechanism also play a major role. Fundamentally important is just which masses of stars eject their mantles of heavy elements and which collapse to black holes while ejecting little new elements. For a given presupernova structure, a shock that imparts $\sim 10^{51}$ erg of kinetic energy to the base of the ejecta, none of which fall back, will give a robust pattern of nucleosynthesis whether that energy is imparted by a piston or as a thermal "bomb". The approximation used by many, however, that the explosion across all masses can be parametrized by a constant kinetic energy at infinity is too crude and needs revisiting. Stars of different masses have different binding energies, compactness parameters, and iron core masses. Rotation probably has a major effect on the explosion, especially of the more massive stars. The next stage of modeling will need to take into account these dependencies.

7.2.3 Stars 30–80 $M_\odot$; Black Hole Progenitors

While the jury is still out regarding the mass-dependent efficiency of an explosion mechanism that includes realistic neutrino transport, rotation, magnetic fields, and relativity in three dimensions, the existence of stellar mass black holes and the absence of observable supernova progenitors with high mass implies that at least some stars do not explode and eject all of their heavy element inventory. Until such time as credible models exist, a reasonable assumption is that the success of the explosion is correlated with the compactness (O'Connor and Ott 2011; Ugliano et al. 2012). By this criterion, one expects the central regions of stars with *helium cores* much larger than about $10 M_\odot$ and lighter than $35 M_\odot$ to collapse (Sukhbold and Woosley 2014) to black holes. Above $10 M_\odot$ of helium, or about $30 M_\odot$ on the main sequence, the iron core is large, typically over $2.0 M_\odot$ and the compactness parameter is large. Above $35 M_\odot$, or about $80 M_\odot$ on the main sequence, one encounters the pulsational pair instability (Sect. 7.3).

For solar metallicity stars, mass loss may reduce the presupernova mass of the star to a level where it can frequently explode. If it does and the entire envelope has been lost, the explosion will be some sort of Type Ib or IC supernova. Because of the large mass, the light curve would be broad, and not as bright as most observed

SN Ibc. The remnant would probably be a neutron star. It is unclear if such events have been observed, though Cas A might be a candidate.

Even if the core of the star collapses to a black hole, its death is not necessarily nucleosynthetically barren or unobservable. The black hole could result from fall back and the envelope may still be ejected. Even if the presupernova star does not explode at all, its evolution will still have contributed to nucleosynthesis by its wind, which may be appreciable (Hirschi et al. 2005). If only the hydrogenic layers are ejected, these winds can be a rich source of ^{12}C, ^{16}O and, at low metallicity, ^{14}N (Meynet 2002). If the wind eats deeply into the helium core, ^{18}O and ^{22}Ne can also be ejected, but the winds of such stars are devoid of heavier elements like silicon and iron.

If the star rotates sufficiently rapidly, a gamma-ray burst may result (Sect. 7.6.2) or a magnetar-powered supernova. Even for non-rotating stars, it is debatable whether the star can simply disappear without a trace. The sudden loss of mass energy from the protoneutron star can trigger mass ejection and a very subluminous supernova (Lovegrove and Woosley 2013). Pulsations or gravity waves generated in the final stages of evolution may partly eject the envelope (Quataert and Shiode 2012). Even a weak explosion might produce a potentially observable bright spike as its shock wave erupts through the surface of the star (Piro 2013). In a tidally locked binary or a low metallicity blue supergiant with diminished mass loss, sufficient angular momentum may exist in the outermost layers of the star to pile up in an accretion disk around the new black hole producing some sort of x-ray and gamma-ray transient (Woosley and Heger 2012; Quataert and Kasen 2012).

7.2.4 Yesterday's Metal Poor Stars

Stars with lower metallicity, as may have predominated in the early universe, can have different presupernova structures for a variety of reasons (Sukhbold and Woosley 2014). Most importantly, metallicity affects mass loss, especially for the more massive stars. If the amount of mass lost is low or zero, the presupernova star including its helium core, is larger, and that has a dramatic effect on its compactness and explodability. A vastly different outcome is expected for e.g., a $60\,M_\odot$ star that retains most of its hydrogen envelope and dies with a helium core of $24\,M_\odot$, and one that loses all of its envelope as well as most of its helium core to die with a total mass of $7.3\,M_\odot$. This small mass is obtained with current estimates of mass loss for solar metallicity stars (Woosley et al. 2002). Indirect effects can also come into play. Because a low metallicity star loses less mass, it loses less angular momentum and thus dies rotating more rapidly. Indeed, there is some suggestion from theory that massive stars are all born rotating near break up and only slow as a consequence of evolution (expansion) and mass loss (Rosen et al. 2012).

Very low metallicity may also enhance the probability of forming more massive stars (Abel et al. 2002). Whether this results in much more massive stars than are being born today is being debated. While this is an important issue for the frequency of first generation stars with masses over 80 $M_\odot$ (Sect. 7.3), an equally important question is whether the IMF for the first generation stars might have been "bottom-light", that is producing a deficiency of stars below some characteristic mass, say ~30 $M_\odot$ (Tan and McKee 2004). Since this would remove the range of masses responsible for most supernovae and nucleosynthesis today, the early universe would have been quite a different place.

Even assuming the exact same masses of stars and explosions as today, nucleosynthesis would be distinctly different in low metallicity stars. The amount of neutrons available to produce all isotopes except those with Z = N depends on the "neutron excess", $\eta = \Sigma(N_i - Z_i)(X_i/A_i)$, where Z_i, N_i, and A_i are the proton number, neutron number and atomic weight of the species "i" and X_I is its mass fraction. At the end of hydrogen burning all CNO (essentially the metallicity of the star) has become ^{14}N. Early in helium burning this becomes ^{18}O by the reaction sequence ^{14}N$(\alpha, \gamma)^{18}$F$(e^+\nu)^{18}$O. The weak interaction here is critical as it creates a net neutron excess that persists throughout the rest of the star's life and limits the production of neutron rich isotopes (like ^{22}Ne, ^{26}Mg, ^{30}Si etc.) and odd-Z elements (like Na, Al, P). Other weak interactions in later stages of evolution also increase η, so that by the time one reaches calcium, the dependence on initial metallicity is not so great, but one does expect an effect on the isotopes from oxygen through phosphorus.

Assuming that the IMF was unchanged and using the same explosion model as for solar metallicity stars (but suppressing mass loss) gives an abundance set that agrees quite well with observations of metal-deficient stars in the range $-4 < [Z/Z_\odot] < -2$ (Lai et al. 2008). All elements from C through Zn are well fit without the need for a non-standard IMF or unusually high explosion energy.

Below $[Z/Z_\odot] = -4$, one becomes increasing sensitive to individual stellar events and to the properties of the first generation stars. If the stars below 30 $M_\odot$ are removed from the sample, the nucleosynthesis is set by (a) the pre-collapse winds of stars in the 30–80 $M_\odot$ range; (b) the results of rotationally powered explosions with uncertain characteristics; and (c) the contribution of pulsational pair- and pair-instability supernovae (see below). If only (a) and (c) contribute appreciably, the resulting nucleosynthesis could be CNO rich and very iron poor.

The light curves of metal deficient supernovae below 80 $M_\odot$ are likely to be different – some of the time. If the stars die as red supergiants, then very similar Type IIp supernovae will result, but more of the stars are expected to die as blue supergiants with light curves like SN 1987A (Heger and Woosley 2010). Rotation can alter this conclusion, however, as it tends to increase the number of red supergiants compared with blue (Maeder and Meynet 2012). To the extent that the massive stars retain their hydrogenic envelope, Type Ib and Ic supernovae will be suppressed, though of course a binary channel remains a possibility.

7.3 Pulsational Pair Instability Supernovae (80–150 $M_\odot$)

The pair instability occurs during the advanced stages of massive stellar evolution when sufficiently high temperature and low density lead to a thermal concentration of electron-positron pairs sufficient to have a significant effect on the equation of state. Only the most massive stars have sufficiently high entropy to encounter this instability. Making the rest mass of the pairs in a post-carbon burning star takes energy that might have otherwise contributed to the pressure. As a result, for a time, the pressure does not rise rapidly enough in a contracting stellar core to keep pace with gravity. The structural adiabatic index of the core dips below 4/3 and, depending on the strength of the instability, the core contracts more or less rapidly to higher temperature, developing considerable momentum as it does so. As temperature rises, carbon, oxygen and, in some cases, silicon burn rapidly. The extra energy from this burning, plus the eventual partial recovery from the instability when the pairs become highly relativistic, causes the pressure to rebound fast enough to slow the collapse. If enough burning occurs before the infall momentum becomes too great, the collapse is reversed and an explosion is possible. For stars that are too big though, specifically for helium non-rotating cores above $133\,M_\odot$, the collapse continues to a black hole.

When an explosion happens, it can be of two varieties. If enough burning occurs to unbind the star in a single pulse, a "pair-instability supernova" results (Sect. 7.4). If not, the core of the star expands violently for a time and may kick off its outer layers, including any residual hydrogen envelope. It then slowly contracts until the instability is encountered again and the core pulses once more. The process continues until enough mass has been ejected and entropy lost as neutrinos that the pair instability is finally avoided and the remaining star evolves smoothly to iron core collapse. Typically this requires a reduction of the helium and heavy element core mass to below $40\,M_\odot$. These repeated thermonuclear outbursts can have energies ranging from "mild", barely able to eject even the loosely bound hydrogen envelope of a red supergiant, to extremely large, with over 10^{51} erg in a single pulse. On the high energy end, collisions of ejected shells can produce very bright transients. The observational counterpart is "pulsational pair-instability supernovae" (PPSN).

Depending upon rotation, the electron-positron pair instability begins to have a marked effect on the post-carbon burning evolution of massive stars with negligible mass loss when their main sequence mass exceeds about 70–$80\,M_\odot$. (Extremely efficient rotationally-induced mixing leading to chemically homogeneous chemical evolution can reduce the threshold main sequence mass still further to approximately the threshold helium core mass (Chatzopoulos and Wheeler 2012)). For solar metallicity, stars this massive are usually assumed to lose all their hydrogen envelope and part of their cores along the way and thus avoid the instability. Suffice it to say that if the combined effects of mass loss and rotation allow the existence of a helium core mass in excess of $34\,M_\odot$ at carbon depletion, the pair instability will have an effect. To get a full-up pair instability supernova, one needs a helium core mass of about 63 $M_\odot$ which might correspond, depending upon the treatment of convection physics, to a main sequence star around 150 $M_\odot$. In between, lies the PPSN. As we shall see, the final evolution of such stars can be quite complicated because of the many pulses, but they have the merit that the explosion hydrodynamics is simple.

7.3.1 Pulsationally Unstable Helium Stars

While the observable display is quite sensitive to whether the presupernova star retains its hydrogen envelope or not, the number, energies, and duration of the pulses driven by the pair instability is determined entirely by the helium core mass. One can thus sample the broad properties of PPSN using only a grid of bare helium cores. This has the appealing simplicity of removing the uncertain effects of convective dredge up and rotational mixing during hydrogen burning and reducing the problem to a one parameter family of outcomes. Table 7.1 and Fig. 7.2 summarize some recent results for helium cores of various masses.

Initially, the instability is quite mild and only happens very close to the end of the star's life, after it has already completed core oxygen burning and is burning oxygen in a shell. For larger helium core masses, a few pulses contribute sufficient energy (about 10^{48} erg), that starting at around 34 $M_\odot$, the hydrogen envelope is ejected, but little else. The low energy ejection of the envelope produces very faint,

Table 7.1 Pulses from helium core explosions of different masses ($M_\odot$)

Mass	N Pulse	Duration	Energy	Rem. Mass
32	Weak	4.0(3)	1.6(45)	32
34	12	6.5(3)	1.5(48)	33.93
36	Many	1.4(4)	9.2(48)	35.81
38	Many	8.7(4)	1.1(50)	37.29
40	Many	2.8(5)	2.7(50)	38.24
42	18	3.3(5)	2.4(50)	39.72
44	10	9.0(5)	5.8(50)	39.94
46	10	2.2(6)	6.6(50)	41.27
48	7	6.4(6)	9.2(50)	41.52
50	4	7.1(7)	8.1(50)	42.80
52	4	4.3(8)	8.1(50)	45.87
54	2	5.4(10)	1.6(51)	43.35
56	2	1.3(11)	1.6(51)	40.61
58	2	3.0(11)	3.7(51)	17.06
60	2	1.3(11)	2.7(51)	36.60
62	2	5.3(11)	7.1(51)	5.33
64	1	–	4.7(51)	0
66	1	–	6.8(51)	0

7 The Deaths of Very Massive Stars

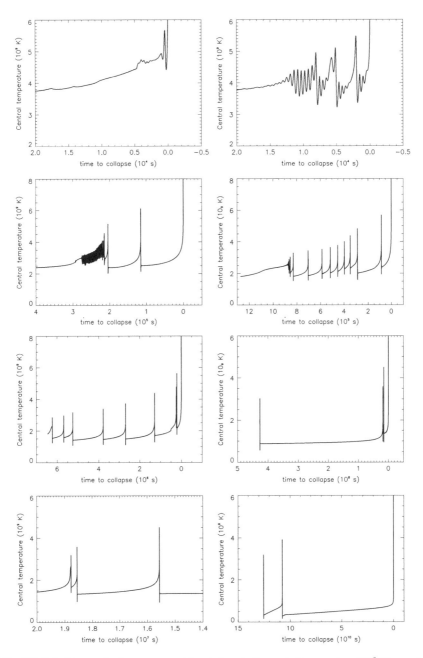

Fig. 7.2 Pair-driven pulsations cause rapid variations in the central temperature (10^9 K) near the time of death for helium cores of 32, 36, 40, 44, 48, 52 (on two different time scales) and 56 M$_\odot$ (*left* to *right*; *top* to *bottom*). The log base 10 of the time scales (s) in each panel are respectively 4, 4, 5, 5, 6, 8, 7, and 10. The last rise to high temperature marks the collapse of the iron core to a compact object. More massive cores have fewer, less frequent, but more energetic pulses. All plots begin at central carbon depletion

long lasting Type IIp supernovae. The continued evolution of such stars yields an iron core of about 2.5 $M_\odot$ that almost certainly collapses to a black hole with a mass nearly equal to the helium core mass. Thus the ejection of the envelope and its nucleosynthesis are the only observables for a distant event.

Moving on up in mass, the pulses have more energy, start earlier, and increase in number until, above 42 $M_\odot$, their number starts to decline again. Figure 7.2 shows that in the mass range 3 $M_\odot$ to about 44 $M_\odot$ a major pulse is typically preceded by a string of smaller ones that grow in amplitude until a single violent event causes a major change in the stellar structure. Recovery from this violent event requires a Kelvin-Helmholtz time scale ($\tau_{KH} \sim GM^2/RL$) for the core to contract back to the unstable temperature, around 2×10^9 K. If the pulse is a weak one, the luminosity in the Kelvin-Helmholtz time scale is the neutrino luminosity and is large, making the time scale short. If the pulse decreases the central temperature below a half-billion degrees however, radiation transport enters in and the time scale becomes long. On the heavier end of this mass range, the total energy of pulses is a few times 10^{50} erg, but their overall duration is less than a week. Since this is less than the time required for the ejected matter to become optically thin, the collisions are usually finished before any supernova becomes visible. Depending upon the presence of an envelope, one expects, for these cases, a rather typical Type Ib or IIp light curve, with some structure possible in the case of the bare helium core because of its short shock transversal time (Sect. 7.3.2). When the pulses are over, a large iron core is again produced, and, some time later, the remaining core of helium and heavy elements probably becomes a black hole.

For still heavier helium core masses, 44–52 $M_\odot$, the total energy of the pulses becomes that of a typical supernova, but spread over several pulses that require from weeks to years to complete. An important alignment of time scales occurs in this mass range. For the masses and energies ejected, average shell speeds for the first pulse are a few thousand km s^{-1} (much less if a hydrogen envelope is in the way). At this speed, a radius of $\sim 10^{16}$ cm is reached in about a year, which is comparable to the interval between pulses. Repeated supernovae and supernovae with complex light curves are thus possible. The photospheric radii of typical supernovae in nature are a few times 10^{15} cm, this being the distance where the expanding debris most efficiently radiate away their trapped energy on an explosive time scale. Since the ejecta of a given pulse will consist of material moving both slower than and faster than the average, and because each pulse is typically more energetic than its predecessor, shells collide at radii 10^{15}–10^{16} cm (Fig. 7.3).

These collisions convert streaming kinetic energy to optical light with high efficiency. In principle, a substantial fraction of the total kinetic energy of the pulses can be radiated, especially if the shells all run into a slowly moving hydrogen shell ejected in the first pulse. Stars in this mass range, in the most extreme cases, can thus give repeated supernovae with up to 10^{51} erg of light.

Still more energetic and less frequent pulsations happen at higher mass, but now the presence of the envelope becomes critical. Without a hydrogen envelope, the time between pulses is so long that the collisions happen at very large radii, 10^{17}–10^{18} cm. For these very large radii, the result would not be so different from

7 The Deaths of Very Massive Stars

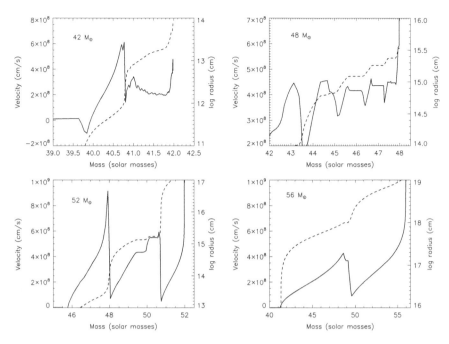

Fig. 7.3 Velocities (*solid lines*) and radii (*dashed lines*) of ejected shells for four helium cores producing mass ejection by the pulsational pair mechanism. The velocities are evaluated at various times when the collision between shells is underway. For the 42 and 48 $M_\odot$ models. this was near iron core collapse. For 52 $M_\odot$, it was at central silicon depletion, and for 56 $M_\odot$, after a strong silicon flash, but before the re-ignition of silicon. Some merging of pulses has already occurred. Regions of flat velocity imply spatially thin, high density shells that may be unstable in two or three dimensions

an ordinary 10^{51} erg supernova running into an unusually dense interstellar medium. Both the very large radii and long time scales preclude any resemblance to ordinary optical supernovae, but the events might instead present as bright radio and x-ray transients.

In the presence of an envelope, the first pulse does not eject matter with such high speed and, given the large variation in speed from the inner part of the moving shell to its outer extremity, substantial energy could still be emitted by explosions in this mass range by shells colliding inside of 10^{16} cm making a bright Type II supernova.

Pulses continue until the helium core has lost enough mass to be stable again. This gives a range of remnant masses typically around 34–46 $M_\odot$ (Table 7.1). The iron core masses and compactness parameters for these stars are both very large, so it seems very likely that black holes will result for the entire range of stars making PPSN, all having typical masses around 40 $M_\odot$.

7.3.2 Light Curves for Helium Stars

Light curves for a sample of helium core explosions are shown in Fig. 7.4 and illustrate the characteristics discussed in the previous section. For the lighter helium cores, the pulses only eject a small amount of matter with low energy. Shell collisions are over before light escapes from the collision region. The light curve for the 26 $M_\odot$ helium core is typical for this mass range – a subluminous "supernova" of less than 10^{42} erg s^{-1} lasting only a few days. These might be looked for in the case of stars that have lost their envelopes prior to exploding. In a star with an envelope, as we shall see later, the situation would be very different. Even the small (10^{49} erg) kinetic energy would unbind the envelope producing a long, faint Type IIp supernova.

For the 42 $M_\odot$ helium core, a brighter, longer lasting transient is produced, but still only a single event, albeit a structured one. The total duration of pulses is about 2 days, followed by a 2 day wait until the core collapse. The last pulse is a particularly violent one. The light curve (Fig. 7.4) shows a faint outburst occurring as many smaller pulses merge and the first big of mass is ejected, followed by a longer more

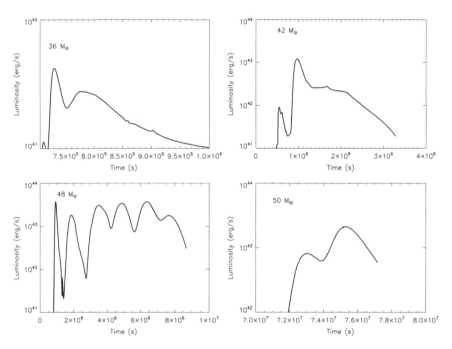

Fig. 7.4 Bolometric light curves from pulsational pair instability supernovae derived from bare helium cores of 36, 42, 48 and 50 $M_\odot$. A wide variety of outcomes is possible. For the 36 and 42 $M_\odot$ models the photospheric radius is inside 10^{15} cm and the transients will be *blue*. For the higher two masses, the photosphere is near 10^{15} cm and the transients might have colors more like an ordinary supernova

luminous peak as that main pulse runs into the prior ejecta. Both of these transients are quite blue since the collisions are occurring at small radius, a few times 10^{14} cm.

By 48 $M_\odot$, the shell collisions are becoming sufficiently energetic and infrequent that the light curve fractures into multiple events. The collisions are now happening at around 10^{15} cm and should be quite bright optically. At 52 $M_\odot$, one sees repeated individual supernovae. Figure 7.4 merely shows the brightest one from this object. Activity at the 10^{41} erg level started two years before.

It should be noted, though, that all these 1D light-curve calculations are quite approximate and need to be repeated in a multi-dimensional code with the appropriate physics, especially for cases where the shells collide in an optically thin regime. KEPLER, a one dimensional implicit hydrodynamics code with flux-limited radiative diffusion does an admirable job in a difficult situation. In 1D however, the snowplowing of a fast-moving shell into a slower one generates a large spike in density, with variations of many orders of magnitude in density between one zone and an adjacent one. For a time this thin shell corresponds to the photosphere. The "linearized" equations of hydrodynamics do not behave well in such clearly non-linear circumstances and the outcome of a multi-dimensional calculation may be qualitatively different. This is an area of active research.

7.3.3 Type II Pulsational Pair Instability Supernovae

The retention of even a small part of the original hydrogen envelope significantly alters the dynamics and appearance of PPSN. For example, what wold have been a brief, faint transient for a 36 $M_\odot$ helium core (Fig. 7.4), provides more than enough energy to eject the entire envelope of a red supergiant. A great diversity of outcomes is possible depending upon the mass of the envelope and helium core and the radius of the envelope

Most striking are the "ultra-luminous supernovae" of Type IIn that happen when very energetic pulses from the edge of the helium core strike a slowly moving, previously ejected hydrogen envelope. A similar (Type I) phenomenon could happen for bare helium cores, but probably with a shorter-lived, less luminous light curve owing to the smaller masses involved. An example is shown in Fig. 7.6 based upon the evolution of a 110 $M_\odot$ star (Woosley et al. 2007). By the end of its life this star had shrunk to 74.6 $M_\odot$ (using a wholly artificial mass loss rate), of which 49.9 $M_\odot$ was the helium core. This core experienced three violent pulsations. The first ejected almost all of the hydrogen envelope, leaving 50.7 $M_\odot$ behind. This envelope ejection produced a rather typical Type IIp supernova although with a slower than typical speed and luminosity (Fig. 7.5). By 6.8 years later, the stellar remnant had contracted to the point that it experienced the pair instability again. Two more pulses, occurring in rapid succession, ejected an additional 5.1 $M_\odot$ with a total kinetic energy of 6×10^{50} erg. Pulses 2 and 3 quickly merged and then run into the ejected envelope (Figs. 7.5 and 7.6).

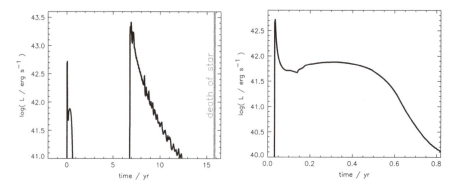

Fig. 7.5 Light curves of the two supernovae produced by the 110 M$_\odot$ PPSN (Woosley et al. 2007). The first pulse ejects the envelope and produces the faint supernova shown in greater detail on the *right*. 6.8 years later the collision of pulses 2 and 3 with that envelope produces another brighter outburst (see Fig. 7.6)

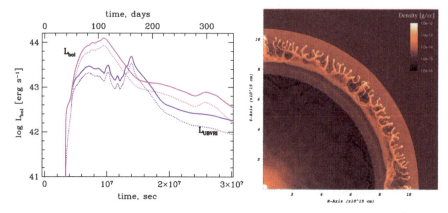

Fig. 7.6 *Left:* Light curve for the second very luminous outburst of the 110 M$_\odot$ model (see Fig. 7.5) of the two supernovae produced by the 110 M$_\odot$ PPSN (Woosley et al. 2007). The brighter set of curves results hen the collision speed is artificially increased by a factor of 2 and resembles SN 2006 gy. *Right:* 2D calculation of the explosion of a 110 M$_\odot$ star as a PPSN. The dense shell produced in 1D by the collision of the ejecta from two pulse is Rayleigh-Taylor unstable. The resulting density contrast is much smaller

These light curves were calculated using 1D codes in which the collision of the shells again produced a very large density spike. When the calculation was run again in 2D, but without radiation transport (Fig. 7.6), a Rayleigh-Taylor instability developed that led to mixing and a greatly reduced density contrast. The combined calculation of multi-D hydro coupled to radiation transport has yet to be carried out, so the light curves shown here are to be used with caution, but a multi-dimensional study would probably give a smoother light curve.

7 The Deaths of Very Massive Stars

Table 7.2 Nucleosynthesis in ejected shells ($M_\odot$) from helium core pulsational explosions

Mass	Total	He	C	O	Ne	Mg	Si	S	Ar	Ca
34	0.071	0.071	–	–	–	–	–	–	–	–
36	0.19	0.19	–	–	–	–	–	–	–	–
38	0.71	0.32	0095	0.17	0.096	0.032	–	–	–	–
40	1.76	0.50	0.29	0.53	0.32	0.11	–	–	–	–
42	2.28	0.60	0.43	0.70	0.41	0.14	–	–	–	–
44	4.06	0.85	0.79	1.36	0.80	0.26	–	–	–	–
46	4.73	1.02	0.94	1.61	0.90	0.27	–	–	–	–
48	6.48	1.34	1.40	2.30	1.15	0.30	–	–	–	–
50	7.20	1.58	1.60	2.61	1.16	0.26	–	–	–	–
52	6.13	1.55	1.33	2.29	0.81	0.16	0.001	–	–	–
54	10.64	1.65	1.83	5.32	1.35	0.41	0.074	–	–	–
56	15.38	1.74	2.06	9.41	1.52	0.50	0.15	–	–	–
58	40.93	1.85	2.87	30.5	2.64	1.42	1.49	0.17	0.020	0.015
60	23.39	1.89	3.10	15.0	2.49	0.60	0.28	0.058	0.008	0.005
62	56.67	1.95	2.87	37.5	2.60	1.43	6.39	2.99	0.51	0.44
64	64	1.92	3.62	44.1	3.60	2.12	5.35	2.41	0.43	0.38
66	66	1.79	3.60	42.8	3.99	2.07	7.11	3.49	0.60	0.53

7.3.4 Nucleosynthesis

The nucleosynthesis from PPSN is novel in that it is heavily weighted towards the light species that are ejected in the shells. For present purposes, given the large iron cores, we assume that all matter not ejected by the pulsations becomes a black hole. This assumption could be violated if rapid rotation energized some sort of jet-like outflows (e.g., a gamma-ray burst), but otherwise it seems reasonable.

Table 7.2 gives the approximate bulk nucleosynthesis, in solar masses, calculated for our standard set of helium cores models. For the lightest cores, the pulses lack sufficient energy to eject more than a small amount of surface material, which by assumption here is pure helium. It should be noted, however, that even these weak explosions would eject at least part of the hydrogen envelope of any red supergiant (typical binding energy less than 10^{48} erg). Since these envelopes often produce primary nitrogen by mixing between the helium core and hydrogen burning shell, an uncertain but possibly large yield of carbon, nitrogen, and oxygen (and of course hydrogen and helium) would accompany these explosions in a star that had not lost its envelope.

Moving up in mass, the violence of the pulses increases rapidly and more material is ejected, eventually reaching the deeper shells rich in heavier elements. In Table 7.2, total yields of less than $0.01\,M_\odot$ have not been included with the single exception of the $66\,M_\odot$ model which made $0.037\,M_\odot$ of ^{56}Ni. The 64 and $66\,M_\odot$ models are actually full up pair instability supernovae and leave no remnants, so perhaps including their yields here with the PPSN is a bit misleading.

If one folds these yields with an IMF to get an overall picture of the nucleosynthesis from a generation of PPSN, it is clear that the production (and the typical spectra of PPSN) will be dominated by H, (He), C, N, O, (Ne) and Mg and little else. In particular, PPSN make no iron-group elements. Given the dearth of strong He and Ne lines, one might expect that the generation of stars following a putative "first generation" of PPSN would show enhancements of C, N, O, and Mg and be "ultra-iron poor". Of course *some* heavier elements could be made by stars sufficiently light (main sequence mass less than $20\,M_\odot$?) to explode by the neutrino-transport process, or sufficiently heavy to make iron in a pair-instability supernova (helium core mass over $65\,M_\odot$).

7.4 $150-260\,M_\odot$; Pair Instability Supernovae

The physics of pair instability supernovae (PISN) is sufficiently well understood that they can be accurately modeled in 1D on a desktop computer. A major question though is their frequency in the universe. PISN come from a range of masses somewhat heavier than we expect for presupernova stars today. This is not to say that stars of over $150\,M_\odot$ are not being born. See e.g., the review by Crowther reported in Vink et al. (2013) which gives $320\,M_\odot$ as the current observational limit. The issue is whether such large masses can be retained in a star whose luminosity hovers near the Eddington limit (Vink et al. 2011). Still observers claim to have discovered at least one PISN event (Galyam et al. 2009). Because the critical quantity governing whether a star becomes PISN is the helium core mass of the presupernova star (greater than $65\,M_\odot$), they are favored by diminished mass loss, i.e., at low metallicity, and may have been more abundant in the early universe.

A common misconception is that all PISN make a lot of ^{56}Ni and therefore are always very bright. As Fig. 7.7 shows, large ^{56}Ni production and very high kinetic energies are limited to a fairly narrow range of exceptionally heavy and rare PISN. Most events will either present as a particularly energetic Type IIp supernova or a *subluminous* SN I. For an appreciable range of masses, less ^{56}Ni is produced than in, e.g., a SN Ia (about $0.7\,M_\odot$).

The nucleosynthesis of very low metallicity PISN is quite distinctive because they lack the excess neutrons needed to make odd-Z elements during the explosion. This is because the initial metallicity of the star, mostly CNO, is turned into ^{14}N during hydrogen burning. During helium burning, ^{14}N captures an alpha particle experiencing a weak decay to make ^{18}O which has two extra neutrons. Subsequent burning stages rearrange these neutrons using them to make isotopes and elements that require an excess of neutrons over protons, like almost all odd Z elements do. During the collapse phase, the time is too short for additional weak interactions so the ejected matter ends up deficient in things like Na, Al, P, Cl, K, Sc, V, and Mn. Very metal poor stars show no such anomalies and this suggests that the contribution of PISN to very early nucleosynthesis was small.

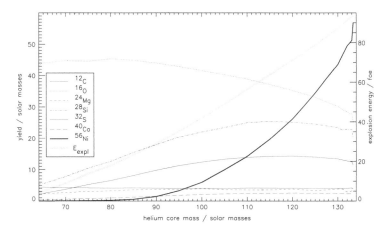

Fig. 7.7 Nucleosynthesis in pair-instability supernovae as a function of helium core mass. Also given is the explosion energy in units of 10^{51} erg (*broad grey line*) which rises steadily with mass. The *dark solid line* is ^{56}Ni synthesis which is not particularly large below 90 M$_\odot$ (Heger and Woosley 2002)

7.5 Above 260 M$_\odot$

Stars heavier than 260 M$_\odot$, or more specifically non-rotating helium cores greater than 133 M$_\odot$, are expected to produce black holes, at least up to about 10^5 M$_\odot$. Starting around 10^5 M$_\odot$, hydrogenic stars encounter a post-Newtonian instability on the main sequence and collapse (Fowler and Hoyle 1964). If these stars have near solar metallicity (above Z = 0.005) then titanic explosions of 10^{56}–10^{57} erg, powered by explosive hydrogen burning, can result for masses in the range 10^5–10^6 M$_\odot$ (Fuller et al. 1986). Lacking a large initial concentration of CNO, stars in this mass range, collapse to black holes.

For lighter stars, $\sim 10^3$–10^5 M$_\odot$, hydrogen burns stably, but helium burning encounters the pair instability, and on the upper end, the post-Newtonian instability. Again black hole formation seems the most likely outcome, though this mass range has not been fully explored.

7.6 The Effects of Rotation

Rotation alters stellar evolution in two major ways. During presupernova evolution it leads to additional mixing processes that can stir up either regions of the star or the whole star. Generally the helium cores of rotating stars are larger and, since the nucleosynthesis and explosion physics of massive stars depends sensitively upon the helium core mass, the outcome of a smaller mass main sequence star with rotation can resemble that of a larger one without rotation. The mixing can also increase

Table 7.3 Pulsar rotation rate predicted by models (Heger et al. 2005)

Mass ($M_\odot$)	Baryon ($M_\odot$)	Gravitational ($M_\odot$)	J (10^{47} erg s)	BE (10^{53} erg)	Pulsar P (ms)
12	1.38	1.26	5.2	2.3	15
15	1.47	1.33	7.5	2.5	11
20	1.71	1.52	14	3.4	7.0
25	1.88	1.66	17	4.1	6.3
35	2.30	1.97	41	6.0	3.0

the lifetime of the star and its luminosity and bring abundances to the surface that might have otherwise remained hidden. In extreme cases, rotation can even lead to the complete mixing of the star on the main sequence, thus avoiding the formation of a supergiant and producing a very rapidly rotating presupernova star that might serve as a gamma-ray burst progenitor (Sect. 7.6.2).

The other way rotation changes the evolution is by affecting how the star explodes and the properties of the compact remnant it leaves behind. Calculations that use reasonable amounts of rotation and approximate the effects of magnetic torques in transporting angular momentum show that rotation may play an increasingly dominant role in the explosion as the mass of the star increases (Heger et al. 2005). This is in marked contrast to the neutrino transport model which shows the opposite behavior (Sect. 7.2.1); heavier stars are *more* difficult to explode with neutrinos.

Table 7.3 shows the expected rotation rates of pulsars derived from the collapse of rotating stars of various main sequence masses. The rotational energy of these neutron stars is given approximately by $10^{51}(5\,\mathrm{ms}/P)^{-2}$ erg, where it is assumed that the neutron star moment of inertia is $80\,\mathrm{km}^2\,M_\odot$ (Lattimer and Prakash 2007). This implies that supernova over about $20\,M_\odot$ or so have enough rotational energy to potentially power a standard supernova. Rapidly rotating stellar cores are also expected to give birth to neutron stars with large magnetic fields (Duncan and Thompson 1992), thus providing a potential means of coupling the large rotation rate to the material just outside the neutron star. Calculations so far are encouraging (e.g. Akiyama et al. 2003; Burrows et al. 2007; Janka 2012). No calculation has yet modeled the full history, of a rotational, or rotational plus neutrino powered supernova all the way through from the collapse to explosion phase including all the relevant neutrino and MHD physics, but probably this will happen in the next decade.

In principle, the outcomes of rotationally powered supernovae and those powered by neutrinos should be very similar, though only rotation offers the prospect of making the explosion hyper-energetic (much greater than 10^{51} erg). To the extent that nucleosynthesis, light curves and spectra only depend upon the prompt deposition of $\sim 10^{15}$ erg at the center of a highly evolved red or blue supergiant, they will be indistinguishable. Rotation breaks spherical symmetry and may produce jets, but except in the case of gamma-ray bursts, it may be hard to disentangle effects essential to the explosion from those that simply modify an already successful explosion. There are interesting constraints on time scales, however, and hence on field strengths. Rotation or neutrinos must overcome a ram pressure from accretion that, in the case of high compactness parameter, may approach a solar mass per second. At a radius of 50 km, roughly typical of a young hot protoneutron star, it would take a field strength of over 10^{15} gauss to impede the flow. A similar estimate comes from nucleosynthesis. In order to synthesize ^{56}Ni, material must be heated to at least 4 and preferably 5×10^9 K. In a hydrodynamical model in which radiation dominates and 10^{51} erg is deposited instantly, this will only occur in a region smaller than 3,000 km. It takes the shock, moving at typically 20,000 km s^{-1}, about 0.1 s to cross that region, after which it begins to cool off. To deposit 10^{51} erg in that time with a standard dipole luminosity (Lang 1980) the field strength would need to exceed about 10^{16} gauss. This probably exceeds the *surface* fields generated by collapse alone. Whether the magneto-rotational instability can generate such fields is unclear, but it may take an exceptionally high rotation rate for this to all work out.

Perhaps the most common case is a neutrino-powered initial explosion amplified by rotation at later times. If that is the case though, a successful outgoing shock must precede any significant pulsar input. That starting point be difficult to achieve in stars with high compactness (Fig. 7.1). In any case we do know that *some* massive stars do make black holes.

7.6.1 Magnetar Powered Supernova Light Curves

If magnetic fields and rotation can provide the $\sim 10^{51}$ erg necessary for the kinetic energy of a supernova, they might, with greater ease, deliver the 10^{48} or even 10^{50} erg needed to make a bright – or a really bright – light curve (Woosley 2010; Kasen and Bildsten 2010). At the outset, one must acknowledge the huge uncertainty in applying the very simple pulsar power formula (Lang 1980),

$$\frac{dE}{dt} \approx 10^{49} \, B_{15}^2 P_{\text{ms}}^{-4} \text{ erg s}^{-1}, \qquad (7.1)$$

to a situation where the neutron star is embedded in a dense medium and that is still be rapidly evolving. Doing this blindly, however, yields some interesting results (Fig. 7.8). Since the energy is deposited late, it is less subject to adiabatic losses and is emitted as optical light with high efficiency. For reasonable choices of magnetic field and initial rotation rate, the supernova can be "ultra-luminous", brighter than a typical SN Ia for a much longer time.

The magnetic fields required are not all that large and are similar to what has been observed for modern day magnetars (Mereghetti 2008). In fact, too large a field results in the rotational energy being deposited too early. That energy then contributes to the explosion kinetic energy, but little to the light curve because, by the time the light is leaking out, the magnetar has already deposited most of

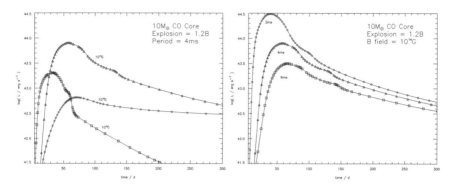

Fig. 7.8 Magnetar powered light curves for (*left*) different values of field strength (10^{14}, 10^{15}, and 10^{16} G at 4 ms) and (*right*) initial rotation periods (2, 4, 6 ms at 10^{14} G). The base event is the 1.2×10^{51} erg explosion of a $10\,M_\odot$ carbon-oxygen core. (Sukhbold and Woosley 2014, in preparation)

its rotational energy. The rotation rates, though large, are also not extreme, not very different, in fact, from the predictions for quite massive stars (Heger et al. 2005). If gamma-ray bursts are to be powered by millisecond magnetars with fields $\sim 10^{15} - 10^{16}$ G, and if ordinary pulsars have fields and rotational energies 100–1,000 times less, one expects somewhere, sometime to make neutron stars with fields and rotational energies that are just ten times less. The long tails on the light curves are interesting and, lacking spectroscopic evidence or very long duration observations, might easily be confused with ^{56}Co decay (Woosley 2010).

Depending upon the mass and radius of the star, the presence or absence of a hydrogenic envelope, and the supernova explosion energy, the resulting magnetar-illuminated transients can be quite diverse. The brighter events will tend to be of Type I because the supernova becomes transparent at an earlier time when greater rotational energy is being dissipated. The upper bound to the luminosity is a few times 10^{51} erg emitted over several months, or $\sim 10^{44.5}$ erg s^{-1}, but much fainter events are clearly possible. For Type II supernovae in red supergiants, the magnetar contribution may present as a rapid rise in brightness after an extended plateau (Maeda et al. 2007). The rise could be even more dramatic and earlier in a blue supergiant.

An interesting characteristic of 1D models for magnetar powered supernovae is a large density spike caused by the pile up of matter accelerated from beneath by radiation. In more than one dimension, this spike will be unstable and its disruption will lead to additional mixing that might have consequences for both the spectrum and the appearance of the supernova remnant.

7.6.2 Gamma-Ray Bursts (GRBs)

In the extreme case of very rapid rotation and the complete loss of its hydrogenic envelope, the death of a massive star can produce a common (long-soft) GRB. For a recent review see Woosley (2013). There are two possibilities for the "central engine" – a "millisecond magnetar" and a "collapsar". The former requires that the product of a successful supernova explosion be, at least for awhile, a neutron star, and that the power source is its rotational energy. The latter assumes the formation of a black hole with a centrifugally supported accretion disk. The energy source can be either the rotational energy of that black hole or of the disk, which is, indirectly, energized by the black hole's strong gravity.

Both models require that the progenitor star have extremely high angular momentum in and around the iron core. Loss of the hydrogen envelope could occur though a wind, binary mass exchange, or because extensive rotationally-induced mixing on the main sequence kept a red giant from ever forming. Loss of the envelope by a wind is disfavored because the existence of a lengthy red giant phase would probably break the rotation of the core to the extent that the necessary angular momentum was lost. One is this left with the possibility of a massive star that lost its envelope quite early in to a companion or a single star that experienced chemically homogeneous evolution (Maeder 1987; Woosley and Heger 2006; Yoon and Langer 2005, 2006). The resulting Wolf-Rayet star must also not lose much mass or its rotation too will be prohibitively damped. This seems to exclude most stars of solar metallicity, so GRBs are relegated to a low metallicity population. The relevant mass loss rate depends upon metallicity (specifically the iron abundance) as $Z^{0.86}$ (Vink and de Koter 2005), and even mild reduction is sufficient to provide the necessary conditions for a millisecond magnetar.

The collapsar model is capable, in principle, of providing much more energy (up to $\sim 10^{54}$ erg) than the magnetar model (up to 3×10^{52} erg). The former is limited only by the efficiency of converting accreted mass into energy, which can be quite high for a rotating black hole, while the latter is capped by a critical rotation rate where the protoneutron star deforms and efficiently emits gravitational radiation. So far, there is no clear evidence for total (beaming corrected) energies above $10^{52.5}$ in any GRB, so both models remain viable. It is interesting that there may be some pile up of the most energetic GRBs and their associated supernovae around a few times 10^{52}. That might be taken as (mild) evidence in favor of the magnetar model. On the other hand, black hole production is likely in the more massive stars and it may be difficult to arrange things such that all the matter always accretes without forming a disk (Woosley and Heger 2012).

Since angular momentum is in short supply, it is definitely easier to produce a millisecond magnetar which requires a mass averaged of angular momentum of only 2×10^{15} erg s (for a moment of inertia $I = 10^{45}$ g cm^2), or a value at its equator of 6×10^{15} erg s (for a neutron star radius of 10 km). For comparison, the angular momentum for the last stable orbit of a Kerr black hole is $1.5 \times 10^{16} \frac{M_{BH}}{3 M_\odot}$ erg s and about three times larger for a Schwarzschild hole. The same sorts of systems that make collapsars thus also seem likely to make, at least briefly, neutron stars with millisecond rotation periods. How these rapid rotators make their fields and how the fields interact with the rapidly accreting matter in which they are embedded is a very difficult problem in 3D, general relativistic magnetohydrodynamics. Analytic arguments suggest however that large fields will be created (Duncan and Thompson 1992) and that the rotation and magnetic fields will play a major role in launching an asymmetric explosion (Akiyama et al. 2003; Burrows et al. 2007).

Just which mass and metallicity stars make GRBs is an interesting issue. Even when the effects of beaming are included, the GRB event rate is a very small fraction of the supernova rate and thus the need for special circumstances is a characteristic of all successful models. These special circumstances include, as mentioned, the lack of any hydrogenic envelope and very rapid rotation. Without magnetic torques, the cores of most massive stars would rotate so rapidly at death that millisecond magnetars, collapsar, and presumably GRBs would abound. Any realistic model thus includes the effects of magnetic braking, even though the theory (Spruit 2002; Heger et al. 2005) is highly uncertain. In fact, most massive stars may be born with extremely rapid rotation, corresponding to 50 % critical in the equatorial plane, because of their magnetic coupling to an accretion disk (Rosen et al. 2012). The fact that most massive stars are observed to be rotating more slowly on the main sequence is a consequence of mass loss which would be reduced in regions with low metallicity. Since these large rotation rates are sufficient, again with uncertain parameters representing the inhibiting effect of composition gradients, to provoke efficient Eddington-Sweet mixing on the main sequence, GRBs should be abundant (too abundant?) at low metallicity. It is noteworthy that models for GRBs that invoke such efficient mixing on the main sequence do not require that the star be especially massive since, for low metallicity, the zero age main sequence mass is not much greater than the presupernova helium core mass (Woosley and Heger 2006). A low metallicity star of only $15\,M_\odot$ could become a GRB and a star of $45\,M_\odot$ could become a pulsational pair instability supernova.

Using a standard set of assumptions, the set of massive stars that might make GRBs by the collapsar mechanism has been surveyed for a grid of masses and metallicities by Yoon and Langer (2006). Averaged over all redshifts they find a GRB to supernova event ratio of 1/200 which declines at low redshift to 1/1,250. Half of all GRBs are expected to be beyond redshift 4. Given that magnetars might also make GRBs, or even most of them, these estimates need to be reexamined. In particular, the mean redshift for bursts may be smaller and the theoretical event rate higher.

7.7 Final Comments

As is frequently noted, we live in interesting times. Most of the basic ideas invoked for explaining and interpreting massive star death are now over 40 years old. This includes supernovae powered by neutrinos, pulsars, the pair-instability, and the pulsational pair instability. Yet lately, the theoretical models and observational data have both experienced exponential growth, fueled on the one hand by the rapid expansion of computer power and the shear number of people running calculations, and on the other, by large transient surveys. Ideas that once seemed "academic", like pair-instability supernovae and magnetar-powered supernovae are starting to find counterparts in ultra-luminous supernovae.

"Predictions" in such a rapidly evolving landscape quickly become obsolete or irrelevant. Still, it is worth stating a few areas of great uncertainty where rapid progress might occur. These issues have been with us a long time, but problems do eventually get solved.

- What range(s) of stellar masses and metallicities explode by neutrino transport alone. The community has hovered on the brink of answering this for a long time. Today some masses explode robustly and others show promise (Janka et al. 2012; Janka 2012), but a comprehensive, parameter-free understanding is still lacking. The computers, scientists, and physics may be up to the task in the next five years. The compactness of the progenitor very likely plays a major role. It would be really nice to know.
- What is the relation between the initial and final (presupernova) masses of stars of all masses and metallicities. Suppose we knew the *initial* mass function at all metallicities (a big given). What is the *final* mass function for presupernova stars? We can't really answer questions about the explosion mechanism of stars of given main sequence masses without answering this one too. Our theories and observations of mass loss are developing, but still have a long way to go.
- What is the angular momentum distribution in presupernova stars? To answer this the effects of magnetic torques and mass loss must be included throughout all stages of the evolution – a tough problem. Approximations exist, but they are controversial and more 3D modeling might help.
- Are the ultra-luminous supernovae that are currently being discovered predominantly pair instability, pulsational pair instability, or magnetar powered (or all three)? Better modeling might help, especially with spectroscopic diagnostics.
- Is the most common form of GRB powered by a rotating neutron star or by an accreting black hole? What are the observational diagnostics of each?
- Does "missing physics", e.g., neutrino flavor mixing or a radically different nuclear equation of state play a role in answering any of the above questions?

This small list of "big theory issues" of course connects to a greater set of "smaller issues" – the treatment of semiconvection, convective overshoot, and rotational mixing in the models; critical uncertain nuclear reaction rates; opacities; the complex interplay of neutrinos, magnetohydrodynamics, convection and general relativity in 3D in a real core collapse – well maybe that is not so small.

Obviously there is plenty for the next generation of stellar astrophysicists to do.

Acknowledgements We thank Tuguldur Sukhbold and Ken Chen for permission to include here details of their unpublished work, especially Figs. 7.1, 7.6, and 7.8. This work has been supported by the National Science Foundation (AST 0909129), the NASA Theory Program (NNX09AK36G), and the University of California Lab Fees Research Program (12-LR-237070).

References

Abel, T., Bryan, G. L., & Norman, M. L. (2002). *Science, 295*, 93.
Akiyama, S., Wheeler, J. C., Meier, D. L., & Lichtenstadt, I. (2003). *Astrophysical Journal, 584*, 954.
Burrows, A., Dessart, L., Livne, E., Ott, C. D., & Murphy, J. (2007). *Astrophysical Journal, 664*, 416.
Brown, J. M., & Woosley, S. E. (2013). *Astrophysical Journal, 769*, 99.
Chandrasekhar, S. (1939). *An introduction to the study of stellar structure*. Chicago: The University of Chicago press.
Chevalier, R. A., & Soderberg, A. M. (2010). *Astrophysical Journal Letters, 711*, L40.
Chieffi, A., & Limongi, M. (2004). *Astrophysical Journal, 608*, 405.
Chieffi, A., & Limongi, M. (2013). *Astrophysical Journal, 764*, 21.
Chatzopoulos, E., & Wheeler, J. C. (2012). *Astrophysical Journal, 748*, 42.
Dessart, L., Hillier, D. J., & Livne, E., et al. (2011). *Monthly Notices of the Royal Astronomical Society , 414*, 2985.
Dessart, L., Hillier, D. J., Li, C., & Woosley, S. (2012). *Monthly Notices of the Royal Astronomical Society, 424*, 2139.
Duncan, R. C., & Thompson, C. (1992). *Astrophysical Journal Letters, 392*, L9.
Fowler, W. A., & Hoyle, F. (1964). *Astrophysical Journal Supplement, 9*, 201.
Fuller, G. M., Woosley, S. E., & Weaver, T. A. (1986). *Astrophysical Journal, 307*, 675.
Gal-Yam, A., Mazzali, P., & Ofek, E. O., et al., (2009). *Nature, 462*, 624.
Heger, A., & Woosley, S. E., (2002). *Astrophysical Journal, 567*, 532.
Heger, A., Woosley, S. E., & Spruit, H. C. (2005). *Astrophysical Journal, 626*, 350.
Heger, A., & Woosley, S. E. (2010). *Astrophysical Journal, 724*, 341.
Hirschi, R., Meynet, G., & Maeder, A. (2005). *Astronomy and Astrophysics, 433*, 1013.
Hoyle, F., & Fowler, W. A. (1960). *Astrophysical Journal, 132*, 565.
Janka, H.-T., Hanke, F., Hüdepohl, L., Marek, A., Müller, B., & Obergaulinger, M. (2012). *Progress of Theoretical and Experimental Physics , 01A309*, 33p.
Janka, H.-T. (2012). *Annual Review of Nuclear and Particle Science , 62*, 407.
Kasen, D., & Woosley, S. E. (2009). *Astrophysical Journal, 703*, 2205.
Kasen, D., & Bildsten, L. (2010). *Astrophysical Journal, 717*, 245.
Lai, D. K., Bolte, M., & Johnson, J. A., et al. (2008). *Astrophysical Journal, 681*, 1524.
Lang, K. (1980). *Astrophysical formulae*. Berlin: Springer.
Lattimer, J. M., & Prakash, M. (2007). *Physics Reports, 442*, 109.
Limongi, M., Straniero, O., & Chieffi, A. (2000). *Astrophysical Journal Supplement, 129*, 625.
Lovegrove, E., & Woosley, S. E. (2013). *Astrophysical Journal, 769*, 109.
Maeda, K., Tanaka, M., & Nomoto, K., et al. (2007). *Astrophysical Journal, 666*, 1069.
Maeder, A. (1987). *Astronomy and Astrophysics, 178*, 159.
Maeder, A., & Meynet, G. (2012). *Reviews of Modern Physics , 84*, 25.
Mereghetti, S. (2008). *Astronomy and Astrophysics Review, 15*, 225.
Meynet, G. (2002). *Astrophysics and Space Science, 281*, 183.
Müller, B., Janka, H.-T., & Heger, A. (2012). *Astrophysical Journal, 761*, 72.
Nomoto, K., Tominaga, N., Umeda, H., Kobayashi, C., & Maeda, K. (2006). *Nuclear Physics A , 777*, 424.
Nomoto, K., Kobayashi, C., & Tominaga, N. (2013). *Annual Review of Astronomy and Astrophysics, 51*, 457.
O'Connor, E., & Ott, C. D. (2011). *Astrophysical Journal, 730*, 70.
Özel, F., Psaltis, D., Narayan, R., & McClintock, J. E. (2010). *Astrophysical Journal, 725*, 1918.
Piro, A. L. (2013). *Astrophysical Journal Letters, 768*, L14.
Quataert, E., & Shiode, J. (2012). *Monthly Notices of the Royal Astronomical Society, 423*, L92–96.
Quataert, E., & Kasen, D. (2012). *Monthly Notices of the Royal Astronomical Society, 419*, L1.
Rosen, A. L., Krumholz, M. R., & Ramirez-Ruiz, E. (2012). *Astrophysical Journal, 748*, 97.

Smartt, S. J. (2009). *Annual Review of Astronomy and Astrophysics, 47*, 63.
Smartt, S. J., Eldridge, J. J., Crockett, R. M., & Maund, J. R. (2009). *Monthly Notices of the Royal Astronomical Society, 395*, 1409.
Spruit, H. C. (2002). *Astronomy and Astrophysics, 381*, 923.
Sukhbold, T., & Woosley, S. E. (2014). *Astrophysical Journal, 783*, 10.
Tan, J. C., & McKee, C. F. (2004). *Astrophysical Journal, 603*, 383.
Thielemann, F.-K., Nomoto, K., & Hashimoto, M.-A. (1996). *Astrophysical Journal, 460*, 408.
Timmes, F. X., Woosley, S. E., & Weaver, T. A. (1996). *Astrophysical Journal, 457*, 834.
Ugliano, M., Janka, H.-T., Marek, A., & Arcones, A. (2012). *Astrophysical Journal, 757*, 69.
Vink, J. S., & de Koter, A. (2005). *Astronomy and Astrophysics, 442*, 587.
Vink, J. S., Muijres, L. E., & Anthonisse, B., et al. (2011). *Astronomy and Astrophysics, 531*, A132.
Vink, J. S., Heger, A., Krumholz, M. R., et al. (2013). To be published in *Highlights of Astronomy*. arXiv:1302.2021.
Wiktorowicz, G., Belczynski, K., & Maccarone, T. J. (2014, submitted). *Astrophysical Journal*. arXiv:1312.5924.
Woosley, S. E. (2010). *Astrophysical Journal Letters, 719*, L204.
Woosley, S. E. (2013). C. Kouveliotou, R. A. M. J. Wijers & S. E. Woosley (Eds.), *Gamma-ray Bursts* (p. 191). Cambridge: Cambridge University Press.
Woosley, S. E., & Weaver, T. A. (1995). *Astrophysical Journal Supplement, 101*, 181.
Woosley, S. E., Heger, A., & Weaver, T. A. (2002). *Reviews of Modern Physics, 74*, 1015.
Woosley, S. E., & Heger, A. (2006). *Astrophysical Journal, 637*, 914.
Woosley, S. E., Blinnikov, S., & Heger, A. (2007). *Nature, 450*, 390.
Woosley, S. E., & Heger, A. (2007). *Physics Reports, 442*, 269.
Woosley, S. E., & Heger, A. (2012). *Astrophysical Journal, 752*, 32.
Yoon, S.-C., & Langer, N. (2005). *Astronomy and Astrophysics, 443*, 643.
Yoon, S.-C., & Langer, N. (2006). *Astronomy and Astrophysics, 460*, 199.

Chapter 8
Observed Consequences of Preupernova Instability in Very Massive Stars

Nathan Smith

Abstract This chapter concentrates on the deaths of very massive stars, the events leading up to their deaths, and how mass loss affects the resulting death. The previous four chapters emphasized the theory of wind mass loss, eruptions, and core collapse physics, but here we emphasize mainly the observational properties of the resulting death throes. Mass loss through winds, eruptions, and interacting binaries largely determines the wide variety of different types of supernovae that are observed, as well as the circumstellar environments into which the supernova blast waves expand. Connecting these observed properties of the explosions to the initial masses of their progenitor stars is, however, an enduring challenge and is especially difficult for very massive stars. Superluminous supernovae, pair instability supernovae, gamma ray bursts, and "failed" supernovae are all end fates that have been proposed for very massive stars, but the range of initial masses or other conditions leading to each of these (if they actually occur) are still very uncertain. Extrapolating to infer the role of very massive stars in the early universe is essentially unencumbered by observational constraints and still quite dicey.

8.1 Introduction

As discussed in previous chapters (Vink, Owocki), two critical aspects in the evolution of very massive stars (VMSs) are that their high luminosities cause strong mass loss in radiation-driven winds, and that high luminosities can also cause severe instabilities in the stellar envelope and interior as the star approaches the Eddington limit. These features become increasingly important as the initial stellar mass increases, but especially so as the star evolves off the main sequence and approaches its death. Moreover, the two are interconnected, since mass loss will increase the star's luminosity/mass ratio, possibly leading to more intense instabilities over time.

It should not be surprising, then, that VMSs show clear empirical evidence of this instability, and this chapter discusses various observational clues that we have.

N. Smith (✉)
Steward Observatory, 933 N. Cherry Ave., Tucson, AZ 85721, USA
e-mail: nathans@as.arizona.edu

This is a particularly relevant topic, as time-domain astronomy is becoming an increasingly active field of observational research. Throughout, the reader should remember that we are focussed on observed phenomena, and that working backward to diagnose possible underlying physical causes is not always straightforward. Hence, this interpretation is where most of the current speculation and debate rests among researchers working in the field. Stellar evolution models make predictions for the appearance of single massive stars late in their lives, but the influence of binary interaction may be extremely important or even dominant (Langer 2012), and the assumptions about mass-loss that go into the single-star models are not very reliable (Smith 2014). In particular, the eruptive instabilities discussed in this chapter are not included in single-star evolution models, and as such, these models provide us with little perspective for understanding the very latest unstable phases of VMSs or their final fates. The loosely bound envelopes that result from a star being close to the Eddington limit may be an important factor in directly causing outbursts, but having a barely bound envelope may also make it easier for other mechanisms to be influential, such as energy injection from non-steady nuclear burning, precursor core explosions, or binary interactions (see e.g., Smith and Arnett 2014 for a broader discussion of this point).

In the sections to follow, we discuss the observed class of eruptive luminous blue variables (LBVs) that have been linked to the late evolutionary phases of VMSs, various types of very luminous supernovae (SNe) or other explosions that may come from VMSs, and direct detections of luminous progenitors of SNe (including a few actual detections of pre-SN eruptions) that provide a direct link between VMSs and their SNe.

8.2 LBVs and Their Giant Eruptions

Perhaps the most recognizable manifestation of the instability that arises in the post-main-sequence evolution of VMSs is the class of objects known as luminous blue variables (LBVs). These were recognized early-on as the brightest blue irregular variables in nearby galaxies like M31, M33, and NGC 2403 (Hubble and Sandage 1953; Tammann and Sandage 1968), and these classic examples were referred to as the "Hubble-Sandage variables". Later, Conti (1984) recognized that many different classes of hot, irregular variable stars in the Milky Way and Magellanic clouds were probably related to these Hubble-Sandage variables, and probably occupy similar evolutionary stages in the lives of massive stars, so he suggested that they be grouped together and coined the term "LBV" to describe them collectively. The LBVs actually form a rather diverse class, consisting of a wide range of irregular variable phenomena associated with evolved massive stars (see reviews by Humphreys and Davidson 1994; van Genderen 2001; Smith et al. 2004, 2011a; Van Dyk and Matheson 2012; Clark et al. 2005).

8.2.1 Basic Observed Properties of LBVs

In addition to their high luminosities, some of the key observed characteristics of LBVs are as follows (although beware that not all LBVs exhibit all these properties):

- **S Doradus eruptions.** Named after the prototype in the LMC, S Dor eruptions are seen as a brightening that occurs at visual wavelengths resulting from a change in apparent temperature of the star's photosphere; this causes the peak of the energy distribution to shift from the UV to visual wavelengths at approximately constant bolometric luminosity. The increase in visual brightness (i.e. 1–2 mag, typically for more luminous stars) corresponds roughly to the bolometric correction for the star, so that hotter stars exhibit larger amplitudes in their S Dor events. LBVs have different temperatures in their quiescent state, and this quiescent temperature increases with increasing luminosity. The visual maximum of S Dor eruptions, on the other hand, usually occurs at a temperature around 7500 K regardless of luminosity, causing the star to resemble a late F-type supergiant with zero bolometric correction (see Fig. 8.1). While these events are defined to occur at constant bolometric luminosity (Humphreys and Davidson 1994), in fact quantitative studies of classic examples like AG Car do reveal some small variation in L_{Bol} through the S Dor cycle (Groh et al. 2009). Similarly, the traditional explanation for the origin of the temperature change was that the star

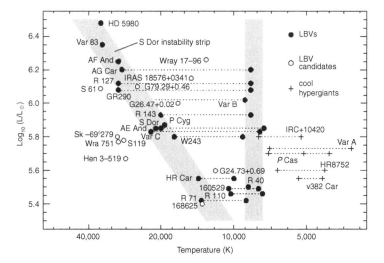

Fig. 8.1 The upper HR Diagram of LBVs and some LBV candidates (from Smith, Vink, and de Koter 2004). The most massive LBVs and LBV candidates like η Car and the Pistol star are off the top of this diagram. The diagonal strip where LBVs reside at quiescence is the S Dor instability strip discussed in the text. Note that LBVs are recognized by their characteristic photometric variability down to luminosities where the S Dor instability strip meets the eruptive temperature

increases its mass-loss rate, driving the wind to very high optical depth and the creation of a pseudo photosphere (Humphreys and Davidson 1994; Davidson 1987). Quantitative spectroscopy reveals, however, that the measured mass-loss rates do not increase enough to cause a pseudo photosphere in classic S Dor variables like AG Car (de Koter et al. 1996), and that the increasing photospheric radius is therefore more akin to a true expansion of the star's photosphere (i.e., a pulsation). Possible causes of this inflation of the star's outer layers is discussed elsewhere in this book (see Owocki's chapter). LBVs that experience these excursions are generally thought to be very massive stars, but their mass range is known to extend down to around 25 $M_\odot$ (Smith et al. 2004).

- **Quiescent LBVs reside on the S Dor instability strip.** As noted in the previous point, LBVs all show roughly the same apparent temperature in their cool/bright state during an outburst, but they have different apparent temperatures in their hot/quiescent states. These hot temperatures are not random. In quiescence, most LBVs reside on the so-called "S Dor instability strip" in the HR Diagram (Wolf 1989). This is a diagonal strip, with increasing temperature at higher luminosity (see Fig. 8.1). Notable examples that do not reside on this strip are the most luminous LBVs, like η Car and the Pistol star, so the S Dor instability strip may not continue to the most massive and most luminous stars, for reasons that may be related to the strong winds in these VMSs (see Vink chapter). Many of the stars at the more luminous end of the S Dor instability strip are categorized as Ofpe/WN9 or WNH stars in their hot/quiescent phases, with AG Car and R127 being the classic examples where these stars are then observed to change their spectral type and suffer bona-fide LBV outbursts. There are also many Ofpe/WN9 stars in the same part of the HR Diagram that have not exhibited the characteristic photometric variability of LBVs in their recent history, but which have circumstellar shells that may point to previous episodes of eruptive mass loss (see below). Such objects with spectroscopic similarity to quiescent LBVs, but without detection of their photometric variability, are sometimes called "LBV candidates".

- **Giant eruptions.** The most dramatic variability attributed to LBVs is the so-called "giant eruptions", in which stars are observed to increase their radiative luminosity for months to years, accompanied by severe mass loss (e.g., Humphreys et al. 1999; Smith et al. 2011a). The star survives the disruptive event. The best studied example is the Galactic object η Carinae, providing us with its historically observed light curve (Smith and Frew 2011), as well as its complex ejecta that contain 10–20 $M_\odot$ and $\sim 10^{50}$ ergs of kinetic energy (Smith et al. 2003; Smith 2006). Besides the less well-documented case of P Cygni's 1600 AD eruption, our only other examples of LBV-like giant eruptions are in other nearby galaxies. A number of these have been identified, with peak luminosities similar to η Car or less (Van Dyk and Matheson 2012; Smith et al. 2011a). Typical expansion speeds in the ejecta are 100–1,000 km s^{-1} (Smith et al. 2011a). These events are discussed more below.

- **Strong emission-line spectra.** Most, but not all, LBVs exhibit strong emission lines (especially Balmer lines) in their visual-wavelength spectra. This is a consequence of their very strong and dense stellar winds (see Vink chapter), combined with their high UV luminosity and moderately high temperature. The wind mass-loss rates implied by quantitative models of the spectra range from 10^{-5} to 10^{-3} $M_\odot$ yr^{-1}; this is enough to play an important role in the evolution of the star (see Smith 2014), and eruptions enhance the mass loss even more. The emission lines in LBVs are, typically, much stronger than the emission lines seen in main-sequence O-type stars of comparable luminosity, and all of the more luminous LBVs have strong emission lines. Other stars that exhibit similar spectra but are not necessarily LBVs include WNH stars, Ofpe/WN9 stars, and B[e] supergiants, some of which occupy similar parts of the HR Diagram.

- **Circumstellar shells.** Many LBVs are surrounded by spatially resolved circumstellar shells. These fossil shells provide evidence of a previous eruption. Consequently, some stars that resemble LBVs spectroscopically and have massive circumstellar shells, but have not (yet) been observed to exhibit photometric variability characteristic of LBVs, are often called LBV candidates. Many authors prefer to group LBVs and LBV candidates together (the logic being that a volcano is still a volcano even when it is dormant). LBV circumstellar shells are extremely important, as they provide the only reliable way to estimate the amount of mass ejected in an LBV giant eruption. The most common technique for measuring the mass is by calculating a dust mass from thermal-IR radiation, and then converting this to a total gas mass with an assumed gas:dust mass ratio (usually taken as 100:1 for the Milky Way, although this value is uncertain[1]). To calculate a dust mass from the IR luminosity, one must estimate the dust temperature from the spectral energy distribution (SED), and then adopt some wavelength-dependent grain opacities in order to calculate the emitting mass. The technique can be quite sensitive to multiple temperature components, and far-IR data have been shown to be very important because most of the mass can be hidden in the coolest dust, which is often not detectable at wavelengths shorter than 20 μm. One can also measure the gas mass directly by various methods, usually adopting a density diagnostic like line ratios of [Fe II] or [S II] and multiplying by the volume and filling factor, or calculating a model for the density needed to produce the observed ionization structure using codes such as CLOUDY (Ferland et al. 1998). The major source of uncertainty here is the assumed ionization fraction. Masses of LBV nebulae occupy a very large range from $\sim$20 $M_\odot$ at the upper end down to 0.1 $M_\odot$ (Smith and Owocki 2006), although even smaller masses become difficult to detect around bright central stars.
- **Wind speeds and nebular expansion speeds.** LBV winds and nebulae typically have expansion speeds of 50–500 km s^{-1}, due to the fact that the escape speed of

---

[1] If this value is wrong, it is probably a conservative underestimate. This is because a gas:dust mass ratio of 100:1 assumes that all refractory elements at $Z_\odot$ are in grains, whereas in reality, the dust formation may be less efficient or UV and shocks may destroy some dust, leaving some of these elements in the gas phase (and thus raising the total mass). In general, nebular gas masses inferred from thermal-IR dust emission should be considered lower limits, especially at $Z < Z_\odot$.

the evolved blue supergiant is lower than for the more compact radii of O-type stars and WR stars that have faster speeds of order $1,000 \, \text{km s}^{-1}$. In many cases, the shell nebulae are expanding with an even slower speed than the underlying wind, but this is not always the case. The slower nebular speeds may suggest that the nebulae were ejected in a state when the star was close to the Eddington limit (lower effective gravity) or that the LBV eruption ejecta have decelerated after colliding with slow CSM or high-pressure ISM.
- **N-rich ejecta.** Lastly, LBVs typically exhibit strong enhancements in their N abundance, measured in the circumstellar nebulae or in the wind spectrum. The most common measurement involves the analysis of visual-wavelength spectra, using nebular [S II] lines to derive an electron density, using the [N II] ($\lambda 6583 + \lambda 6548)/\lambda 5755$ ratio to derive an electron temperature, and then using the observed intensity of the [N II] lines compared to H lines for a relative N^+/H ratio (and then doing a similar analysis of O and C lines in order to estimate N/O and N/C ratios). One must make assumptions about the ionization levels of N and other elements, but if UV spectra are available, one can constrain the strength of a wide range of ionization levels of each atom. In the case of η Carinae, for example, strong lines of N I, II, III, IV, and V are detected, but O lines of all ionization levels are extremely faint (Davidson et al. 1986). The observed levels of N enrichment in LBVs suggest that the outer layers of the stars include large quantities of material processed through the CNO cycle and mixed to the surface, requiring that LBVs are post-main-sequence stars.

8.2.2 The Evolutionary State of LBVs

While evidence for N enrichment and C+O depletion suggest that LBVs are massive post-main-sequence stars, their exact evolutionary status within that complex and possibly non-monotonic evolution has been controversial – moreso in recent years.

The traditional view of LBVs, which emerged in the 1980s and 1990s, is that they correspond to a very brief transitional phase of massive star evolution, as the star moves from core H burning when it is seen as a main sequence O-type star, to core He burning when it is seen as a Wolf-Rayet (WR) star. A typical monotonic evolutionary scheme for a VMS is as follows:

$$100 \, M_\odot : \text{O star} \rightarrow \text{Of/WNH} \rightarrow \text{LBV} \rightarrow \text{WN} \rightarrow \text{WC} \rightarrow \text{SN Ibc}$$

In this scenario, the strong mass-loss experienced by LBVs is important for removing what is left of the star's H envelope after the main sequence, leaving a hydrogen-poor WR star following the end of the LBV phase. The motivation for thinking that this is a very brief phase comes from the fact that LBVs are extremely rare, even for very massive stars: taking the relative numbers of LBVs and O-type stars at high luminosity, combined with the expected H-burning lifetime of massive

O-type stars, would imply a duration for the LBV phase of only a few 10^4 years or less. This view fits in nicely with a scenario where the observed population of massive stars is dominated by single-star evolution.

However, a number of problems and inconsistencies have arisen with this standard view of LBVs. For one thing, the very short transitional lifetime depends on the assumption that the observed LBVs are representative of the whole transitional phase. In fact, there is a much larger number of blue supergiant stars that are not bona fide LBVs seen in eruption, but which are probably related—these are the LBV candidates discussed earlier. Examining populations in nearby galaxies, for example, Massey et al. (2007) find that there are more than *an order of magnitude* more LBV candidates than there are LBVs confirmed by their photometric variability. (For example, there are several hundred LBV candidates in M31 and M33, compared to the 8 LBVs known by their photometric variability.) If the LBV candidates are included with LBVs, then the average lifetime of the LBV phase must rise from a few 10^4 year to several 10^5 year. Now we have a problem, because this is a significant fraction of the whole He burning lifetime, making it impossible for LBVs to be fleeting *transitional* objects. There is not enough time in core-He burning to link them to both WR stars and LBVs. Should we include the LBV candidates and related stars? Are they dormant LBVs? If indeed LBVs go through dormant phases when they are not showing their instability (or when they have temporarily recovered from the instability after strong mass loss), then it would be a mistake not to include the duty cycle of instability in the statistics of LBVs. Massey (2006) has pointed to the case of P Cygni as a salient example: its 1600 A.D. giant LBV eruption was observed and so we consider it an archetypal LBV, but it has shown no eruptive LBV-like behavior since then. If the observational record had started in 1700, then we would have no idea that P Cygni was an LBV and we would be wrong. So how many of the other LBV candidates are dormant LBVs? The massive circumstellar shells seen around many LBV candidates imply that they have suffered LBV giant eruptions in the previous 10^3 year or so.

Another major issue is that we have growing evidence that LBVs or something like them (massive stars with high mass loss, N enrichment, H rich, slow $\sim$100 km s^{-1} winds, massive shells) are exploding as core-collapse SNe while still in an LBV-like phase (see below). This could not be true if LBVs are only in a brief transition to the WR phase, which should last another 0.5–1 Myr before core collapse to yield a SN Ibc. Pre-supernova eruptive stars that resemble LBVs are discussed in more detail in following sections.[2]

Last, the estimates for lifetimes in various evolutionary phases in the typical monotonic single-star scenario (see above) ignore empirical evidence that binary evolution dominates the evolution of a large fraction of massive stars. Many massive O-type stars (roughly 1/2 to 2/3) are in binary systems whose orbital separation is small enough that they should interact and exchange mass during their lifetime (Kobulnicky and Fryer 2007; Kiminki and Kobulnicky 2012; Kiminki et al. 2012;

---

[2] See Smith & Tombleson (2014), in press.

Chini et al. 2012; Sana et al. 2012). These binary systems *must* make a substantial contribution to the observed populations of evolved massive stars and SNe, so to find agreement between predictions of single-star evolutionary models and observed populations indicates that something is wrong with the models. Unfortunately, solutions to these problems are not yet readily apparent; some current effort is focussed here, and these topics are still a matter of debate among massive star researchers.

8.2.3 A Special Case: Eta Carinae

The enigmatic massive star η Carinae is perhaps the most famous and recognizable example of an evolved and unstable VMS. It is sometimes regarded as the prototype of eruptive LBVs, but at the same time it has a long list of peculiarities that make it seem unique and very atypical of LBVs. In any case, it is by far the *best studied* LBV, and (for better or worse) it has served as a benchmark for understanding LBVs and the physics of their eruptions.

Several circumstances conspire to make η Car such a fountain of information. It is nearby (about 2.3 kpc; Smith 2006) and bright with low interstellar extinction, so one is rarely photon-starved when observing this object at any wavelength. It is one of the most luminous and massive stars known, with rough values of $L \simeq 5 \times 10^6 \, L_\odot$ and a present-day mass for the primary around $100 \, M_\odot$ (its ZAMS mass is uncertain, but was probably a lot more than this). Its giant eruption in the nineteenth century was observed at visual wavelengths so that we have a detailed light curve of the event (Smith and Frew 2011), and η Car is now surrounded by the spectacular expanding Homunculus nebula that provides us with a fossil record of that mass loss. This nebula allows us to estimate the ejected mass and kinetic energy of the event, which are $\sim 15 \, M_\odot$ and $\sim 10^{50}$ erg, and we can measure the geometry of the mass ejection because the Homunculus is still young and in free expansion (Smith et al. 2003; Smith 2006).

Davidson and Humphreys (1997) provided a comprehensive review of the star and its nearby ejecta in the mid-1990s, but there have been many important advances in the subsequent 16 years. It has since been well established that η Car is actually in a binary system with a period of 5.5 year and $e \simeq 0.9$ (Daminelli et al. 1997), which drastically alters most of our ideas about this object. Accordingly, much of the research in the past decade has been devoted to understanding the temporal variability in this colliding-wind binary system (see Madura et al. 2012, and references therein). Detailed studies of the Homunculus have constrained its 3D geometry and expansion speed to high precision (Smith 2006), and IR wavelengths established that the nebula contains almost an order of magnitude more mass than was previously thought (Smith et al. 2003; Morris et al. 1999; Gomez et al. 2010). The larger mass and kinetic energy force a fundamental shift in our understanding of the physics of the Great Eruption (see below). Observations with *HST* have dissected the detailed ionization structure of the nebula and measured its expansion proper

8 Observed Deaths of Very Massive Stars

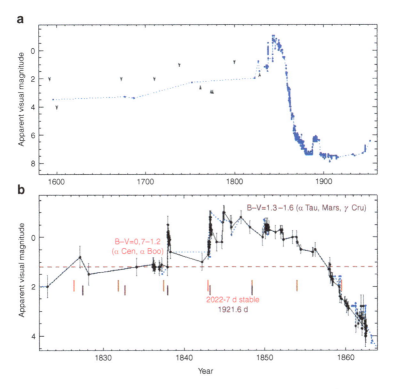

Fig. 8.2 The historical light curve of the nineteenth century Great Eruption of η Carinae, from Smith and Frew (2011)

motion (e.g., Gull et al. 2005; Morse et al. 2001). Spectra have revealed that the Great Eruption also propelled extremely fast ejecta and a blast wave outside the Homunculus, moving at speeds of 5,000 km s^{-1} or more (Smith 2008). We have an improved record of the nineteenth century light curve from additional released historical documents (Smith and Frew 2011; Fig. 8.2), and perhaps most exciting, we have now detected light echoes from the nineteenth century eruption, allowing us to obtain spectra of the outburst itself after a delay of 160 years (Rest et al. 2012).

Altogether, the outstanding observational record of η Car suggests a picture wherein a VMS suffered an extremely violent, $\sim 10^{50}$ erg explosive event comparable to a weak supernova, which ejected much of the star's envelope – but the star apparently survived this event. This gives us a solid example of the extreme events that can result from the instability in a VMS, but the underlying physics is still not certain. Interactions with a close companion star are critical for understanding its present-day variability; the binary probably played a critical role in the behavior of the nineteenth century Great Eruption as well, although the details are still unclear.

While η Car is the best observed LBV, it may not be very representative of the LBV phenomenon in general. In what ways is η Car so unusual among LBVs?

Its nineteenth century Great Eruption reached a similar peak absolute magnitude (−14 mag) to those of other so-called "SN impostor" events in nearby galaxies (see Smith et al. 2011a), but unlike most extra-galactic examples, its eruption persisted for a decade or more, whereas most extra-galactic examples of similar luminosity last only 100 days or less. Among well-studied LBVs in the Galaxy and Magellanic Clouds, only η Car is known to be in a wide colliding-wind binary system that shows very pronounced, slow periodic modulation across many wavelengths (HD 5980 in the SMC is in a binary, but with a much shorter period). Its 500 km s^{-1} and 10^{-3} $M_\odot$ year^{-1} wind is unusually fast and dense compared to most LBVs, which are generally an order of magnitude less dense. Its Homunculus nebula is the youngest LBV nebula, and together with P Cygni these are the only sources for which we have both an observed eruption event and the nebula it created. Thus, it remains unclear if η Car represents a very brief (and therefore rarely observed) violent eruption phase that most VMSs pass through at some time in their evolution, or if it really is so unusual because of its very high mass and binary system parameters.

In any case, the physical parameters of η Car's eruption are truly extreme, and they push physical models to limits that are sometimes hard to meet. The nineteenth century event has long been the prototype for a super-Eddington wind event, but detailed investigation of the physics involved shows that this is quite difficult to achieve (see Owocki's chapter in this volume). At the same time, we now have mounting observational evidence of an explosive nature to the Great Eruption: (1) A very high ratio of kinetic energy to integrated radiated energy, exceeding unity; (2) Brief spikes in the light curve that occur at times of periastron; (3) evidence for a small mass of very fast moving (∼5,000 km s^{-1}) ejecta and a blast wave outside the Homunculus, which requires a shock-powered component to the eruption, and (4) behavior of the spectra seen in light echoes (Rest et al. 2012), which do not evolve as expected from an opaque wind. These hints suggest that some of the phenomena we associate with LBVs (and their extra-galactic analogs) are driven by explosive physics (i.e. hydrodynamic events in the envelope) rather than (or in addition to) winds driven from the surface by high luminosity. This is discussed in more detail in the following subsection.

8.2.4 Giant Eruptions: Diversity, Explosions, and Winds

Giant eruptions are simultaneously the most poorly understood, most puzzling, and probably the physically most important of the observed phenomena associated with LBVs. They are potentially the most important aspect for massive stars because of the very large amounts of mass (as much as 10–20 $M_\odot$) that are ejected in a short amount of time, and consequently, because of their dramatic influence on immediate pre-SN evolution (next section). Although the giant eruptions themselves are rarely observed because they are infrequent and considerably fainter than SNe, a large number of LBVs and spectroscopically similar stars in the Milky Way and

Magellanic Clouds are surrounded by massive shell nebulae, indicating previous eruptions with a range of ejecta masses from 1–20 $M_\odot$ (Clark et al. 2005; Smith and Owocki 2006; Wachter et al. 2010; Gvaramadze et al. 2010). Thus, eruptive LBV mass loss is inferred to be an important effect in late evolution of massive stars, and perhaps especially so in VMSs.

Originally the class of LBV giant eruptions was quite exclusive, with only four approved members: η Car's 1840s eruption, P Cygni's 1600 AD eruption, SN 1954J (V12 in NGC 2403), and SN 1961V (see Humphreys, Davidson, and Smith 1999). Due the the advent of dedicated searches for extra-galactic SNe from the late 1990s onward, the class of giant eruptions has grown to include a few dozen members (see recent summaries by Smith et al. 2011a; Van Dyk and Matheson 2012). Because of their serendipitous discovery in SN searches, they are also referred to as "SN impostors". Other names include "Type V" supernovae (from F. Zwicky), "η Car analogs", and various permutations of "intermediate luminosity transients".

Although the total number of SN impostors is still quite small (dozens) compared to SNe (thousands), the actual rates of these events could potentially be comparable to or even exceed those of core-collapse SNe. The difference is due to the fact that by definition, SN impostors are considerably fainter than true SNe, and are therefore much harder to detect. Since they are $\sim$100 times less luminous than a typical Type Ia SN, their potential discovery space is limited to only 1/1,000 of the volume in which SNe can be discovered with the same telescope. Their discovery is made even more difficult because of the fact that their contrast compared to the underlying host galaxy light is lower, and because in some cases they have considerably longer timescales and much smaller amplitudes of variability than SNe. Unfortunately, there has not yet been any detailed study of the rates of SN impostors corrected for the inherent detection bias in SN searches. We are limited to small numbers, but one can infer that the rates of LBV eruptions and core-collapse SNe are comparable based on a local guesstimate: in our nearby region of the Milky Way there have been 2 giant LBV eruptions (P Cyg & η Car) and 3 SNe (Tycho, Kepler, and Cas A; and only 1 of these was a core-collapse SN) in the past $\sim$400 year.

The increased number of SN impostors in the past decade has led to recognition of wide diversity among the group, and correspondingly, increased ambiguity about their true physical nature. It is quite possible that many objects that have been called "SN impostors" are not LBVs, but something else. The SN impostors have peak absolute magnitudes around -14 mag, but there is actually a fairly wide spread in peak luminosity, ranging from -15 mag down to around -10 mag. At higher luminosity, transients are assumed to be supernovae, and at lower luminosity we call them something else (novae, stellar mergers, S Dor eruptions, etc.)— but these dividing lines are somewhat arbitrary. Most of their spectra are similar, the most salient characteristic being bright, narrow H emission lines (so they are all "Type IIn") atop either a smooth blue continuum or a cooler absorption-line spectrum. Since the outbursts all look very similar, many different types of objects might be getting grouped together by observers. When more detailed pre-eruption information about the progenitor stars is available, however, we find a range of cases. Some are indeed very luminous, blue, variable stars; but some are not so luminous

($<10^5$ $L_\odot$), and are sometimes found among somewhat older stellar populations than one expects for a VMS (Prieto et al. 2008a,b; Thompson et al. 2009). Some well-studied extra-galactic SN impostors that are clearly massive stars suffering LBV-like giant eruptions are SN 1997bs, SN 2009ip, UGC 2773-OT, SN 1954J, V1 in NGC 2366, SN 2000ch; some well-studied objects that appear to be lower-mass stars (around 6–10 $M_\odot$) are SN 2008S, NGC 300-OT, V838 Mon, and SN 2010U. There are many cases in between where the interpretation of observational data is less straightforward or where the data are less complete. In any case, it is interesting that even lower mass stars (8–15 $M_\odot$) may be suffering violent eruptive instabilities similar to those seen in the most massive stars. If the physical cause of the outbursts is at all related, it may point to a deep-seated core instability associated with nuclear burning or some binary collision/merger scenario, rather than an envelope instability associated with the quiescent star being near the Eddington limit.

Physically, the difference between a "SN impostor"/giant eruption and a true (but under-luminous) SN is that the star does not survive the latter type of event. Observationally, it is not always so easy to distinguish between the two. Even if the star survives, it may form dust that obscures the star at visual wavelengths, while IR observations may not be available to detect it. On the other hand, even if the star dies, there may appear to be a "surviving" source at the correct position if it is a host cluster, a companion star, an unrelated star superposed at the same position, or ongoing CSM interaction from the young SN remnant. It is often difficult to find decisive evidence in the faint, noisy, unresolved smudges one is forced to interpret when dealing with extra-galactic examples. Consider the extremely well-observed case of SN 1961V. This object was one of the original "Type V" SNe and a prototype of the class of LBV giant eruptions (Humphreys et al. 1999). However, two recent studies have concluded that it was most likely a true core-collapse Type IIn SN, and for two different reasons: Smith et al. (2011a) point out that all of the observed properties of the rather luminous outburst are fully consistent with the class of Type IIn SNe, which did not exist in 1961 and was not understood until recently. If SN 1961V were discovered today, we would undoubtedly call it a true SN IIn since its high peak luminosity (-18 mag) and other observed properties clearly make it an outlier among the SN impostors. On the other hand, Kochanek et al. (2011) analyzed IR images of the site of SN 1961V and did not find an IR source consistent with a surviving luminous star that is enshrouded by dust, like η Car. Both studies conclude that since the source is now $\sim$6 mag fainter than the luminous blue progenitor star, it probably exploded as a core-collapse event. Although there is an Hα emission line source at the correct position (Van Dyk et al. 2002), this could be due to ongoing CSM/shock interaction, since no continuum emission is detected. It is hard to prove definitively that the star is dead, however (for an alternative view, see Van Dyk and Matheson 2012). This question is very important, though, because the progenitor of SN 1961V was undoubtedly a very luminous star with a likely initial mass well exceeding 100 $M_\odot$. If it was a true core-collapse SN, it would prove that some very massive stars do explode and make successful SNe.

What is the driving mechanism of LBV giant eruptions? What is their source of luminosity and kinetic energy? Even questions as simple and fundamental as these have yet to find answers. Two broad classes of models have developed: super-Eddington winds, and explosive mass loss. Both may operate at some level in various objects.

Traditionally, LBV giant eruptions have been discussed as super-Eddington winds driven by a sudden and unexplained increase in the star's bolometric luminosity (Humphreys and Davidson 1994; Shaviv 2000; Owocki et al. 2004; Smith and Owocki 2006), but there is growing evidence that some of them are non-terminal hydrodynamic ejections (see Smith 2008, 2013). Part of the motivation for this is based on detailed study of η Carinae, which as noted above, has shown several signs that the 1840s eruption had a shock-driven component to it. One normally expects sudden, hydrodynamic events to be brief (i.e., a dynamical time), which may seem incongruous with the 10 year long Great Eruption of η Car. However, as in some very long-lasting core-collapse SNe, it is possible to power the observed luminosity of the decade-long Great Eruption with a shock wave plowing through dense circumstellar gas (Smith 2013). In this model, the duration of the transient brightening event is determined by how long it takes for the shock to overrun the CSM (this, in turn, depends on the relative speeds of the shock and CSM, and the radial extent of the CSM). Since shock/CSM interaction is such an efficient way to convert explosion kinetic energy into radiated luminosity, it is likely that many of the SN impostors with narrow emission lines are in fact powered by this method. The catch is that even this method requires something to create the dense CSM into which the shock expands. This may be where super-Eddington winds play an important role. The physical benefit of this model is that the demands on the super-Eddington wind are relaxed to a point that is more easily achievable; instead of driving 10 $M_\odot$ in a few years (as for η Car), the wind can provide roughly half the mass spread over several decades or a century. The required mass-loss rates are then of order 0.01–0.1 $M_\odot$ year^{-1}, which is more reasonable and physically plausible than a few to several $M_\odot$ year^{-1}. Also, the wind can be slow (as we might expect for super-Eddington winds; Owocki et al. 2004), whereas the kinetic energy in observed fast LBV ejecta can come from the explosion.

In any case, the reason for the onset of the LBV eruption remains an unanswered question. In the super-Eddington wind model, even if the wind can be driven at the rates required, we have no underlying physical explanation for why the star's bolometric luminosity suddenly increases by factors of 5–10 or more. In the explosion model, the reason for an explosive event preceding core collapse is not known, and the cause of explosive mass loss at even earlier epochs is very unclear. It could either be caused by some instability in late nuclear burning stages (see e.g., Smith and Arnett 2014), or perhaps by some violent binary interaction like a collision or merger (Smith 2011; Smith and Arnett 2014; Podsiadlowski et al. 2010). Soker and collaborators have discussed an accretion model to power the luminosity in events like η Car's Great Eruption, but these assume that an eruption occurs to provide the mass that is then accreted by a companion, and so there is no explanation for what triggers the mass loss from the primary in the first place. In any case, research on these eruptions is actively ongoing; it is a major unsolved problem in astrophysics, and in the study of VMSs in particular.

8.3 Very Luminous Supernovae

8.3.1 Background

The recognition of a new regime of SN explosions has just occurred in the last few years—this includes SNe that are observed to be substantially more luminous than a standard bright Type Ia SN (the brightest among "normal" SNe). Although this is still a young field, the implications for and connections to the evolution and fate of VMSs is exciting. Here we discuss these luminous SNe as well as gamma ray bursts (GRBs), and their connection to the lives and deaths of the most massive stars.

This field of research on the most luminous SNe took on a new dimension with the discovery of SN 2006gy (Smith et al. 2007; Ofek et al. 2007), which was the first of the so-called "super-luminous SNe" (SLSNe). The surprising thing about this object was that with its high peak luminosity (−21.5 mag) and long duration (70 days to rise to peak followed by a slow decline), the integrated luminous energy E_{rad} was a few times 10^{51} erg, more than any previous SN. A number of other SLSNe have been discovered since then (see below). Why were these SLSNe not recognized previously? There may be multiple reasons, but clearly one reason is that earlier systematic SN searches were geared mainly toward maximizing the number of Type Ia SN discoveries in order to use them for cosmology. This meant that these searches, which usually imaged one galaxy per pointing due to the relatively small field of view, mainly targeted large galaxies to maximize the chances of discovering SNe Ia each night. Since it appears that SLSNe actually seem to prefer dwarf galaxy hosts (either because of lower metallicity, or because dwarf galaxies have higher specific star-formation rates), these searches may have been biased against discovering SLSNe. More recent SN searches have used larger fields of view and therefore search large areas of the sky, rather than targeting individual large galaxies; this is probably the dominant factor that led to the increased discovery rate of SLSNe (see Quimby et al. 2011). Additionally, even if SLSNe were discovered in these earlier targeted searches, precious followup resources for spectroscopy on large telescopes are limited, and so SNe that were not Type Ia were given lower priority.

8.3.2 Sources of Unusually High Luminosity

So what can make SLSNe 10–100 times more luminous than normal SNe? There are essentially two ways to get a very luminous explosion. One is by having a relatively large mass of ^{56}Ni that can power the SN with radioactive decay; a higher luminosity generally requires a larger mass of synthesized ^{56}Ni. While a typical bright Type Ia SN might have 0.5–1 $M_\odot$ of ^{56}Ni, a super-luminous SN must have 1–10 $M_\odot$ of ^{56}Ni to power the observed luminosities. Currently, the only proposed explosion mechanism that can do this is a pair instability SN (see Chapter 7 by Woosley & Heger). It is interesting to note that most normal SNe are powered by radioactive decay – were it not for the synthesis of ^{56}Ni in these explosions, we wouldn't ever see most SNe.

The synthesized mass of ^{56}Ni needed to supply the luminosity of a PISN through radioactivity is estimated from observations the same way as for normal SNe:

$$L = 1.42 \times 10^{43} \text{ergs s}^{-1} e^{-t/111d} \, M_{\text{Ni}}/M_\odot \qquad (8.1)$$

(Sutherland and Wheeler 1984) where L is the bolometric luminosity at time t after explosion (usually measured at later times when the SN is clearly on the radioactive decay tail). Important uncertainties here are that L must be the *bolometric* luminosity, which is not always easily obtained without good multiwavelength data (otherwise this provides only a lower limit to the ^{56}Ni mass), and the time of explosion t must be known (this is often poorly constrained observationally, since most SNe have been discovered near maximum luminosity). An additional cause of ambiguity is that in very luminous SNe, it is often difficult to determine if the source of luminosity is indeed radioactivity, since other mechanisms (see below) may be at work.

The other way to generate an extraordinarily high luminosity is to convert kinetic energy into heat, and to radiate away this energy before the ejecta can expand and cool adiabatically. This mechanism fails for many normal SNe, since the explosion of any progenitor star with a compact radius (a white dwarf, compact He star, blue supergiant) must expand to many times its initial radius before the photosphere is large enough to provide a luminous display. These SNe are powered primarily by radioactivity, as noted above. Red supergiants, on the other hand, have larger initial radii, and so their peak luminosity is powered to a much greater extent by radiation from shock-deposited thermal energy. However, even the bloated radii of red supergiants (a few AU) are far smaller than a SN photosphere at peak ($\sim 10^{15}$ cm), and so the most common Type II-P SNe from standard red supergiants never achieve an extraordinarily high luminosity. Most of the thermal energy initially deposited in the envelope is converted to kinetic energy through adiabatic expansion. This inefficiency (and relatively low ^{56}Ni yields of only $\sim 0.1\, M_\odot$) is why the total radiated energy of a normal SN II-P (typically 10^{49} erg) is only about 1 % of the kinetic energy in the SN ejecta.[3]

Smith and McCray (2007) pointed out that this shock-deposition mechanism could achieve the extremely high luminosities of SLSNe like SN 2006gy if the initial "stellar radius" was of order 100 AU, where this radius is not really the hydrostatic photospheric radius of the star, but is instead the radius of an opaque CSM shell ejected by the star before the SN. The key in CSM interaction is that something else (namely, pre-SN mass loss) has already done the work against gravity to put a large mass of dense and slow-moving material out at large radii ($\sim 10^{15}$ cm) away from the star. When the SN blast wave crashes into this material, already located at

---

[3]Of course, most of the energy from a core collapse SN escapes in the form of neutrinos ($\sim 10^{53}$ erg).

a large radius, the fast SN ejecta are decelerated and so the material is heated far from the star, where it can radiate away its thermal energy before it expands by a substantial factor. By this mechanism, large fractions (∼50 % or more) of the total ejecta kinetic energy can be converted to thermal energy that is radiated away. In a hydrogen-rich medium, the photosphere tends to an apparent temperature around 6000–7000 K, and so a large fraction of the radiated luminosity escapes as visual-wavelength photons. Since this mechanism of optically thick CSM interaction is very efficient at converting ejecta kinetic energy into radiation, this process can yield a SLSN without an extraordinarily high explosion energy or an exotic explosion mechanism. What makes this scenario extraordinary (and a challenge to understand) is the requirement of ejecting ∼10 $M_\odot$ in just the few years before core collapse. This is discussed more below.

A variant of this conversion of kinetic energy into light is powering a SLSN with the birth of a magnetar (Woosley 2010; Kasen and Bildsten 2010). In this scenario, a normal core-collapse SN explodes the star and sends its envelope (10 s of $M_\odot$) expanding away from the star. For the SN itself, there is initially nothing unusual compared to normal SNe. But in this case a magnetar is born instead of a normal neutron star or black hole. The rapid spin-down of the magnetar subsequently injects ∼10^{51} ergs of energy into the SN ejecta (which have now expanded to a large radius of ∼100 AU). Similar to the opaque CSM interaction model mentioned previously, this mechanism reheats the ejected material at a large radius, so that it can radiate away the energy before the heat is lost to adiabatic expansion, providing an observer with a SLSN. It would be very difficult to tell the difference observationally between the magnetar model and the opaque shocked shell model during the early phases around peak when photons are diffusing out through the shell or ejecta. It may be possible to see the difference at late times if late-time data are able to see the signature of the magnetar (Inserra et al. 2013).

In summary, there are three proposed physical mechanisms for powering SLSNe. For each, there are also reasons to suspect a link to VMSs.

1. **Pair instability SNe.** This is a very powerful thermonuclear SN explosion. To produce the observed luminosity and radiated energy, one requires of order 10 $M_\odot$ of synthesized ^{56}Ni. These explosions are only expected to occur in VMSs with initial masses of >150 $M_\odot$, because those stars are the only ones with a massive enough CO core to achieve the high temperatures needed for the pair-instability mechanism. The physics of these explosions is discussed more in the chapter by Woosley & Heger. So far, there is only one observed example of a SN that has been suggested as a good example of a PISN, and this is SN 2007bi (Gal-Yam et al. 2009). However, this association with a PISN is controversial. Dessart et al. (2012) have argued that SN 2007bi does not match predictions for a PISN; it has a very blue color with a peak in the UV, whereas the very large mass of Fe-group elements in a PISN should cause severe line blanketing, leading to very red observed colors and deep absorption features. Thus, it is unclear if we have ever yet observed a PISN.

2. **Opaque shocked shells.** Here we have a normal SN explosion that collides with a massive CSM shell, providing a very efficient way of converting the SN ejecta kinetic energy into radiated luminosity when the SN ejecta are decelerated. The reason that this mechanisms would be linked to VMS progenitors is because one requires a very large mass of CSM (10–20 $M_\odot$) in order to stop the SN ejecta. Given expectations for the minimum mass of SN ejecta in models and the fact that stars also suffer strong mass loss during their lifetimes, a high mass progenitor star is needed for the mass budget. Also, sudden eruptive mass loss in non-terminal events that eject ∼10 $M_\odot$ is, so far, a phenomenon exclusively associated with VMSs like LBVs. Although lower-mass stars do appear to be suffering eruptions that look similar (see above), these do not involve the ejection of 10 $M_\odot$.
3. **Magnetar-powered SNe.** In principle, the mechanism is quite similar to the opaque shocked shell model, in the sense that thermal energy is injected at a large radius, although here we have magnetar energy being dumped into a SN envelope, rather than SN ejecta colliding with CSM. Although the SN explosion that leads to this SLSN may be normal, the potential association with VMSs comes from the magnetar. Some magnetars have been found in the Milky Way to be residing in massive young star clusters that appear to have a turnoff mass around 40 $M_\odot$, suggesting that the progenitor of the magnetar had an initial mass above 40 $M_\odot$.

8.3.3 Type IIn SLSNe

Since massive stars are subject to strong mass loss, it is common that there is CSM surrounding a massive star at the time of its death, into which the fast SN ejecta must expand. The collision between the SN blast wave and this CSM is referred to as "CSM interaction", which is commonly observed in core-collapse SNe in the form of X-ray or radio emission (Chevalier and Fransson 1994). However, only about 8–9 % of core-collapse SNe (Smith et al. 2011b) have CSM that is dense enough to produce strong visual-wavelength emission lines and an optically thick continuum. In these cases, the SN usually exhibits a smooth blue continuum with strong narrow H emission lines, and is classified as a Type IIn SN.

Intense CSM interaction can occur in two basic regimes: (1) If the interaction is optically thick so that photons must diffuse out through the material in a time that is comparable to the expansion time, or (2) an effectively optically thin regime, where luminosity generated by CSM interaction escapes quickly. This is equivalent to cases where the outer boundary of the CSM is smaller or larger, respectively, than the "diffusion radius" (see Chevalier and Irwin 2011). The former case will yield an observed SN without narrow lines, resembling a normal broad-lined SN spectrum. The latter will exhibit strong narrow emission lines with widths comparable to the speed of the pre-shock CSM, emitted as the shock continues to plow through the extended CSM. In most cases, the SN will transition from the optically thick case to the optically thin case around the time of peak luminosity (see Smith et al. 2008). If the CSM is hydrogen rich, the narrow H lines earn the SN the designation of Type IIn. (If the CSM is H-poor and He-rich, it will be seen as a Type Ibn, but these are rare and no SLSNe have yet been seen of this type.)

Although narrow H emission lines are the defining characteristic of the Type IIn class, the line widths and line profiles can be complex with multiple components. They exhibit wide diversity, and they evolve with time during a given SN event as the optical depth drops and as the shock encounters density and speed variations in the CSM. These line profiles are therefore a powerful probe of the pre-SN mass loss from the SN progenitor star. Generally, the emission line profiles in SNe IIn break down into three sub-components: narrow, intermediate-width, and broad.

- The narrow (few 10^2 km s^{-1}) emission lines arise from a photo-ionized shock precursor, when hard ionizing photons generated in the hot post-shock region propagate upstream and photo-ionize much slower pre-shock gas. The width of the narrow component, if it is resolved in spectra, gives an estimate of the wind speed of the progenitor star in the years leading up to core collapse. Since these speeds are generally between about 200–600 km s^{-1}, this seems to suggest blue supergiant stars or LBVs for the progenitors of SNe IIn, because the escape speeds are about right (see Smith et al. 2007, 2008, 2010a). Bloated red supergiants or compact WR stars have much slower or faster wind speeds, respectively. In some cases when relatively high spectral resolution is used, one can observe the narrow P Cygni absorption profile. This gives an even more precise probe of the wind speed of the pre-shock gas along the line-of-sight, which in some cases has multiple velocity components showing that the wind speed has been changing (see below, and Groh and Vink 2011). Since the absorption occurs in the densest gas immediately ahead of the shock, one can potentially use the time variation in the P Cyg absorption to trace out the radial velocity law in the wind. A dramatic example of this was the case of SN 2006gy (Fig. 8.4; Smith et al. 2010a), where the velocity of the P Cyg absorption increased with time as the shock expanded, indicating a Hubble-like flow in the CSM (i.e. $v \propto R$). In this case, the Hubble-like law in the pre-shock CSM indicated that the dense CSM was ejected only about 8 years before the SN (Smith et al. 2010a). The rather close synchronization between the pre-SN eruptions and the SN has important implications, and is discussed more below.
- Intermediate-width ($\sim 10^3$ km s^{-1}) components usually accompany the narrow emission-line cores. Generally these broader components exhibit a Lorentzian profile at early times and gradually transition to Gaussian, asymmetric, or irregular profiles at late times. This is thought to be a direct consequence of dropping optical depth (see Smith et al. 2008). At early times in very dense CSM, line photons emitted in the ionized pre-shock CSM must diffuse outward through optically thick material outside that region. The multiple electron scatterings encountered as the photons escape produces the Lorentzian-shaped wings to the narrow line cores. For these phases, it would therefore be a mistake to fit multiple components to the Hα line profile, for example, and to adopt the

broader component as indicative of some characteristic expansion speed in the explosion. At later times when the pre-shock density is lower and we see deeper into the shock, the intermediate-width components can trace the kinematics of the post-shock region more directly. These generally indicate shock speeds of a few 10^3 km s^{-1} or less.
- Sometimes, in special cases of lower-density CSM (or at late times), clumpy CSM, or CSM with non-spherical geometry, one can also observe the broad-line profiles from the underlying fast SN ejecta. In these cases one can estimate the speed of the SN ejecta directly. This usually does not occur in the most luminous SNe IIn, however, simply because the lower-density CSM or the small solid angle for CSM interaction (i.e. a disk) needed to allow one to see the broad SN ejecta lines also limits the luminosity of the CSM interaction, making it hard to have both transparency and high luminosity in the same explosion. A recent case of this is the 2012 SN event of SN 2009ip (Smith et al. 2014).

8.3.4 CSM Mass Estimates for SLSNe IIn

Cold Dense Shell (CDS) Luminosity: Armed with empirical estimates of the speed of the CSM and the speed of the advancing shock, one can then calculate a rough estimate for the density and mass-loss rate of the CSM required to power the observed luminosity of the SN. Dense CSM slows the shock, and the resulting high densities in the post-shock region allow the shock to become radiative. With high densities and optical depths, thermal energy is radiated away primarily as visual-wavelength continuum emission. This loss of energy removes pressure support behind the forward shock, leading to a very thin, dense, and rapidly cooling shell at the contact discontinuity (usually referred to as the "cold dense shell", or CDS; see Chugai et al. 2004; Chugai and Danziger 1994). This CDS is pushed by ejecta entering the reverse shock, and it expands into the CSM at a speed V_{CDS}. In this scenario, the maximum emergent continuum luminosity from CSM interaction is given by

$$L_{CSM} = \frac{1}{2} \dot{M} \frac{V_{CDS}^3}{V_W} = \frac{1}{2} w V_{CDS}^3 \qquad (8.2)$$

where V_{CDS} is the outward expansion speed of the CDS derived from observations of the intermediate-width component, V_W is the speed of the pre-shock wind derived from the narrow emission line widths or the speed of the P Cygni absorption trough, $\dot{M}$ is the mass-loss rate of the progenitor's wind, and $w = \dot{M}/V_W$ is the so-called wind density parameter (see Chugai et al. 2004; Chugai and Danziger 1994; Smith et al. 2008, 2010a). The wind density parameter is a convenient way to describe the CSM density, because it does not assume a constant speed (for the highest mass-loss rates, it may be a poor assumption to adopt a constant wind with a standard

R^{-2} density law, since the huge masses involved are more likely to be the result of eruptive/explosive mass loss).

In general, this suggests that more luminous SNe require either higher density in the CSM, faster shocks, or both. Thus, a wide range of different CSM density (resulting from different pre-SN eruption parameters or different wind mass-loss rates) should produce a wide variety of luminosities in SNe IIn. This is, in fact observed. Figure 8.3 shows several examples of light curves for well-studied SNe IIn, which occupy a huge range in luminosity from the most luminous SNe down to the lower bound of core-collapse SNe (below peaks of about -15.5 mag, we would generally refer to a SN IIn as a SN impostor).

To derive a CSM mass, it is common to re-write the previous equation with an efficiency factor ϵ as:

$$L_{CSM} = \epsilon \, \frac{1}{2} \, \dot{M} \, \frac{V_{CDS}^3}{V_W} = \epsilon \, \frac{1}{2} \, w \, V_{CDS}^3. \tag{8.3}$$

With representative values, this can be rewritten as:

$$\dot{M} = 0.3 \, M_\odot \, \text{yr}^{-1} \times \frac{L_9}{\epsilon^{-1}} \frac{V_w/200}{(V_{CDS}/2000)^3} \tag{8.4}$$

where L_9 is the bolometric luminosity in units of $10^9 \, L_\odot$, $V_W/200$ is the CSM expansion speed relative to $200 \, \text{km s}^{-1}$, and $V_{CDS}/2{,}000$ is the expansion speed of the post-shock gas in the CDS relative to $2{,}000 \, \text{km s}^{-1}$. These velocities are representative of those observed in SNe IIn, although there is variation from one object to the next. L_9 corresponds roughly to an absolute magnitude of only -17.8 mag, which is relatively modest for SNe IIn (Fig. 8.3). Thus, we see that even for relatively normal luminosity SNe IIn, extremely high pre-SN mass-loss rates are required, much higher than is possible for any normal wind. For SLSNe that are ~ 10 times more luminous, extreme mass-loss rates of order $\sim 1 \, M_\odot \, \text{year}^{-1}$ are needed. Moreover, this mass-loss rate is really a lower limit, due to the efficiency factor ϵ, which must be less than 100 %. In favorable cases (fast SN ejecta, slow and dense CSM) the efficiency can be quite high (above 50 %; see van Marle et al. 2010). However, for lower densities and especially non-spherical geometry in the CSM, the efficiency drops and CSM mass requirements rise.

In cases where the post-shock Hα emission is optically thin, one can, in principle, also estimate the CSM mass in a similar way, by replacing the bolometric luminosity with $L_{H\alpha}$, and the efficiency ϵ with the corresponding Hα efficiency $\epsilon_{H\alpha}$. This is perhaps most appropriate at late times, as CSM interaction may continue for a decade after the SN. During this time the assumption of optically thin post-shock Hα emission may be valid. In practice, however, there are large uncertainties in the value of $\epsilon_{H\alpha}$ (usually assumed to be of order 0.005 to 0.05; e.g. Salamanca et al. 2002), so this diagnostic provides only very rough order of magnitude estimates.

Light Curve Fits: The rough estimate in the previous method provides a mass-loss rate corresponding only to the density overtaken at one moment by the shock

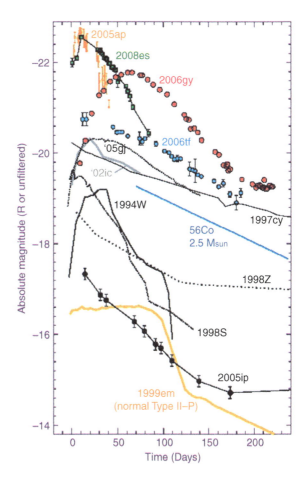

Fig. 8.3 Example light curves of several Type IIn SNe, along with two non-IIn SLSNe (SN 2005ap, a Type Ic) and SN 2008es (Type II) for comparison. SN 1999em is also shown to illustrate a "normal" Type II-P light curve. The fading rate of radioactive decay from ^{56}Co to ^{56}Fe is indicated, although for most SNe IIn this is not thought to be the power source despite a similar decline rate at late times in some objects. Note that SN 2002ic and SN 2005gl are thought to be examples of SNe Ia interacting with dense CSM, leading them to appear as Type IIn (see text)

(assuming the CDS radiation escapes without delay; see below). In reality, the values of V_{CDS}, V_w, and the CSM density can change with time as the shock decelerates while it expands into the CSM, as does the speed of the SN ejecta crashing into the reverse shock. Moreover, pre-SN mass loss is likely to be episodic, so it is unclear for how long that value of $\dot{M}$ was sustained. To get the total CSM mass ejected by the progenitor within some time frame before core collapse (and hence, an average value of $\dot{M}$), one must integrate over time. This means producing a model to fit the observed light curve.

One can calculate a simple analytic model for the CSM mass needed to yield the light curve by demanding that momentum is conserved in the collision between the SN ejecta and the CSM, and that the change in kinetic energy resulting from the deceleration of the fast SN ejecta is lost to radiation. Assuming an explosion energy, a density law for the SN ejecta, and a speed and density law of the CSM, one can calculate the resulting analytic light curve assuming that high densities and H-rich composition lead to a small bolometric correction (see Smith et al. 2008,

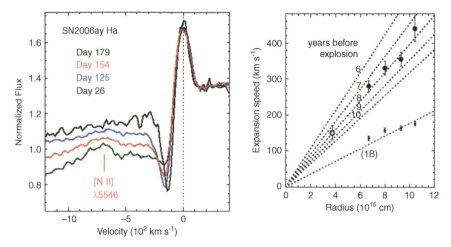

Fig. 8.4 Observations of pre-shock CSM speeds in the SLSN IIn SN 2006gy. The *left panel* shows several tracings of the narrow P Cygni feature. The *right* shows velocities measured for various radii, where dates have been converted to radii based on the observed expansion speed of the cold dense shell (upper points are for the blue edge of the absorption, while the lower points are for the velocity at the minimum of the absorption). The CSM velocity follows a Hubble-like law, indicating a single ejection date for the CSM about 8 year prior to the SN. Both figures are from Smith et al. (2010a)

2010a; Smith 2013a,b; Chatzopoulos et al. 2013; Moriya et al. 2013). One can also do the same from a numerical simulation (e.g., Woosley et al. 2007; van Marle et al. 2010). In general, very high CSM masses of order 10–20 $M_\odot$ are found for SLSNe like SN 2006gy and 2006tf, emitted in the decade or so preceding the explosion. Considering the CSM mass within the radius overtaken by the shock, the uncertainty in this mass estimate is roughly a factor of 2, but should also be considered a lower limit to the total mass since more mass can reside at larger radii. When very high mass and dense CSM is involved, this method is usually more reliable than other methods (emission lines, X-rays, radio) that may severely underestimate the mass due to high optical depths.

Diffusion Time: In extreme cases where the CSM is very dense, the diffusion time $\tau_{diff} \simeq (n\sigma R^2)c$ may be long. If τ_{diff} becomes comparable to the expansion timescale of the shock moving through the CSM $\tau_{exp} \simeq R/V_s$, then the shock-deposited thermal energy can leak out after the shock has broken out of the CSM. Since the radius of the CSM may be very large (of order 10^{15} cm), this may produce an extremely luminous SN display (Smith and McCray 2007). This is essentially the same mechanism as the normal plateau luminosity of a SN II-P (Falk and Arnett 1977), but the radius here is the radius of the CSM, not the hydrostatic radius of the star. This can be simplified to

$$M_{CSM}/M_\odot \simeq R_{15}(\tau_{diff}/23\ days) \qquad (8.5)$$

where R_{15} is the assumed radius of the opaque CSM in units of 10^{15} cm, and τ_{diff} can be estimated from observations of the characteristic fading time of the SN light curve. Applying this to SLSNe like SN 2006gy yields a CSM mass of order 10–20 $M_\odot$ (Smith and McCray 2007; Chevalier and Irwin 2011). This is comparable to the estimates through the previous method. The underlying physical mechanism is the same as normal CSM interaction discussed above, but the optical depths are assumed to be too high for the luminosity to escape quickly. In fact, even lower-luminosity SNe IIn may have diffusion-powered light curves at early times as the shock breaks through the inner and denser parts of the wind; their lower luminosity compared to SLSNe reflects the smaller radius in the CSM where this breakout occurs (see Ofek et al. 2013a).

Hα Emission from Unshocked CSM: When high resolution spectra reveal a narrow P Cygni component to the Hα line (widths of order 100–500 km s^{-1}), one can infer that this emission arises from the pre-shock CSM. (Note that if a narrow P Cyg profile is not seen, but rather a simple emission profile, it is uncertain if this narrow component arises from a distant circumstellar nebula or a nearby H II region.) Following Smith et al. (2007), the mass of emitting ionized hydrogen in the CSM around a SN IIn can be inferred from the total narrow-component Hα luminosity $L_{H\alpha}$ from

$$M_{H\alpha} \simeq \frac{m_H L_{H\alpha}}{h\nu \alpha_{H\alpha}^{eff} n_e} \quad (8.6)$$

where $h\nu$ is the energy of an Hα photon, $\alpha_{H\alpha}^{eff}$ is the Case B recombination coefficient, and n_e is the average electron density. This simplifies to

$$M_{H\alpha} \simeq 11.4 \, M_\odot (L_{H\alpha}/n_e) \quad (8.7)$$

with $L_{H\alpha}$ in units of $L_\odot$ and n_e in cm^{-3} (see Smith et al. 2007). Note that this is only the mass of ionized H at high densities, so it is only a lower limit to the CSM mass if some of the CSM remains neutral. However, as with mass-loss rates of normal O-type stars, the Hα emission depends on the degree of clumping in the wind (see review by Smith 2014), which can lower the total required mass. For more luminous SNe IIn with very dense pre-shock CSM, the narrow Hα component may arise from a relatively thin zone ahead of the shock, and it therefore provides a useful probe of the immediate pre-shock CSM in cases where a narrow P Cyg profile is observed. For a SLSN IIn like SN 2006gy, this method yields a CSM mass of order 10 $M_\odot$ or a mass-loss rate of order 1 $M_\odot$ year^{-1} (Smith et al. 2007). For a more moderate-luminosity SN IIn like SN 2009ip, Ofek et al. (2013a) applied this same method and found a mass-loss rate of order 10^{-2} $M_\odot$ year^{-1}.

X-ray and radio emission: For SLSNe IIn the X-ray and radio emission is of limited utility in diagnosing the pre-SN mass-loss rate, since very high CSM densities cause the X-rays to be self absorbed (the reprocessing of X-rays and their thermalization to lower temperatures is what powers the high visual-wavelength continuum luminosity of SNe IIn) and the CSM is optically thick to radio emission during the main portion of the visual light curve peak.

When X-rays are detected, the X-ray luminosity L_X can be used to infer a characteristic mass-loss rate (see Ofek et al. 2013a; Smith et al. 2007; Pooley et al. 2002):

$$L_X \simeq 3.8 \times 10^{41} \, ergs\,s^{-1} (\dot{M}/0.01)^2 (V_w/500) - 2R_{15} e^{-(\tau + \tau_b f)} \tag{8.8}$$

where $\dot{M}$ is in units of $0.01\,M_\odot$ year^{-1}, the wind speed is relative to $500\,\text{km s}^{-1}$, R_{15} is the shock radius in units of 10^{15} cm, τ is the Thomson optical depth in the wind, and the exponential term is due to wind absorption (see Ofek et al. 2013a for further detail). Caution must be used when inferring global properties, however. If the CSM is significantly asymmetric (as most nebulae around massive stars are), X-rays may indeed escape from less dense regions of the CSM/shock interaction, while much denser zones may yield high optical depths and a strong visual-wavelength continuum. Thus, one could infer both low and high densities simultaneously, which might seem contradictory at first glance. This was indeed the case in SN 2006gy, where the CSM density indicated by X-rays was not nearly enough to provide the observed visual luminosity (Smith et al. 2007).

Radio synchrotron emission is quashed for progenitor mass-loss rates much higher than about $10^{-5}\,M_\odot$ year^{-1} in the first year or so after explosion, and as a result, radio emission is rarely seen from SNe IIn at early times. (In order for the CSM interaction luminosity to compete with the normal SN photosphere luminosity, the mass-loss rate of a SN IIn progenitor must generally be higher than $10^{-4}\,M_\odot$ year^{-1}. Moreover, very massive stars almost always have normal winds in this range anyway, due to their high luminosity.) Radio emission can be detected at later times when the density drops, but this emission is then tracing the mass-loss rate that occurred centuries before the SN, rather than the eruptions in the last few years before explosion. For a discussion of how to use radio emission as a diagnostic of the progenitor's mass-loss rate, we refer the reader to Chevalier and Fransson (1994).

8.3.5 Connecting SNe IIn and LBVs

There are several lines of evidence that suggest a possible connection between LBVs and the progenitors of luminous SNe IIn. While each one is not necessarily conclusive on its own, taken together they clearly favor LBVs as the most likely known type of observed stars that fit the bill. If the progenitors of SNe IIn are not actually LBVs, they do a very good impersonation. Here is a list of the different lines of evidence that have been suggested:

1. Super-luminous SNe IIn, where the demands on the amount of CSM mass are so extreme (10–20 $M_\odot$ in some cases) that unstable massive stars are required for

the mass budget, and the inferred radii and expansion speeds of the CSM require that it be ejected in an eruptive event within just a few years before core collapse (Smith et al. 2007, 2008, 2010a; Smith and McCray 2007; Woosley et al. 2007; van Marle et al. 2010). So far, the only observed precedent for stars known to exhibit this type of extreme, eruptive mass loss is LBV giant eruptions. (In fact, one could argue that since LBV is an observational designation, if any such pre-SN event were to be observed, we would probably call it an LBV-like eruption.)
2. Direct detections of progenitors of SNe IIn that are consistent with massive LBV-like stars (Gal-Yam and Leonard 2009; Gal-Yam et al. 2007; Smith et al. 2010b, 2011a, 2012; Kochanek et al. 2011). This is discussed in the next section (Sect. 8.4).
3. Direct detections of non-terminal LBV-like eruptions preceding a SN explosion. This is seen by some as a smoking gun for an LBV/SN connection. So far there are only two clear cases of this, and two more with less complete observations, discussed later (Sect. 8.5).
4. The narrow emission-line components from the CSM indicate H-rich ejecta surrounding the star. H-rich CSM is obviously not exclusive to LBVs, but it argues against most WR stars as the progenitors. If SNe IIn (especially SLSNe IIn) indeed require very massive progenitors, this is a pretty severe problem for standard models of massive-star evolution. In any case, among massive stars with very strong mass loss, LBVs are the only ones with the combination of H-rich ejecta and high densities comparable to those required.
5. Wind speeds consistent with LBVs. As noted above, the observed line widths for narrow components in luminous SNe IIn suggest wind speeds of a few 10^2 km s^{-1}. This is consistent with the expected escape velocities of blue supergiants and LBVs (Salamanca et al. 2002; Smith 2006; Smith et al. 2007, 2008, 2010a; Trundle et al. 2008). While it doesn't prove that the progenitors are in fact LBVs, it is an argument against red supergiants or WR stars as the likely progenitors. Wind speeds alone are not conclusive, however, since radiation from the SN itself may accelerate pre-shock CSM to these speeds.
6. Wind variability that seems consistent with LBVs. Modulation in radio light curves indicates density variations that suggest a connection to the well-established variability of LBV winds (Kotak and Vink 2006). Also, multiple velocity components along the line of sight seen in blue-shifted P Cygni absorption components of some SNe IIn resemble similar multi-component absorption features seen in classic LBVs like AG Car (Trundle et al. 2008). This may hint that some SN IIn progenitors had winds that transitioned across the bi-stability jump, as do LBVs (see Vink chapter; Groh and Vink 2011). As with the previous point (wind speed), this is not a conclusive connection to LBVs, since other stars do experience density and speed variations in their winds, and the sudden impulse of radiation driving from the SN luminosity itself might give the impression of multiple wind speeds seen in absorption along the line of sight. Nevertheless, the variability inferred does hint at a possible connection to LBVs, and is consistent with that interpretation.

We must note, however, that not all SNe IIn are necessarily tied to LBVs and the most massive stars. Some SNe IIn may actually be Type Ia explosions with dense CSM (e.g., Silverman et al. 2013 and references therein), some may be electron-capture SN explosions of stars with initial masses around 8–10 $M_\odot$ (Smith 2013b; Mauerhan et al. 2013b; Chugai et al. 2004), and some may arise from extreme red supergiants like VY CMa with very dense winds (Smith et al. 2009; Mauerhan & Smith 2012; Chugai and Danziger 1994). The argument for a connection to LBVs and VMSs is most compelling for the SLSNe IIn because of the required mass budget, which is hard to circumnavigate (Smith and McCray 2007; Smith et al. 2007, 2008, 2010a; Woosley et al. 2007; Rest et al. 2011).

8.3.6 Requirements for Pre-SN Eruptions and Implications

In order for the characteristic Type IIn spectrum to be observed, and to achieve a high luminosity from CSM interaction, the collision between the SN shock and the CSM must occur immediately after explosion. This places a strong constraint on the location of the CSM and the time before the SN when it must have been ejected. Given the luminosity of SLSNe, the photosphere must be at a radius of a few 10^{15} cm, which must also be the location of the CSM if interaction drives the observed luminosity. Another way to arrive at this same number is to require that a SN shock front (the cold dense shell or CDS, as above) expands at a few 10^3 km s^{-1} in order to overtake the CSM in the first $\sim$100 days. Then we have $D = v \times t = (2{,}000 \text{ km s}^{-1}) \times (100 \text{ days}) = 2\times10^{15}$ cm. Note that the observed blue-shifted P Cygni absorption profiles in narrow line components indicate that the CSM is *outflowing*. This observed expansion rules out possible scenarios where the CSM is primordial (i.e. disks left-over from star formation).

How recently was this CSM ejected by the progenitor star? From the widths of narrow lines observed in spectra we can derive the speed of the pre-SN wind, and these show speeds of typically 100–600 km s^{-1} (Smith et al. 2008, 2010a; Kiewe et al. 2012), as noted earlier. In order to reach radii of 1–2$\times 10^{15}$ cm, then, the mass ejection must have occurred only a few years before the SN. Since the lifetime of the star is several Myr and the time of He burning is 0.5–1 Myr, a timescale of only 2–3 year is very closely synchronized with the time of core collapse. This is a strong hint that something violent (i.e., hydrodynamic) may be happening to these stars very shortly before core collapse, apparently as a *prelude* to the core collapse event.

As noted earlier, the CSM mass must be substantial in order to provide enough inertia to decelerate the fast SN ejecta and extract the kinetic energy. This is especially true for SLSNe, where high CSM masses of order 10 $M_\odot$ are required. Combined with the expansion speeds of several 10^2 km s^{-1} derived from narrow emission lines in SNe IIn, we find that whatever ejected the CSM must have been provided with an energy of order 10^{49} ergs. Since the mass loss occurred in only a few years before core collapse, it is necessarily an eruptive event that is short in duration.

The H-rich composition, high mass, speed, and energy of these pre-SN eruptions are remarkably similar to the physical conditions derived for LBV giant eruptions. This is the primary basis for the connections between LBVs and SNe IIn, as noted earlier. Consequently, we are left with the same ambiguity about the underlying physical mechanism of pre-SN outbursts as we have for LBVs. The SN precursors seem to be some sort of eruptive or explosive mass-loss event, but the underlying cause is not yet known. Unlike many of the LBVs, however, the pre-SN eruptions provide a telling clue—i.e. for some reason they appear to be synchronized with the time of core collapse. This is interesting, since we do know that core evolution proceeds rapidly through several different burning stages as a massive star approaches core collapse. It is perhaps natural to associate these pre-SN eruptions with Ne and O burning, each of which lasts roughly a year (see Quataert & Shiode 2011; Smith and Arnett 2014). Carbon burning lasts at least several centuries (too long for the immediate SN precursors, but possibly important in some SNe IIn), while Si burning lasts only a day or so (too short). A number of possible instabilities that may occur in massive stars during these phases is discussed in more detail by Smith and Arnett (2014), as well the specific case of wave-driven mass loss by Quataert & Shiode (2011) and Shiode and Quataert (2013).

In extremely massive stars with initial masses above $\sim$100 $M_\odot$, a series of precursor outbursts can occur as a result of the pulsational pair instability (PPI; see chapter 7 by Woosley & Heger). These eruptions are thought to occur in a range of initial masses (roughly 100–150 $M_\odot$) where explosive O burning events are insufficient to completely disrupt the star as a final SN, but which can give rise to mass ejections with roughly the mass and energy required for conditions observed in luminous SNe IIn precursors. The PPI should occur far too rarely ($\sim$1 % or less of all core-collapse SNe) to explain all of the SNe IIn (which are about 8–9 % of ccSNe; Smith et al. 2011b). It may, however, provide a plausible explanation for the much more rare cases of SLSNe of Type IIn.

8.3.7 Type Ic SLSNe and GRBs

Not all SLSNe are Type IIn, and not all SLSNe have H in their spectra. The progenitors of SNe IIn are required to eject a large mass of H in just a few years before core collapse, so they must retain significant amounts of H until the very ends of their lives. This fact is in direct conflict with stellar evolution models, as noted above. There are also, however, a number of SNe that may be associated with the deaths of VMSs which have shed all of their H envelopes and possibly their He envelopes as well before finally exploding. Recall that SNe with no visible sign of H, but which do show strong He lines are Type Ib, and those which show neither H nor He are Type Ic (see Filippenko 1997 for a review of SN classification). (Type IIb is an intermediate category that is basically a Type Ib, but with a small mass of residual H left, and so the SN is seen as a Type II in the first few weeks, but then transitions to look like a Type Ib.) Together, Types Ib, Ic, and IIb are sometimes referred to as "stripped envelope" SNe. The stripped envelope SNe most closely related to the deaths of VMSs are the SLSN of Type Ic, and the broad-lined Type Ic supernovae that are observed to be associated with GRBs.

SLSN Ic. The most luminous SNe known to date turn out to be of spectral Type Ic. The prototypes for this class are objects like SN 2005ap (Quimby et al. 2007) and a number of other cases discussed by Quimby et al. (2011). Although these SNe were discovered around the same time as SN 2006gy, their true nature as the most luminous Type Ic SNe wasn't recognized until a few years later. This is because they were actually located at a fairly substantial redshift ($z \simeq 0.2$ to 0.3), causing their visual-wavelength spectra to appear unfamiliar. It turns out that these objects are closest to Type Ic spectra, with no H and little if any He visible in their spectra.[4] Once their redshifts were recognized, it became clear that these SNe were the most luminous of any SNe known, having peak absolute magnitudes around -22 to -23. These SNe are also hotter than normal Type Ic SNe, however, with the peak of their spectral energy distribution residing in the near-UV; this enhances their visual-wavelength apparent brightness (and detectability) because of the redshifts at which they are found. The hotter photospheric temperatures are likely related to their lack of H and He. Unlike SNe IIn, these object do not have narrow lines in their spectra; their spectra exhibit broad absorption lines that are more like normal SNe. More detailed information about these objects is available in two recent reviews (Quimby et al. 2011; Gal-Yam 2012).

The three possible physical driving mechanisms for these explosions are the same as those mentioned above for all SLSNe: (1) Interaction between a SN shock and an opaque CSM shell, (2) Magnetar birth, or (3) Pair instability SN. Even though these objects do not have narrow lines in their spectra (and therefore lack tell-tale signatures of CSM interaction), the first is a possible power source if the opaque CSM shell has a sharp outer boundary that is smaller than the diffusion radius in the CSM. If this is the case, then the shock will break out of the CSM and photons will diffuse out afterward, producing a broad-lined spectrum (Smith and McCray 2007; Chevalier and Irwin 2011). Magnetar-driven SNe (Kasen and Bildsten 2010; Woosley 2010) provide another possible power source for SLSNe Ic, and so far appear to be consistent with all available observations. Recently, Inserra et al. (2013) have presented evidence that favors the magnetar model for these SLSNe Ic, seen in the late-time data. The third mechanism of a pair instability SN (PISN) is perhaps the oldest viable idea for making SLSNe from very massive stars (Barkat et al. 1967; Bond et al. 1984), but so far evidence for this type of explosion remains unclear. Most of the SLSNe Ic fade too quickly to be PISNe; for their observed peak luminosities they would require $\sim 10\,M_\odot$ of ^{56}Ni in order to be powered by radioactive decay, but the rate at which they are observed to fade from peak is much

---

[4]There is so far only one exception to this, which is SN 2008es (Miller et al. 2010; Gezari et al. 2010), whose light curve is shown in Fig. 8.3. This object is a SLSN of Type II, with broad H lines in its spectra, and is not a Type IIn. The total mass of H in its envelope is not well constrained, however.

faster than the ^{56}Co rate (Quimby et al. 2011). So far only one object among the SLSNe Ic, SN 2007bi, has a fading rate that is consistent with radioactivity (Gal-Yam et al. 2009), but the suggestion that this is a true PISN is controversial, as noted earlier (Dessart et al. 2012). It remains unclear if any PISN have yet been directly detected. Originally these SNe were predicted to occur only for extremely massive stars in the early universe (with little mass loss), as discussed more extensively in the chapter by Woosley & Heger.

SNe Ic-BL associated with GRBs. Gamma Ray Bursts (GRBs) represent another example of the possible deaths of VMSs. The detailed observed properties of GRBS, the variety of GRBs (short vs. long duration, etc.), and their history is too rich to discuss here (see Woosley and Bloom 2006 for a review). Instead we focus on the observable SNe that are associated with long-duration GRBs, which are thought to result from core collapse to a black hole in the death of a massive star.

So far, the only type of SN explosion seen to be associated with GRBs are the so called "broad-lined" Type Ic, or SN Ic-BL. Here we must be careful in terminology. While earlier in this chapter we referred to the fact that normal SNe have broad lines, at least compared to the narrow and intermediate-width lines seen in SNe IIn, the class of SN Ic-BL have extremely broad absorption lines in their spectra. A normal SN typically has lines that indicate outflow speeds of $\sim$10,000 km s^{-1}, but SNe Ic-BL exhibit expansion speeds closer to 30,000 km s^{-1}, or 0.1c. These trans-relativistic speeds are related to the fact that a GRB has a highly relativistic jet that is seen as the GRB, if we happen to be observing it nearly pole-on. Since kinetic energy goes as velocity squared, these very fast expansion speeds in SNe Ic-BL imply large explosion energy, and have led them to be referred to as "hypernovae" by some researchers. The reason to associate these SNe Ic-BL and GRBs with the possible deaths of VMSs is that the favored scenario for producing the relativistic jet (the "collapsar", see Woosley & Heger chapter) involves a collapse to a black hole that is thought to occur in stars with initial masses above 30 $M_\odot$. Although the GRBs and the afterglows are extremely luminous, the SN explosion seen as SNe Ic-BL that follow the GRBs are not extremely luminous (they are near the top end of the luminosity distribution for normal SNe, with peaks of -19 or -20 mag), and certainly not as luminous as the class of SLSN Ic discussed above.

Host Galaxies. An interesting commonality is found between SLSNe Ic and the class of SN Ic-BL associated with GRBs. In addition to sharing the Ic spectral type, indicating a progenitor stripped of both its H and He layers, the two groups seem to arise preferentially in similar environments. Namely, both classes of Ic occur preferentially in relatively low-mass host galaxies with low metallicity (Neill et al. 2011; Modjaz et al. 2008). This may hint that these two classes of SNe are the endpoints of similar evolution in massive stars at low metallicity, but that some additional property helps to determine if the object is a successful GRB or not. Since one normally associates stronger mass loss and stripping of the H and He layers with stronger winds (and therefore higher metallicity), the low-metallicity hosts of these SNe may hint that binary evolution plays a key role in the angular momentum that is needed (especailly for the production of GRB jets), with an alternative explanation relying upon chemically homogeneous evolution of rapidly rotating stars (see Yoon and Langer 2005). In this vein, it is perhaps interesting to note that magnetars have been suggested as another possible driving source for GRBs, while magnetar birth is also a likely explanation for SLSNe Ic as noted above. This is still an active topic of current research.

8.4 Detected Progenitors of Type IIn Supernovae

While the previous section described inferred connections between very luminous SNe and VMSs, these connections are however indirect, based primarily on circumstantial evidence. For example, they rely upon the large mass of CSM needed in SNe IIn, the observed wind speeds, and the requirement of extreme eruptive variability only demonstrated (to our knowledge) by evolved massive stars like LBVs, and the possible association with magnetars or collapsars. However, our most direct way to draw a connection between a SN and the mass of the star that gave rise to it is to directly detect the progenitor star itself in archival images of the explosion site taken before the SN occurred. The increase of successful cases of this in recent years is thanks in large part to the existence of archival *HST* images of nearby galaxies, and this has now been done for a number of normal SNe and for a small collection of Type IIn explosions (only one of which qualifies as a superluminous SN). For this technique to work in identifying the progenitor star, one must be lucky[5] enough to have a high quality, deep image of the explosion site in a public archive.

The first cases of a direct detection of a SN progenitor star were the very nearby explosions of SN 1987A in the Large Magellanic Cloud and SN 1993J in M 81, using archival ground-based data. With the advent of *HST*, this technique could be pushed to host galaxies at larger distances, and a number of such cases up until 2008 were reviewed by Smartt (2009). New examples continue to be added since the Smartt (2009) review, including the very nearby SN IIb in M101, SN 2011hd (Van Dyk et al. 2013). Most of the progenitor detections so far are for SNe II-P and IIb, all with relatively low implied initial masses ($<20\,M_\odot$).

The technique for identifying SN progenitors requires very precise work. Once a nearby SN is discovered, one must determine if an archival image of sufficient quality exists (it is frustrating, for example, to find that your SN occurred at a position that is right at the very edge of a CCD chip in an archival image, or just past that edge). Then one must obtain an *HST* image or high-quality ground-based

---

[5]Another somewhat less direct technique for estimating the mass of a SN progenitor star is to analyze the stellar population in the nearby SN environment. The age of the surrounding stellar population provides a likely (although not necessarily conclusive) estimate of the exploded star's lifetime and initial mass. While this information can only be obtained for the nearest SNe, it can be performed after the SN fades and therefore does not require the lucky circumstance of having a pre-existing high-quality archival image.

image (with either excellent seeing or adaptive optics) of the SN itself, in order to perform very careful and precise astrometry to pinpoint the exact position of the SN (usually the precision is a few percent of an HST pixel). The exact position of the SN must then be identified on the pre-explosion archival image of the SN site, using reference stars in common to both images (preferably at the same wavelengths), and then finally one can determine if there is a detected point source at the SN's location. If not, one can derive an upper limit to the progenitor star's luminosity and mass, which is most useful in the nearest cases where the upper limit can be quite restrictive. If there is a source detected, then it becomes a "candidate" progenitor, because it could also be a chance alignment, a companion star in a binary or triple system, or a host cluster. The way to tell is to wait several years and verify that this candidate progenitor source has disappeared after the SN fades beyond detectability.

Once a secure detection is made, one can then use the pre-explosion image to estimate the apparent and absolute magnitudes of the star, and to estimate colors if there are multiple filters. After correcting for the effects of extinction and reddening of the progenitor (which might include the effects of unknown amounts of CSM dust that was vaporized by the SN), one can place the progenitor star on an HR diagram. One can then use single-star evolution tracks to infer a rough value for the star's initial mass, by comparing the progenitor's position on the HR diagram to the expected luminosity and temperatures at the endpoints of evolution models (note, however, that trajectories of these evolution tracks are highly sensitive to assumptions about mass loss and mixing in the models, and the 1D models do not include possible instabilities in late burning phases; see Smith and Arnett 2014). The technique favors types of progenitor stars that are luminous in the filters used for other purposes (usually nearby galaxy surveys, using R and I-band filters), allowing them to be more easily detected. For example, WR stars are the expected progenitors of at least some SNe Ibc, but while these stars are luminous, they are also hot and therefore emit most of their flux in the UV. Compared to a red supergiant at the same distance, they are therefore less easily detected in the red I-band filters that are often used in surveys of nearby galaxies that populate the HST archive. Similarly, very luminous progenitors that emit much of their luminosity at visual wavelengths, like LBVs, should be relatively easy to detect at a given distance. This probably explains why we have multiple cases of LBV-like progenitors, despite the relatively small numbers of very massive stars.

A central issue for understanding VMSs is whether they make normal SNe when they die, rare and unusual types of SNe (like SLSNe or Type IIn), or if instead they have weak/failed SNe as core material and ^{56}Ni falls back into a black hole (making them difficult or impossible to observe). A common expectation from single-star evolution models combined with core collapse studies (e.g., Heger et al. 2003 and references therein; see also the chapter by Woosley & Heger) is that stars with initial masses above some threshold (for example, 30 $M_\odot$, although the exact value differs from one study to the next) will collapse to a black hole and will fail to make a successful bright SN explosion, unless special conditions such as very rapid rotation and envelope stripping can lead to a collapsar and GRB.

Observationally, there are at least four cases where stars more massive than 30 $M_\odot$ do seem to have exploded successfully, and all of these are Type IIn (recall that these cases may be biased because LBV-like progenitors are very bright and easier to detect than hotter stars of the same bolometric luminosity). The four cases are listed individually below.

SN 2005gl. SN 2005gl was a moderately luminous SN IIn (Gal-Yam et al. 2007). Pre-explosion images showed a source at the SN position that faded below detection limits after the SN had faded (Gal-Yam and Leonard 2009). Its high luminosity suggested that the progenitor was a massive LBV similar to P Cygni, with an initial mass of order 60 $M_\odot$ and a mass-loss rate shortly before core-collapse of $\sim$0.01 $M_\odot$ year^{-1} (Gal-Yam et al. 2007).

SN 1961V. Another example of a claimed detection of a SN IIn progenitor, SN 1961V, has a more complicated history because it is much closer to us and more highly scrutinized. For decades SN 1961V was considered a prototype (although the most extreme case) of giant eruptions of LBVs, as noted above, and an analog of the nineteenth century eruption of η Carinae (Goodrich et al. 1989; Filippenko et al. 1995; Van Dyk et al. 2002). However, two recent studies (Smith et al. 2011a; Kochanek et al. 2011) argue for different reasons that SN 1961V was probably a true core-collapse SN IIn. Both studies point out that the pre-1961 photometry of this source's variability was a detection of a very luminous quiescent star, as well as a possible precursor LBV-like giant eruption in the few years before the supposed core collapse. While the explosion mechanism of SN 1961V is still debated (e.g., Van Dyk and Matheson 2012), the clear detection and post-outburst fading of its LBV progenitor is at least as reliable as the case for SN 2005gl. SN 2005gl was shown to have faded to be about 1.5 mag fainter than its progenitor star, whereas SN 1961V is now at least 6 mag fainter than its progenitor. In any case, the luminosity of the progenitor of SN 1961V suggests an initial mass of at least 100–200 $M_\odot$.

In the previous two cases, the SN has now faded enough that it is fainter than its detected progenitor star. The implication is that the luminous progenitor stars detected in pre-explosion images are no longer there, and are likely dead. This provides the strongest available evidence that these detected sources were indeed the stars that exploded to make the SNe we saw, and not simply a chance alignment of another unrelated star, a star cluster, or a companion star in a binary. This is not true for the next two sources, which are still in the process of fading from their explosion. We will need to wait until they fade to be sure that the candidate sources are indeed the star that exploded.

SN 2010jl. Of the four progenitor detections discussed here, SN 2010jl is the only explosion that qualifies as a SLSN, with a peak absolute magnitude brighter than -20 mag. Smith et al. (2011c) identified a source at the location of the SN in pre-explosion *HST* images. The high luminosity and blue colors of the candidate progenitor suggested either an extremely massive progenitor star or a very young and massive star cluster; in either case it seems likely that the progenitor had an initial mass well above 30 $M_\odot$. In this case, however, the SN has not yet faded (it is still bright after 3 years), so we will need to wait to solve the issue of whether the source was the progenitor or a likely host cluster.

SN 2009ip. Although its name says "2009", SN 2009ip is the most recent addition to the class of direct SN IIn progenitor detections, because while the 2009 discovery event was a SN impostor, the same object now appears to have suffered a true SN in 2012 (Mauerhan et al. 2013; Smith et al. 2014). SN 2009ip is an exceptional case, and is discussed in more detail below (Sect. 8.5). For now, the relevant point to mention is that archival *HST* images obtained a decade before the initial discovery revealed a luminous point source at the precise location of the transient. If this was the quiescent progenitor star, the implied initial mass is 50–80 $M_\odot$ (Smith et al. 2010b) or >60 $M_\odot$ (Foley et al. 2011), depending on the assumptions used to calculate the mass. Thus, the case seems quite solid that the progenitor was indeed a VMS.

Altogether, all four of these cases of possible progenitors of SNe IIn suggest progenitor stars that are much more massive than the typical red supergiant progenitors of SNe II-P (Smartt 2009).

8.5 Direct Detections of Pre-SN Eruptions

SNe IIn (and SNe Ibn) require eruptive or explosive mass loss in just the few years preceding core collapse in order to have the dense CSM needed for their narrow-line spectra and high luminosity from CSM interaction. As noted above, the timescale is constrained to be within a few years beforehand, based on the observed expansion speed of the pre-shock gas and the derived radius of the shock and photosphere.

Until recently, these pre-SN eruptions were mostly hypothetical, limited to conjectures supported by the circumstantial evidence that *something* must deposit the outflowing CSM so close to the star. However, we now have examples of SN explosions where a violent outburst was detected in the few years before a SN, and in all cases the SN had bright narrow emission lines indicative of CSM interaction. The two most conclusive detections of an outburst are SN 2006jc and SN 2009ip, and they deserve special mention. SN 1961V and SN 2010mc also had pre-peak detections, although the data are less complete, and the interpretations are more controversial.

SN 2006jc. - SN 2006jc was the first object clearly recognized to have a brief outburst 2 years before a SN. The precursor event was discovered in 2004 and noted as a possible LBV or SN impostor. It had a peak luminosity similar to that of η Car (absolute magnitude of -14), but was fairly brief and faded after only a few weeks (Pastorello et al. 2007). No spectra were obtained for the precursor transient source, but the SN explosion 2 years later was a Type Ibn with strong narrow emission lines of He, indicating moderately slow (1,000 km s^{-1}) and dense H-poor CSM (Pastorello et al. 2007; Foley et al. 2007). There is no detection of the quiescent progenitor, but the star is inferred to have been a WR star based on the H-poor composition of the CSM.

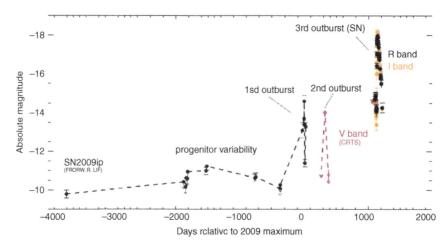

Fig. 8.5 The pre-SN light curve of SN 2009ip, from Mauerhan et al. (2013)

SN 2009ip. A much more vivid and well-documented case was SN 2009ip, mentioned earlier (see Fig. 8.5). It was initially discovered and studied in detail as an LBV-like outburst in 2009, again with a peak absolute magnitude near −14. This time, however, several spectra of the pre-SN eruptions were obtained, and these spectra showed properties similar to LBVs (Smith et al. 2010b; Foley et al. 2011). Also, a quiescent progenitor star was detected in archival *HST* data taken 10 year earlier, which as noted above, indicated a VMS progenitor. In the 5 year preceding its discovery as an LBV-like eruption, the progenitor also showed slow variability consistent with an S Dor-like episode without a major increase in bolometric luminosity, characteristic of LBVs. The object then experienced several brief luminosity peaks over 3 years that looked like additional LBV eruptions (unlike SN 2006jc, detailed spectra of these progenitor outbursts were obtained), culminating in a final SN explosion in 2012 (Mauerhan, et al. 2013a; Smith et al. 2014). The SN light curve was double-peaked, with an initially fainter bump (−15 mag) that had very broad (8,000 km s^{-1}) emission lines probably formed in the SN ejecta photosphere, and it rose quickly 40 days later to a peak of −18 mag, when it looked like a normal SN IIn (caused by CSM interaction, as the SN crashed into the slow material ejected 1–3 years earlier; see Mauerhan, et al. 2013a and Smith et al. 2014). A number of detailed studies of the bright 2012 transient have now been published, although there has been some controversy about whether the 2012 event was a true core-collapse SN (Mauerhan, et al. 2013a; Prieto et al. 2013; Ofek et al. 2013a; Smith et al. 2013, 2014) or not (Pastorello et al. 2013; Fraser et al. 2013; Margutti et al. 2014). More recently, Smith et al. (2014) have shown that the object continues to fade and its late-time emission is consistent with late-time CSM interaction in normal Type IIn supernovae. In any case, SN 2009ip provides us with the most detailed information about any SN progenitor for a decade preceding the SN, with a detection of a quiescent progenitor, several LBV-like precursor eruptions of two different types, and detailed high-quality spectra of the star. This object paints a very detailed picture of the violent death throes in the final years in the life of a VMS.

SN 2010mc. Ofek et al. (2013b) reported the discovery of a precursor outburst in the ∼40 days before the peak of SN 2010mc, recognized after the SN by analyzing archival data. Smith (2013a,b) showed that the light curve of SN 2010mc was nearly identical to that of the 2012 supernova-like event of SN 2009ip, to a surprising degree. Smith et al. (2014) proposed that the ∼40 day precursor events in both SN 2009ip and SN 2010mc were in fact the SN explosions, since this is when very broad P Cygni features were seen in the spectra, and that the following rise to peak was actually due to additional luminosity generated by intense CSM interaction. In that case, the ∼40 day precursor event of SN 2010mc was not actually a pre-SN eruption, but the SN itself. Nevertheless, the similarity in light curves and spectra between SN 2009ip and SN 2010mc would obviously suggest that SN 2010mc probably did have a series of pre-SN LBV-like eruptions too, although those preceding events were not detected.

SN 1961V. The remarkable object SN 1961V has extensive temporal coverage of its pre-SN phases and solid detections of a luminous and highly variable progenitor, moreso than any other SN. The luminous (−12.2 mag absolute at blue/photographic wavelengths) progenitor is well detected in data reaching back to more than 20 year preceding the SN, which includes some small (∼0.5 mag) fluctuations in brightness that could be S Dor-like LBV episodes. In the year before the SN, there is one detection at an absolute magnitude of roughly −14.5, although since it is only one epoch, we don't know if this was an LBV giant eruption or the beginning of the SN. Then in 1961 there was a ∼100 day plateau at almost −17 mag followed by a brief peak at about −18 mag. After this, the SN faded rapidly and has been fading ever since, except for some plateaus or humps in the declining light curve within ∼5 year after peak. Currently, the suggested source at the same position is about 6 mag fainter than the progenitor, and shows Hα emission. In chronological order, SN 1961V was therefore the first direct detection of a pre-SN eruption. In practice, however, the significance of this has been overlooked because the 1961 event was discussed in terms of LBV eruptions (it was considered a "super-η Car-like event"), and was not thought to be a true SN. It is only the much more recent recognition that SN 1961V could have been a true core-collapse Type IIn supernova (Smith et al. 2011a; Kochanek et al. 2011) that underscores the implications of the pre-1961 photometric evidence.

These direct discoveries of pre-SN transient events provide strong evidence that VMSs suffer violent instabilities associated with the latest phases in a massive star's life. The extremely short timescale of only a few years probably hints at severe instability in the final nuclear burning sequences, especially Ne and O burning (Smith and Arnett 2014; Shiode and Quataert 2013; Quataert and Shiode 2012), each of which lasts about 1 year. These instabilities may be exacerbated in the most massive stars, although much theoretical work remains to be done. The increased instability at very high initial masses is certainly true in cases where the pre-SN eruptions result from the pulsational pair instability (see the chapter by Woosley &

Heger), but it may extend to other unknown nuclear burning instabilities as well (Smith and Arnett 2014). Although the events listed above are just a few very lucky cases, they may also be merely the tip of the iceberg. Undoubtedly, continued work on the flood of new transient discoveries will reveal more of these cases. Future cases will be interesting if high-quality data can place reliable constraints on the duration, number, or luminosity of the pre-SN outbursts that will allow for a meaningful comparison with LBV-like eruptions. The limitation will be the existence of high-quality archival data over long timescales of years before the SNe, but these sorts of archives are always becoming more populated and improved. When LSST arrives, it will probably become routine to detect pre-SN outbursts.

8.6 Looking Forward (or Backward, Actually)

Very massive stars are very bright, and their SLSNe are even brighter. Thus, we can see them at large distances, and there is hope that we may soon be able to see light from the explosions of some of the earliest stars in the Universe. The fact that VMSs appear to suffer pre-SN instability that leads to the ejection of large amounts of mass —which in turn enhances the luminosity of the explosion—helps our chances of seeing the first SNe. There is an expectation that the low metallicity environments in the early Universe may favor the formation of very massive stars because of the difficulty in cooling and fragmentation during the star-formation process.

So then we must ask what happens to these stars and their explosions as we move to very low metallicity? How does the physics of eruptions and explosions in the local universe translate to the low-metallicity environments of the earlier universe?

Traditional expectations for massive star evolution are that lower metallicity means lower mass-loss rates (e.g., Heger et al. 2003), since line-driven winds of hot stars have a strong metallicity dependence. It is somewhat ironic, then, that the SNe associated with VMSs have some of the most extreme mass-loss rates (SNe IIn and SLSNe Ic), *but these appear to favor host galaxies with low metallicity.* This contradicts the simple expectation that lower metallicity means lower mass loss, and the implication is that eruptive mass-loss and mass transfer in binary systems may play an extremely important role. It may, in fact, dominate the observed populations of different types of SNe (Smith et al. 2011b). In that case, extrapolating back to low-metallicity conditions in the early universe is not so easy. Binary evolution is not well understood even in the local universe, so extrapolating to a regime where there is no data remains rather adventurous.

The main theme throughout this chapter is that VMSs seem to suffer violent eruptions that impact their evolution and drastically modify the type of SN seen. These eruptions may be very important and may actually dominate the mass lost by VMSs in the local universe, and it is important to recognize that they are probably much less sensitive to changes in metallicity than line-driven winds. The two leading candidates for the physical mechanism of driving this eruptive mass loss are continuum-driven super-Eddington winds and hydrodynamic explosions. While we are not yet certain of the triggering mechanism(s) for either type of event, which may turn out to depend somehow on metallicity, the *physical mechanisms* that drive the mass loss are not metallicity dependent.

Super-Eddington continuum-driven winds rely on electron scattering opacity to transfer radiation momentum to the gas (see the chapter by Owocki; Owocki et al. 2004; Smith and Owocki 2006), and this is independent of metallicity since it only requires electrons supplied by ionized H. This occurs because absorption lines are saturated for high densities in winds with mass-loss rates much above $10^{-4}\,M_\odot$ year^{-1} (recall that LBV eruptions typically have mass-loss rates of $0.01\,M_\odot$ year^{-1} or more). Non-terminal hydrodynamic explosions are driven by a shock wave, and shock waves can obviously still accelerate gas even with zero metal content. If the shocks are driven by some sort of instability in advanced nuclear burning stages (using the ashes of previous burning stages as fuel), it seems unlikely that this would depend sensitively on the initial metallicity that the star was born with. Since these eruptive mechanisms appear to be important for heavy mass loss of VMS in the local universe, there is a good chance that they will still operate or may even be enhanced at low metallicity (Smith and Owocki 2006). The recent recognition that SLSNe appear to favor low-metallicity hosts (see above) would seem to reinforce this suspicion.

One of the key missions for the *James Webb Space Telescope* (*JWST*) will be to detect the light of the explosions from the first stars. Given the arguments above, we should perhaps be hopeful that *JWST* may be able to see extremely luminous SNe from very massive stars, if they suffer similar types of pre-SN eruptive mass loss.

References

Barkat, Z., et al. (1967). *Physical Review Letters, 18*, 379.
Bond, J. R., Arnett, W. D., & Carr, B. J. (1984). *Astrophysical Journal, 280*, 825.
Chatzopoulos, E, Wheeler, J. C., Vinko, J, Horvath, Z. L., & Nagy, A. (2013). *Astrophysical Journal, 773*, 76.
Chevalier, R. A., & Fransson C. (1994). *Astrophysical Journal, 420*, 268.
Chevalier, R. A., & Irwin, C. M. (2011). *Astrophysical Journal, 729*, L6.
Chini, R., Hoffmeister, V. H., Nasseri, A., Stahl, O., & Zinnecker, H. (2012). *Monthly Notices of the Royal Astronomical Society, 424*, 1925.
Chugai, N. N., & Danziger, I. J. (1994). *Monthly Notices of the Royal Astronomical Society, 268*, 173.
Chugai, N. N., Blinnikov, S. I., Cumming, R. J., et al. (2004). *Monthly Notices of the Royal Astronomical Society, 352*, 1213.
Clark, J. S., Larionov, V. M., & Arkharov, A. (2005). *Astronomy and Astrophysics, 435*, 239.
Conti, P. S. (1984). *Institute of Architecture and Urban & Spatial Planning of Serbia, 105*, 233.
Daminelli, A., et al. (1997). *New Astronomy, 2*, 107.
Davidson, K. (1987). *Astrophysical Journal, 317*, 760.
Davidson, K, & Humphreys, R. M. (1997). *Annual Review of Astronomy and Astrophysics, 35*, 1.
Davidson, K., et al. (1986). *Astrophysical Journal, 305*, 867.
de Koter, A., et al. (1996). *Astronomy and Astrophysics, 306*, 501.
Dessart, L., et al. (2012). *Monthly Notices of the Royal Astronomical Society, 426*, L76.
Falk, S., & Arnett, W. D. (1977). *Astrophysical Journal Supplement, 33*, 515.

Ferland, G. J., et al. (1998). *Publications of the Astronomical Society of the Pacific, 110*, 761.
Filippenko, A. V. (1997). *Annual Review of Astronomy and Astrophysics, 35*, 309.
Filippenko, A. V., et al. (1995). *Astronomical Journal, 110*, 2261.
Foley, R. J., Smith, N., Ganeshalingam, M., et al. (2007). *Astrophysical Journal, 657*, L105.
Foley, R. J., Berger, E., Fox, O., et al. (2011). *Astrophysical Journal, 732*, 32.
Fraser, M., Inserra, C., Jerkstrand, A., et al. (2013). *Monthly Notices of the Royal Astronomical Society, 433*, 1312.
Gal-Yam, A., Leonard, D. C., Fox, D. B., et al. (2007). *Astrophysical Journal, 656*, 372.
Gal-Yam, A., & Leonard, D. C. (2009). *Nature, 458*, 865.
Gal-Yam, A., et al. (2009). *Nature, 462*, 624.
Gal-Yam, A. (2012). *Science, 337*, 927.
Gezari, S., et al. 2010. *Astronomy and Astrophysics, 690*, 1313.
Goodrich, R. W., et al. (1989). *Astrophysical Journal, 342*, 908.
Gomez, H., et al. (2010). *Monthly Notices of the Royal Astronomical Society, 401*, L48.
Groh, J. H., Hillier, D. J., Damineli, A., et al. (2009). *Astrophysical Journal, 698*, 1698.
Groh, J. H., & Vink, J. (2011). *Astronomy and Astrophysics, 531*, L10.
Gull, T., et al. (2005). *Astrophysical Journal, 620*, 442.
Gvaramadze, V. V., Kniazev, A. Y., & Fabrika, S. (2010). *Monthly Notices of the Royal Astronomical Society, 405*, 520.
Heger, A., et al. (2003). *Astrophysical Journal, 591*, 288.
Hubble, E., & Sandage, A. (1953). *Astrophysical Journal, 118*, 353.
Humphreys, R. M., & Davidson K. (1994). *Publications of the Astronomical Society of the Pacific, 106*, 1025.
Humphreys, R. M., Davidson, K., & Smith, N. (1999). *Publications of the Astronomical Society of the Pacific, 111*, 1124.
Inserra, C., et al. (2013). *Astrophysical Journal, 710*, 128.
Kasen, D., & Bildsten, L. (2010). *Astrophysical Journal, 717*, 245.
Kiewe, M., et al. (2012). *Astrophysical Journal, 744*, 10.
Kiminki, D. C., & Kobulnicky, H. A. (2012). *Astrophysical Journal, 751*, 4.
Kiminki, D. C., Kobulnicky, H. A., Erwig, I., et al. (2012). *Astrophysical Journal, 747*, 41.
Kobulnicky, H. A., & Fryer, C. L. (2007). *Astrophysical Journal, 670*, 747.
Kochanek, C. S., Szczygiel, D. M., & Stanek, K. Z. (2011). *Astrophysical Journal, 737*, 76.
Kotak, R., & Vink, J. S. (2006). *Astronomy and Astrophysics, 460*, L5.
Langer, N. (2012). *Annual Review of Astronomy and Astrophysics, 50*, 107.
Madura et al. (2012). *Monthly Notices of the Royal Astronomical Society, 420*, 2064.
Margutti, R., et al. (2014). *Astrophysical Journal, 780*, 21.
Massey, P. (2006). *Astrophysical Journal, 638*, L93.
Massey, P., et al. (2007). *Astronomical Journal, 134*, 2474.
Mauerhan, J. C., & Smith, N. (2012). *Monthly Notices of the Royal Astronomical Society, 424*, 2659.
Mauerhan, J. C., Smith, N., Filippenko, A. V., et al. (2013a). *Monthly Notices of the Royal Astronomical Society, 430*, 1801.
Mauerhan, J. C., Smith, N., Silverman, J. M., et al. (2013b). *Monthly Notices of the Royal Astronomical Society, 431*, 2599.
Miller, AA., et al. 2010. *Monthly Notices of the Royal Astronomical Society, 404*, 305.
Modjaz, M., Kewley, L., Kirshner, R. P., et al. (2008). *Astrophysical Journal, 702*, 226.
Morris, P., et al. (1999). *Nature, 402*, 502.
Moriya, T. J., Blinnikov, S. I., Baklanov, P. V., Sorokina, E. I., & Dolgov, A. P. (2013). *Monthly Notices of the Royal Astronomical Society, 430*, 1402.
Morse, J. A., et al. (2001). *Astrophysical Journal, 548*, L207.
Neill, J. D., Sullivan, M., Gal-Yam, A., et al. (2011). *Astrophysical Journal, 727*, 15.
Ofek, E. O., Cameron, P. B., Kasliwal, M. M., et al. (2007). *Astrophysical Journal, 659*, L13.
Ofek, E. O., Lin, L., Kouveliotou, C., et al. (2013a). *Astrophysical Journal, 768*, 47.
Ofek, E. O., Sullivan, M., Cenko, S. B., et al. (2013b). *Nature, 494*, 65.

Owocki, S. P., Gayey, K. G., & Shaviv, N. J. (2004). *Astrophysical Journal, 616*, 525.
Pastorello, A., Smartt, S. J., Mattila, S., et al. (2007). *Nature, 447*, 829.
Pastorello, A., Cappaellaro, E., Inserra, C., et al. (2013). *Astrophysical Journal, 767*, 1.
Podsiadlowski, P., Ivanova, N., Justham, S., & Pappaport, S. (2010). *Monthly Notices of the Royal Astronomical Society, 406*, 840.
Pooley, D., et al. (2002). *Astrophysical Journal, 572*, 932.
Prieto, J. L., Stanek, K. Z., & Beacom, J. F. (2008a). *Astrophysical Journal, 673*, 999.
Prieto, J. L., Stanek, K. Z., Kochanek, C. S., et al. (2008b). *Astrophysical Journal, 673*, L59.
Prieto, J. L., et al. (2013). *Astrophysical Journal, 763*, L27.
Quataert, E., & Shiode, J. (2012). *Monthly Notices of the Royal Astronomical Society, 423*, L92.
Quimby, R., et al. (2007). *Astrophysical Journal, 668*, L99.
Quimby, R. M., Kulkarni, S. R., Kasliwal, M. M., et al. (2011). *Nature, 474*, 487.
Rest, A., Foley, R. J., Gezari, S., et al. (2011). *Astrophysical Journal, 729*, 88.
Rest, A., Prieto, J. L., Walborn, N. R., et al. (2012). *Nature, 482*, 375.
Salamanca, I., et al. (2002). *Monthly Notices of the Royal Astronomical Society, 330*, 844.
Sana, H., de Mink, S. E., de Koter, A., et al. (2012). *Science, 337*, 444.
Shaviv, N. J. (2000). *Astrophysical Journal, 532*, L137.
Shiode, J., & Quataert, E. (2013). *Astrophysical Journal, 780*, 96.
Silverman, J. M., Nugent, P. E., Gal-Yam, A., et al. (2013). *Astrophysical Journal Supplement, 207*, 3.
Smartt, S. J. (2009). *Annual Review of Astronomy and Astrophysics, 47*, 63.
Smith, N. (2006). *Astrophysical Journal, 644*, 1151.
Smith, N. (2008). *Nature, 455*, 201.
Smith, N. (2011). *Monthly Notices of the Royal Astronomical Society, 415*, 2020.
Smith, N., et al. (2011c). *Astrophysical Journal, 723*, 63.
Smith, N. (2013a). *Monthly Notices of the Royal Astronomical Society, 429*, 2366.
Smith, N. (2013b). *Monthly Notices of the Royal Astronomical Society, 434*, 102.
Smith, N. (2014, in press). *Annual Review of Astronomy and Astrophysics*.
Smith, N., & Arnett, W. D. (2014). *Astrophysical Journal, 785*, 82.
Smith, N., & Frew, D. (2011). *Monthly Notices of the Royal Astronomical Society, 415*, 2009.
Smith, N., & McCray, R. (2007). *Astrophysical Journal, 671*, L17.
Smith, N., & Owocki, S. P. (2006). *Astrophysical Journal, 645*, L45.
Smith, N., Gehrz, R. D., Hinz, P. M., et al. (2003). *Astronomical Journal, 125*, 1458.
Smith, N., Vink, J. S., de Koter, A. (2004). *Astrophysical Journal, 615*, 475.
Smith, N., Li, W., Foley, R. J., et al. (2007). *Astrophysical Journal, 666*, 1116.
Smith, N., Chornock, R., Li, W., et al. (2008). *Astrophysical Journal, 686*, 467.
Smith, N., Hinkle, K. H., & Ryde, N. (2009). *Astronomical Journal, 137*, 3558.
Smith, N., Chornock, R., Silverman, J. M., Filippenko, A. V., & Foley, R. J. (2010a). *Astrophysical Journal, 709*, 856.
Smith, N., Miller, A. A., Li, W., et al. (2010b). *Astronomical Journal, 139*, 1451.
Smith, N., Li, W., Silverman, J. M., Ganeshalingam, M., & Filippenko, A. V. (2011a). *Monthly Notices of the Royal Astronomical Society, 415*, 773.
Smith, N., Li, W., Filippenko, A. V., & Chornock, R. (2011b). *Monthly Notices of the Royal Astronomical Society, 412*, 1522.
Smith, N., Li, W., Miller, A. A., et al. (2011c). *Astrophysical Journal, 732*, 63.
Smith N, Mauerhan JC, Silverman JM, et al. (2012). *Monthly Notices of the Royal Astronomical Society, 426*, 1905.
Smith, N., Mauerhan, J. C., Kasliwal, M., & Burgasser, A. (2013). *Monthly Notices of the Royal Astronomical Society, 434*, 3721.
Smith N, & Tombleson R. (2014). *Monthly Notices of the Royal Astronomical Society*, in press (arXiv:1406.7431).
Smith, N., Mauerhan, J. C., & Prieto, J. L. (2014). *Monthly Notices of the Royal Astronomical Society, 438*, 1191.
Sutherland, P. G., & Wheeler, J. C. (1984). *Astrophysical Journal, 280*, 282.

Tammann, G. A., & Sandage, A. (1968). *Astrophysical Journal, 151*, 825.
Thompson, T. A., Prieto, J. L., Stanek, K. Z., et al. (2009). *Astrophysical Journal, 705*, 1364.
Trundle, C., et al. (2008). *Astronomy and Astrophysics, 483*, L47.
Van Dyk, S. D., & Matheson, T. (2012). *Astrophysical Journal, 746*, 179.
Van Dyk, S. D., et al. (2002). *Publications of the Astronomical Society of the Pacific, 114*, 700.
Van Dyk, S. D., Zheng, W. K., Clubb, K. I., et al. (2013). *Astrophysical Journal, 772* L32.
van Genderen, A. (2001). *Astronomy and Astrophysics, 366*, 508.
van Marle A. J., Smith, N., Owocki, S. P., & van Veelen, B. (2010). *Monthly Notices of the Royal Astronomical Society, 407*, 2305.
Wachter, S., Mauerhan, J. C., Van Dyk, S. D., et al. (2010). *Astronomical Journal, 139*, 2330.
Wolf, B. (1989). *Astronomy and Astrophysics, 217*, 87.
Woosley, S. E. (2010). *Astrophysical Journal, 719*, L204.
Woosley, S. E., & Bloom, J. S. (2006). *Annual Review of Astronomy and Astrophysics, 44*, 507.
Woosley SE, Blinnikov S, Heger A. (2007). *Nature, 450*, 390.
Yoon, S. C., & Langer, N. (2005). *Astronomy and Astrophysics, 443*, 643.

Index

Accretion, 2, 5, 43, 44, 191, 205, 219, 239
Arches cluster, 2, 4, 11, 24, 32, 60, 66, 90

Bi-stability, 82, 94–96, 103, 106, 162, 177
Binary dynamics, 4, 5, 33
Binary evolution, 2, 5, 62, 168, 184, 202, 221, 228, 255

Clumping, 23, 81, 87, 90, 97, 98, 100, 140, 161
Collapsar, 93, 192, 221, 255
Convection, 23, 122, 144, 163, 164, 168, 202

Eddington limit, 2, 5, 77, 86, 90, 93, 96, 114, 178, 216
Eta Car, 5, 10, 93, 94, 96, 97, 149, 153, 234

Gamma-ray burst (GRB), 2, 93, 166, 205, 218, 255

Hertzsprung-Russell diagram (HRD), 14, 116, 125, 170, 230
Humphreys-Davidson limit, 5, 116

Inflation, 5, 85, 124, 126
Initial mass function (IMF), 2, 24, 43, 65, 66, 184, 206
Interstellar medium (ISM), 9, 11

Jeans mass, 10, 45

Luminous Blue Variable (LBV), 5, 10, 13, 91, 93, 116, 125, 134, 143, 151, 228

Magnetar, 6, 192, 205, 221, 242
Magnetic field, 46, 166, 167
Mass loss, 1–3, 5, 6, 9, 55, 63, 77, 116, 126, 129, 204, 227
Merger, 59, 61, 62, 70
Metallicity, 1, 2, 10, 54, 81, 92, 158, 175, 178, 240

NGC 3603, 4, 24, 31, 37, 61

Opacity, 50, 54, 79, 80, 86, 89, 96, 97, 99, 100, 103, 106, 114, 117–120, 122, 130, 141, 162, 168, 179
Overshooting, 23, 202

P Cygni, 93, 98, 105, 230
P Cygni profile, 11, 97, 100, 251
Pair-instability supernova (PISN), 1, 5, 185, 189, 194, 216, 242
Porosity, 90, 96–98, 100, 102, 141
Pulsational pair-instability (PPI), 3, 189, 192

R 136, 2, 4, 10, 56, 61, 66
Radiation pressure, 2, 50, 77, 114
Rayleigh-Jeans tail, 14
Rayleigh-Taylor instability, 51, 140
Red supergiant (RSG), 5, 12, 203, 241

Redshift, 2, 222, 254
Reionization, 2

Schwarzschild criterion, 23, 164
Star formation, 10, 158, 252
Supernova, 2, 11, 96, 151, 190, 200, 228

Tarantula nebula, 2, 67

Toomre instability, 47, 49

Upper-mass limit, 4, 10, 30, 43, 45, 66, 115, 116

WNh, 3, 25, 32, 34, 78, 230
Wolf-Rayet, 2, 5, 10, 16, 24, 31, 85, 126, 162, 221

Lightning Source UK Ltd.
Milton Keynes UK
UKOW06n1144041215

264108UK00003B/23/P